Overtourism, Technology Solutions and Decimated Destinations

Azizul Hassan • Anukrati Sharma
Editors

Overtourism, Technology Solutions and Decimated Destinations

Springer

Editors
Azizul Hassan
Tourism Consultants Network
The Tourism Society
London, UK

Anukrati Sharma
Department of Commerce and Management
University of Kota
Kota, Rajasthan, India

ISBN 978-981-16-2473-5 ISBN 978-981-16-2474-2 (eBook)
https://doi.org/10.1007/978-981-16-2474-2

This Springer imprint is published by the registered company Springer Nature Singapore Pte Ltd.
The registered company address is: 152 Beach Road, #21-01/04 Gateway East, Singapore 189721, Singapore

Introduction

Emerging technologies have created a low-cost and convenient way to travel, but most tourist cities do not keep pace with handling the increased flows of tourists. Overtourism Many of the world's most attractive cities are affected by overtourism, since they are typically unable to deal with huge crowds of visitors encouraged by low prices and easy mobility, but the same Information and Communications Technology (ICT) sector is now called on to provide smart solutions to deal with overtourism and escape the congestion of city centers. Residents are alienated as a result of rising pricing, low-cost housing, congested infrastructure, excessive water and resource use, noise pollution, waste, and environmental and cultural heritage destruction. Although emerging technologies are part of the issue, smart ICT solutions can provide an escape strategy, as peer-to-peer applications and social media have encouraged a focus on a limited number of over-visited destinations. Literature studies evident the application of technologies for dealing overtourism.

Bourliataux-Lajoinie, Dosquet and del Olmo Arriaga (2019) aimed to deliver a three-pronged reflection in major cities such as Barcelona on overtourism. The research outlined how technology can influence overtourism and ultimately how to use technology to fix the issue. The research found that the quick distribution of feedback and reviews about attractions and venues is the most important feature. These modern developments have a dark side, despite the interest in the generalisation of ICT. Closely linked to trends in fashion, some tourist destinations are easily overbooked. According to Camatti, Bertocchi, Carić, and van der Borg (2020), Tourism Carrying Capacity (TCC) of a destination is an important reference point in order to design successful responses to the complex phenomenon of overtourism. This paper presents an in-depth analysis of this connection with Dubrovnik. The research applies a model of TCC estimation that can provide the key effects of overtourism quantitatively. The paper demonstrates how these findings can be used to simplify specific decongestion policies through conceptualizing a real-time intervention automated response framework to minimise the adverse effects of overtourism. World Travel & Tourism Council (WTTC) (2018) also has already spotted the relevancy and necessity of emerging technologies and practices (i.e., Wi-Fi; tourists dispersing; speeding and flow control; digital and augmented reality etc.) to help dealing with overtourism. More research studies are required to cover these aspects comprehensively.

In order to see a new viewpoint on overtourism, technology solutions, and decimated destinations, the book is a rare mix of positive, plentiful, and enduring recommendations. This book meticulously focuses on seeing technical solutions in a new trend of overtourism on decimated destinations. Therefore, it will be a forum for addressing the steps that need to be taken to resolve the issue of the most recent COVID-19 and overtourism crisis. The book also explains the duties of tourists towards their decimated destinations. The book offers in-depth awareness of technical solutions for understanding the overtourism problem in decimated destinations. This book comes with 21 chapters in seven parts (i.e., overtourism and technology; social media images; marketing and promotion; psychology, community and environmental issues; challenges; and possible solutions). Summaries of these chapters are presented below:

In chapter one, Séraphin theorizes emerging innovations to counter overtourism as future incremental management tools. According to the author, overtourism has highlighted three key things as a global problem: (1) Tourism is not the solution for all socio-economic challenges facing destinations (2) Sustainability is still an issue to be tackled by the tourism industry (3) Incremental strategies are not the way forward for the sustainability of the industry (as opposed to drastic strategies). As a consequence, this chapter indicates that new technologies (incremental strategies) could contribute to combating overtourism, but only if a driver of fundamental and drastic changes (radical changes) such as an external shock is substantiated (COVID-19, for instance). In that case, it would be possible to assimilate emerging developments with an amplifier and external shock as a motor. This conceptual chapter, based on secondary data (literature review), offers an overview of overtourism research and new tourism technologies. The chapter then offers proof that new technologies act only as enablers but not as evolving motors. Finally, taking wedding tourism as an example, this chapter demonstrates that COVID-19, as an external shock, is driving the industry towards dramatic changes in which technology might theoretically play a significant role, as opposed to gradual strategies (new technologies). This research provides a strategy (theory) that has not previously been proposed to counter overtourism, namely the combination of gradual (enablers) and radical (drivers) strategies. This technique comes under the methodology of ambidextrous management.

In chapter two, Ray interacts between technology and tourism to encourage responsible practice to avoid overtourism. The author believes that tourism and technology are inextricably linked. The relationship between the two is then critical to tourism's growth-mobility, resulting in changes in production, consumption, and the business ecosystem. Interaction was primarily responsible for transforming dispersed and unorganized travel into orderly and organized tourism. In the post-1950s, it also aided the expansion of organized tourism beyond the elitist fringe. It also contributed significantly to the qualitative and quantitative transformation of tourism from mass to aware, sustainable, and finally responsible activities. Technology-tourism interactions drive qualitative and quantitative shifts from fragmented to coordinated mobility, then mass-mobility beyond carrying ability, regulated or sustainable mobility, and finally responsible mobility. Increased mobility in the tourism sector after 1950 became a visible symbol of economic growth, resulting in socio-economic transformation in many countries. The overtourism phenomenon arose as a logical consequence of tourism's rapid growth. By the early twenty-first century, this phenomenon and its countercultural response had become global, thanks to civil society's increasing reliance on information and communication technology (ICT). As a result, prudent practise emerged, understanding that neither unrestricted development nor regulated mobility would improve a place's quality of life or its ability to attract visitors and visitors. During the 2020 pandemic, the growing use of virtual reality (VR) technology is also causing the world to notice this new understanding. The chapter looks at how the connection of technology and tourism will help people resist the overtourism epidemic by putting new knowledge into effect.

Chapter three by Sharma and Yedla analyses social media to advise visitors to visit unexplored destinations for facing overtourism. The aim of this chapter is to define the status of various tourist destinations in terms of their capacity to accommodate tourists. A kind of imbalance leads to congestion and adversely impacts the destination and its surrounding population. Although the world's popular destinations are overcrowded, neither the government nor the private actors and media are searching for unexplored tourist destinations and related activities. The study was conducted by the two writers of this chapter, who belong to India, during a visit to Sri Lanka in September 2019. To write this chapter, the case study method of research was chosen. The researchers also surveyed residents and visitors from both India and Sri Lanka by means of the questionnaire. Thus, in this segment, the primary data produced were analyzed and presented. One of the key reasons for using the term overtourism in a derogatory way is that neither the host community nor the visited visitor is pleased with the destination's overtourism. Tourism is a service sector, an industry for the prosperity of places with balanced ecosystem by the citizens and for the people. But it is sad that both stability, prosperity, and public health are absent today due to the promotion of tourism beyond its borders. The consumption and mishandling of natural resources by visitors is so high that the inhabitants do not have much left. The situation is increasingly moving towards a campaign against tourism's further growth. The chapter seeks to find the answer to the question 'how to get out of overtourism by exploring the unexplored destinations in India and Sri Lanka, what is the level of tourists' knowledge of the tourist destination, and what role do social media play in increasing overtourism? The chapter also includes information about both countries' unexplored destinations as well as the overcrowded tourist destinations. More respondents from India were unaware of Sri Lanka's tourist destinations.

The fourth chapter by Solanky and Gupta discuss the transformation of tourist conduct through social media: contributing towards overtourism. Following the authors, most of the world's admired and exotic tourist destinations are distressed by challenges raised by overtourism, which in turn affect nature and human wellbeing. There are several causes of overtourism, but social media can be considered as the primary explanation for overtourism in the Web 2.0 period. This chapter aims to discuss the role of social media in transforming tourist behavior by using secondary sources of knowledge gathered from e-resources, papers, articles, and books through a descriptive research approach. It has been inferred from the literature study that due to the craze of user-generated content such as reviews, selfies, photos, visuals, and much more, social media plays a tremendous role in transforming tourist behavior. There was a constructive effort to propose ways to tackle overtourism. Due to the novel coronavirus, visitor behavior has been changed. In order to provide insight, the effect of COVID-19 on overtourism was also affected.

The fifth chapter is contributed by Singh that outlines social networking and customer behavior transformation in tourism. In terms of acceptance and its use for various purposes, social media enjoys an extraordinary achievement. While ample efforts have been made to discover the use of social media in consumer behavior for the purchase of travel services from different viewpoints, the use of social media remains imprecise during various stages of travel from pre-trip to post-trip point. Therefore, by describing the uses of social media on consumer behavior in various stages of travel, the present chapter aims to fill this void. The chapter is focused on qualitative methods to research and discover the various aspects of consumers' usage of social media during the travel process and the effects of such use on consumer behavior. For the fulfillment of the objectives of the chapter, expert discussions and reviews of blogs, social media, and digital channels relevant to the tourism and hospitality sector were accessed. The section noted that during all stages of the travel process, social media had been commonly used by customers to make a purchasing decision starting from reaction and ending in the exchange of information. The research shows that the use of social media in travel decision-making has a major effect on customer behavior and has been used extensively in the sharing of information, versatility in implementing this dynamic technology, searching for and finding travel-related information. Moreover, from the perspective of practitioners in the tourism sector, some practical implications can also be taken from the report.

The sixth chapter is written by Mohanty, Shreya, and Choudhury and explores overtourism management through ICT. Overtourism has emerged as one of the largest and most intensely debated issues worldwide, with growing attention being paid to factors such as carrying capacity and sustainability of destinations. Overtourism has been not only a buzzword in the last decade but also a social tourism phenomenon that has received a lot of media attention. Effective, effective, and creative strategies have been called for to tackle the issue of overtourism at different destinations. One such approach is the use of advanced destination management ICT software that can contribute to reducing stress on overbooked destinations, preparing for alternative destinations, targeting an alternative community of visitors, and, most significantly, by systematic and scheduled destination management strategy, gaining influence over the tourist inflow. Although there is plenty of literature on the use of ICT for tourism promotion, at a comprehensive level, the aspect of destination management by the use of advanced technology has rarely been discussed. By creating a structure for destination marketing organizations (DMOs) in order to mitigate the issues of overtourism through the use of emerging advanced technologies, this paper aims to fill this void. This paper aims to illustrate the issues associated with overtourism based on the study of secondary literature and then discusses the ways in which the problems can be mitigated by the use of advanced and modified technical resources.

The seventh chapter is written by Kafa and Korkmaz that covers the perception of overtourism, tourism impacts, and economic dependence of residents Tourism Impacts Economic Dependence on Gökceada Island. The authors argue that it is a fact that certain obstacles are faced by the most tourist-receiving and tourist-sending nations, as well as enjoying different tourism benefits. Deterioration and loss of social, cultural, and historical values are seen as issues that are evergrowing. Such problems are grouped under five titles in the Coping with Progress, Handling Overcrowding in Tourism Destinations December 2017 study prepared by McKinsey and World Travel and Tourism Council as alienated local people, deteriorated tourist experience, crowded infrastructure, damage to the environment, and threats to culture and heritage. The most significant players impacted by overtourism and its problems are local individuals. Studies on the attitudes of people to tourism have also been formulated in a limited way. The basic reasoning is to get an idea of how visitors are seen by local residents. The development of attitudes, however, does not only consist of direct interaction and personal experience but also depends on whether the local people's economic activities are connected to tourism. Companies, on the other hand, may be afraid that they may become the targets of the anti-tourism sentiment of residents and may be interrupted if demand limits on numbers and ability. As the anti-tourist sentiment of local residents increases, visitors are also potential losers. However, over the last few years, the notion of overtourism has become more and more popular as one of the most debated problems with regard to tourism by academia. Today, it is seen to be among the world's top 10 countries in terms of the number of visitors to Turkey. With regard to the regions of Turkey, the number of tourist destinations facing problems linked to overtourism is growing, especially during certain periods of time (tourism season, religious holidays, and public holidays). The goal of this study is to decide how the problems caused by tourism are viewed based on whether or not local people carry out tourism-related economic activities on Gökceada (Imbros Island). The survey was carried out with a questionnaire focused on a study of the literature. In order to establish the connections between the views of local people regarding overtourism and socio-demographic variables, the findings were statistically analyzed.

Chapter eight, as contributed by Saltık and Turgut explains the growth in the community/anger and overtourism. In the chapter, overtourism is seen as one of the concepts that recently emerged in the literature, meaning the disproportionate number of visitors and the effects on local populations and ecology of their flows. An important indicator of the impact of overtourism on populations is the rise in anti-tourism incidents around the world. The underlying explanation for the problem is focused primarily on decisions made in a rush, without proper assessment, consideration of the necessary issues, and sometimes without active involvement and inclusion of communities at the beginning of the growth of tourism in the destinations. Communities also fear that they are inadequately prepared to cope with tourism's increasingly evolving social, environmental, and economic impacts. Local communities can feel held responsible for significant impacts that reduce their quality of life and contribute to community dissatisfaction that contrasts with tourism's idealized intent. The creation of the Group (based tourism) is one of the main elements for avoiding the needless feeling of anti-tourism and for getting communities and cultures closer together through the development of understanding and touch. Community development seeks to improve the social, economic, environmental, and cultural, and community well-being, especially those made up of people who are vulnerable and marginalized. Community development means finding consensus within the community to jointly launch action not only for developments such as economic growth and changes in infrastructure, etc. but also for collaboration and cooperation between individuals, informal groups, and formal organizations. Growing visitor arrivals in a destination inflict a great deal of strain on the environment and local community and their culture, thereby worsening both the quality of the experiences of visitors and the quality of life of local residents. Rapid tourism growth and undesirable condensation disrupt the viability of a destination causing local community and other stakeholders to no longer show any interest in the further development of tourism and make them feel like they are suffering because they are inhibited from accessing opportunities, comfort, and health, etc. In this context, the intent of this chapter is to clarify the role of community development in tourism and the importance of respecting the dissatisfaction of the community in preventing the adverse effects of tourism, especially overtourism.

The ninth chapter is written by ÇİĞDEMLİ and it is about overtourism problems, risk, and duty. Tourism has become an industry in which, due to its contribution to world revenues, many countries and destinations compete to draw visitors. As a consequence of tourism's competitive climate and marketing activities to captivate more and more visitors, human concentration beyond the carrying capacity in destinations contributes to human concentration. On the one hand, the density of tourist inflows increases the popularity of destinations, but on the other hand, concerns about the crowding of visitors and local communities. The negative effects of tourism such as cultural degradation, disruption to the local language, noise pollution, tensions caused by everyday life crowds, rise in living costs, instances of crime, the widespread use of drugs, stresses due to overcrowding in public spaces are risky problems for both local and tourist communities. This situation would have a negative effect on the quality of life of locals, who are important stakeholders in tourism growth. Overtourism, which has arisen as the definition of unsustainable tourism demand, impacts both visitor experiences and the quality of life. Historical places, cultural heritage, environmental quality, and scenery can be affected by the footprint impact caused by crowding in sensitive areas. Given these components as the key sources of tourism attraction, it is evident that there can be a decrease in the tourist profit from the destination. In view of all these variables, the main aim of this study is to define the overtourism definition, to disclose the problems related to overtourism as capacity carrying, density, and overcrowding, to evaluate the outputs of overtourism in terms of local residents, visitors, climate and socio-cultural aspects. Finally, in the form of marketing, management, and tourism aspects, the steps to be taken to avoid the detrimental consequences of overtourism can be addressed.

Chapter ten is written by Godha that questions whether undertuourism can be an option after COVID-19. The primary disease epidemic technique is non-pharmaceutical treatments without any flu vaccine and a restricted medical capacity to treat the symptoms (NPI). The danger of coming back from traveling abroad with the disease. Anxiety, more fear, more confusion, but tourism has also been viewed as a resilient industry, one of the largest and fast-growing industries. Compared with other professions, the tourism industry is even more affected by any internal or external disturbance or depression. In this context, the recent outbreak of COVID-19 has caused national fear. Due to sudden worldwide travel restrictions and homestay orders, the world's greatest disruption in the world economy since the Second World War. The downturn in the outbreak of COVID-19 disease may provide an unprecedented and extremely valuable opportunity for a good long-term reconsideration and restoration of the tourist industry. "Leading" approaches to tourism alone, however, cannot provide sufficient capacity to support such a reset. COVID-19 is widely regarded as a challenge or maybe even a monumental gamble in travel & leisure. The "tourism academy" served as a basis for serious debate, comparing the effects of COVID-19 on the tourist industry in terms of problems, threats, and responsibilities. It notes that industry advocates for a quick recovery are opposed to much more socially responsible, sustainable, and transparent efforts to change the tourism industry. The war includes the important role of intellectuals from the tourist industry and academics from the tourist industry. The effects of such a debate have consequences for the future of tourism in the disciplinary growth of the tourism industry.

In the eleventh chapter, Karim and Hassan describe overtourism in Cox's Bazar in the pre, during and post COVID-19 time period. Overtourism has been one of the most bustling words in many of the famous tourism destinations in the world since the outbreak of COVID-19. Overtourism has been an issue for decades, and, obviously, local residents' demonstrations against tourists have been frequent. At the same time, due to overcrowding, longer queues and socio-environmental deterioration of the destinations, tourists registered disappointing holiday experiences. Cox's Bazar, Bangladesh's most popular tourist destination, was not an exception during the peak season. Thus, this study aims to explain the causes and consequences of overtourism prior to COVID-19 in Cox's Bazar. The research then examines the general overtourism scenario in Cox's Bazar during and after COVID-19. Based on both primary and secondary data sources, this analysis is carried out. In Cox's Bazar, along with in-depth interviews with residents, visitors, and industry stakeholders, journal articles, organizational and government publications, website materials, and other sources are used as data sources and knowledge generation for the overtourism scenario in Cox's Bazar before, during and after COVID-19. The chapter makes some recommend recommendations to track and minimize the adverse effects of overtourism in Cox's bazaar on the basis of performance. This study, as planned, can help tourism destinations to fix potential overtourism concerns as well as direct authorities to a more beneficial tourism industry.

The twelfth chapter by Shukla and Shamurailatpam maps overtourism opportunities by harnessing knowledge and communication technologies. This chapter offers an insight into the overtourism challenges faced by countries that can be mitigated to support the tourism industry by the use of Information and Community Technology (ICT) frameworks. The fundamental proposition is that through technology-led ambidextrous management policies such as smart technology, the use of mobile apps to detect tourist data, analytics for crowd management, censoring tools to observe tourist behavior, tourist cards, eco-friendly biodiversity conservation apps, and censoring tools to observe tourist behavior, the adverse effects of excessive tourism in destinations can be addressed. The essence of the relationship between overtourism and the propagation of technical advancement as well as various facets of the economy is conceptualized by a model. In mitigating the issues of overtourism, our conceptual model generalizes the value of technology and its feasible solution.

The thirteenth Chapter by Koba presents if alternative tourism can be an option for overtourism. Over time, mass tourism has caused natural, cultural, and environmental resources to be used quickly and unplanned. This has destroyed the wealth of nature, society, and the environment in a way that cannot be renewed and reaches a degree that is harmful to future generations. By creating personalized holidays that address the special needs of customers, these negative effects can be minimized. In addition to the sea-sand-sun trio, alternate forms of tourism consisting of special interest trips have therefore grown insignificance. Changes in customer travel (holiday) preferences have been created by the damage caused by mass tourism to natural, cultural, and environmental resources and by casualization over time. As customers began opting for customized, high-quality goods and services, suppliers of tourism products began developing alternatives that could meet various standards. Local people living in regions where tourism is developing often seek to receive a share of tourism income, which has spread tourism activity across the year, and the diversity of tourism products has begun to increase. To mitigate the detrimental impact of tourism, sustainable tourism has gained significance. As a perfect way to remove the problems raised by mass tourism and to ensure sustainability, alternative tourism is seen. This chapter focuses on alternative tourism as a solution to the problems of mass tourism described earlier.

The fourteenth chapter by Akhtar, Saeed, and Waheed briefs Smart tourism as an alternative solution to overtourism. One of the key sources of recreation, exploration, and social gathering is tourism. Many factors, such as the increasingly global economy, the rise of the middle class in developed economies, technological change, new business models, low travel costs, and relaxation of visa policies, contribute to the growth of this industry. The unique geographical location and endless beauty of nature with natural diversity, covering the range from the seaside to the high Himalayan Mountains, bless Pakistan. People from all over the world visited Pakistan to enjoy endless scenes of nature and to celebrate traditional rituals. Given the coronavirus situation, this confusion poses many questions in the minds of people across the globe. In such a moment of collective uncertainty, a major radical paradigm change has centered on the economic structure of all world economies.

Therefore, the issue of overtourism and COVID-19 needs to be considered holistically so that we can have strong solutions for the government, tourism agencies, visitors, local businesses, health care, infrastructure, etc. The goal of the study is to provide a realistic solution in a holistic way to tackle overtourism and COVID-19. Smart tourism is a solution that solves this industry's multi-faceted issues. While selecting the destination such as confidence, facilities, assistance in case of emergency, and reputation, there are few tourist bookings on their agencies. Smart tourism is a better choice for all stakeholders, such as the government, tourism, and tourism companies, to increase opportunities. This opens the door for the business community to think about a new paradigm in which companies driven by technology are more competitive than conventional ones. For researchers, advertisers, visitors, and society as a whole, this chapter is useful.

The fifteenth chapter by Kumara, Guruge, and Silva showcases the prospects and obstacles in Sri Lanka's promotion all-inclusive holiday tourism of All-Inclusive Holiday with particular reference to the Kalutara resorts. The main objective of this study is to understand the opportunities and challenges in consideration of resort/hotel occupancy, revenue, and profitability in promoting All-Inclusive Holiday Tourism in Sri Lanka. The conceptual structure has been built on the basis of a study of literature. There were four independent variables and one dependent variable in the Conceptual System. To test the conceptual model and understand the relationship between the independent and dependent variables, four hypotheses were developed. In the analysis, null hypotheses presumed that four independent variables had no influence on the dependent variable variance, whereas the alternative hypotheses assumed that the independent variables had an impact on the dependent variable variance. The study population consisted of all resort hotels in Sri Lanka's Kalutara district, while the study sample population consisted of 8 resort hotels situated in Sri Lanka's Kalutara district. The primary data collection instrument used in this study was a self-administered questionnaire consisting of 69 questions. In addition, the relationship between the independent variables and the dependent variable has been evaluated using SPSS version 17.0. Using Pearson's association, the correlation between the different holiday packages provided by resort hotels, occupancy, revenue, and profitability was analyzed, while the influence of independent variables on the dependent variable was analyzed using linear regression. Data analysis results show that the AI holiday package has a clear statistically significant positive association with the occupancy, income, and profitability of resort hotels and has a significant effect on the variance in the dependent variable. Data analysis reveals that B&B and HB holiday packages have a statistically significant weak negative association with dependent variables in terms of the other holiday packages, while FB holiday packages have a weak negative relationship with the dependent variable that is not statistically significant. The contribution of these three holiday packages to the variance in the dependent variable is exceedingly limited in comparison to the unique contribution of these three holiday packages to the variance in the dependent variable (less than 14% in total). It can be inferred on the basis of the data analysis that All-Inclusive Holiday Tourism contributes significantly to the occupancy, profits, profitability of resort/hotels and is crucial for the long success and viability of these hotels.

The sixteenth chapter by Silva, Mubarak and Duminduhewa analyses the effect of tourism accommodation knowledge on responsible tourism for destination development in the Hambantota District. The main objective of this study was to examine the perception of responsible tourism and its benefits in the tourist accommodation sector in the district of Hambantota, to recognize the difficulties and challenges faced by businesses in this sector in implementing responsible tourism practices, and to identify the non-financial and financial benefits of responsible tourism derived from these businesses. In order to examine the research issue, a total of 32 Hambantota tourist accommodation institutions were selected for the study. A semi-structured questionnaire was the key data collection method employed in the analysis. To analyze the quantitative data obtained from the survey, basic statistical methods such as mean, median, and percentages were used, while thematic analysis was used to analyze the qualitative data obtained from the questionnaire. The results of the study show that all businesses in the tourist accommodation sector in the district of Hambantota are aware of the idea of responsible tourism. As regards the understanding of the benefits associated with responsible tourism by enterprises in the sector, 93.7% of enterprises were aware of financial benefits, 75% were aware of social benefits, while all enterprises in the sector were aware of the environmental benefits associated with the idea. Results from the study show that 43.8% of companies in this field encountered obstacles and difficulties in implementing responsible tourism practices with respect to the challenges faced in implementing responsible tourism practices. A lack of resources to enforce these practices, insufficient funding from visitors for these practices, and lack of support from policy-makers and local authorities in this regard were the main challenges faced in this regard. In terms of the financial and non-financial gains, those companies in the tourist accommodation sector gain from responsible tourism activities. Data analysis results indicate that only 37.5% of businesses derive financial benefits, while only 43.8% of businesses derive non-financial benefits.

The seventeenth chapter by Dawayan and Annuar outlines the abilities of homestay operators to use social media marketing to solve problems of overtourism. The aim of this chapter is to provide an overview of the Malaysian tourism sector and to implement Malaysian homestay programs. The chapter sheds light on the current issue of overtourism in Sabah, Malaysia in particular. The use of social media among homestay operators in Sabah, and the impact of social media competence on homestay performance are also defined in this chapter. The key highlight of this chapter is the debate on moving tourists to unexplored destinations and how homestay operators can be empowered to use social media marketing as a tool to shift the attention of tourists from mainstream destinations to less explored destinations. Thus, diverting tourists to these destinations can help to curb overtourism issues currently faced by most tourist destinations. While homestay operators in Sabah are mainly from the rural areas, it is believed that if they have some basic knowledge in using social media and are competent in using this type media to market their homestays. They are able to disseminate information about their homestays and build awareness among potential tourists. Mass tourist destinations can promote themselves via social media among some homestay operators. This can allow potential tourists to better plan their holidays by getting more options to choose from. At the same time, this may also be a step in the drive to divert visitors to less-visited destinations and, thus, seek to fix overtourism problems faced by many mainstream holiday destinations at present.

The eighteenth chapter by İştin and Turpcu discusses the application of technology for managing overtourism in Turkey. In particular, the existence of unique attractions motivates potential visitors to visit certain places. In popular destinations, this situation boosts the density of unsustainable tourist numbers. Overtourism, which has many negative economic, social, cultural, and physical consequences, puts pressure on and affects local people's everyday lives living in the area. To reduce or eradicate these negative effects that affect negative perceptions on visitors, it is, therefore, necessary for sustainability to control carrying capacity and find out and overcome the reasons. Technological advances can be seen in this sense as one of the most useful fields for combating overtourism, as well as movements that raise overtourism. The use of technical applications for overtourism in Turkey is discussed in this section of the book. Finally, in the light of technical applications, both visitors coming to the area and the local people living in the region are faced with simpler solutions and suggestions.

The nineteenth chapter by Ferrari and Iaffaldano narrates the creation of tourism in Italian protected natural areas in terms of conventional and creative tools for coping with overtourism and other potential negative effects. Natural protected areas are also successful tourist destinations around the world nowadays. However, they should establish sustainable forms of tourism because of its existence and functions and avoid attracting mass tourism. For this reason, and also because the segment of green tourism is growing very rapidly, the management of protected areas has to face many challenges linked to various aspects of the sustainability of tourism. In reality, overtourism and its effects, such as environmental disruption, negative local community reactions, or other sustainability issues, may occur along with tourism growth. As for socio-cultural sustainability, it is linked to the effect of tourism growth on the socio-cultural fabric of the place and the optimization of the relationship between hosts and visitors. In terms of the potential negative effects on the natural and social-cultural climate, overtourism is perhaps the most difficult factor to analyze, forecast, and prepare. The authors define and explain the variables that could have a major impact on the sustainability of tourism in green destinations, especially in protected natural areas. The case of the Tremiti Islands Marine Reserve is presented, taking into account the different goals and priorities of the stakeholders concerned. In order to define and evaluate the most successful instruments that can be used to deal with these problems in a natural protected area. Different groups of stakeholders are interviewed in a qualitative sample. The results identify instruments to deter negative behavior and to track the effect of tourist flows and conventional and creative measurement methods to minimize the impact of visitor flows on the Tremiti Islands.

The twentieth chapter by Tiwari, Chowdhary, and Kainthola elaborates the paradoxical view of overtourism and technology. Globally, the exponential development of the tourism and hospitality industry is also linked to the advancement of digital media and technical instruments. At every point of the decision-making process of travelers, the domination of information and communication technology is predominant (i.e., from searching for the right destination until sharing comments on social media sites). Technology not only affects the behavioral habits of travelers, but destinations often use technology to market and improve the experience of consumers. However, for the tourism industry, the development of technology has operated like a double-axed sword. Digital media are also kept responsible for popularizing a destination to the degree that it becomes a platform for mass tourism. Via technology and online social networks, problems such as tourismphobia, anti-tourism campaigns, and tourism etc., are gaining hype. Alternatively, for better management, destination managers use technical instruments to sustain tourism growth and visitor experience. ICT has played a crucial role in influencing visitors to visit common destinations, contributing to the problem of overtourism. Similarly, the implementation of technology is equally critical for handling the influx of visitors and, subsequently, preventing crowding and overtourism. The goal of the chapter is to emphasize the ambidextrous role of technology in open tourism. The study is conceptual and uses briefcases of different common overtourism-affected destinations and how technology has acted as an emancipator to tackle unhealthy patterns of consumption. The chapter explores the practical consequences of the use of technology to tackle tourism issues that contribute to unsustainability. It also highlights technology's evolving position in improving the experience of tourists in the post-COVID-19 scenario. This study provides a holistic viewpoint and a link between tourism and technology. The bright side of technology in the tourism and hospitality industry has been addressed in many reports, but the darker side is less recognized.

The twenty first chapter by Korstanje questions what exactly to say about overtourism. The current chapter synthesizes an interesting debate on the issue of overtourism. Tourism study has hotly explored the dichotomies of tourism use and sustainability from its inception. Although tourism appears to be considered a desirable practice that brings prosperity and stability to nations, if not a privilege of liberal democracy, it is no less valid that the industry generates some collateral harm. In recent years, overtourism has demonstrated not only the anger and animosity of locals toward foreign visitors but also the emergence of emerging innovations that are escaping government control. In the area of overtourism, technology provides fertile ground for new solutions and discussions. Some voices have eloquently cautioned about the issues and restrictions that ICT has generated for tourist destinations. Nevertheless, though resocializing visitors, ICT provides a fertile basis for planning more sustainable destinations. To mitigate the detrimental effects of tourism, the implementation of new ethical codes on tourism consumption appears to be crucial. Scholars should separate overtourism, which is a direct influence of the industry, and touristphobia. While overtourism involves practical solutions, touristphobia denotes a slippery matter that arose from the West's inability to tolerate the "Non-Western Other" who became a "unwanted visitor". This drastic reform means, at least as we know it, the end of hospitality. Undoubtedly, after 9/11 and the War on Terror, the drop in hospitality has become more acute.

This book is an important addition to the already limited knowledge of this identified field of study. The entire world is a potential market for this book. The demand and interest in technology and overtourism is expanding enormously. This book can be a reading companion for students of tourism and hospitality in higher academic institutions. The related policy planners and business experts can read this book as well. Researchers from tourism-related disciplines with a deep interest in the industry can also enjoy reading this book.

Tourism Consultants Network
The Tourism Society,
London, UK
Azizul Hassan

Department of Commerce and Management
University of Kota,
Kota, Rajasthan, India
Anukrati Sharma

References

Bourliataux-Lajoinie, S., Dosquet, F., & del Olmo Arriaga, J. L. (2019). The dark side of digital technology to overtourism: the case of Barcelona. *Worldwide Hospitality and Tourism Themes, 11*(5), 582–593.

Camatti, N., Bertocchi, D., Carić, H., & van der Borg, J. (2020). A digital response system to mitigate overtourism: The case of Dubrovnik. *Journal of Travel & Tourism Marketing, 37*(8–9), 887–901.

World Travel & Tourism Council (WTTC) (2018). *How can new technologies help deal with overcrowding?* Retrieved from: https://worldtraveltourismcouncil.medium.com/how-can-new-technologies-help-deal-with-overcrowding-dda554fb164b. Accessed: 21 Jan 2021.

Contents

Contributors

Muhammad Farooq Akhtar is a PhD student of Consumer and Market Analysis at Othman Yeop Abdullah Graduate School of Business (OYAGSB), Universiti Utara Malaysia (UUM), Malaysia. His research interest includes Consumer Behavior, Consumer Psychology, Green Marketing, Food Marketing, Islamic Marketing, Tourism, and areas related to marketing. He has published one paper as a book chapter entitled "Green Consumer Behaviour: Integration of Theory of Planned Behaviour and Technology Acceptance Model" in a book entitled "Leveraging Consumer Behavior and Psychology in the Digital Economy". He has the experience to work as a research fellow on US-AID funded research project at The University of Agriculture, Faisalabad, Pakistan. He is a Lecturer in Marketing (study leave) at The University of Lahore, Gujrat, Pakistan, where he has experience of teaching marketing subjects (i.e., Seminar in Consumer Behavior, Seminar in Marketing Theory, Marketing Research, Customer Relationship Marketing, Global Marketing, Consumer Behavior, Strategic Marketing).

Rashed Al Karim is currently working as an Associate Professor at the School of Business, East Delta University, Chattogram Bangladesh. He has completed his PhD from Universiti Utara Malaysia in Tourism and Hospitality Management and MBA in Marketing from the University of Wales, UK. His research interests are in tourism and hospitality, service marketing, consumer behavior, sustainable development. He has some publications in reputed peer-reviewed journals. He has also presented his research works at international conferences. He has reviewed several articles from reputed journals such as International Journal of Contemporary Hospitality Management, Journal of Hospitality and Tourism Insights, GeoJournal of Tourism and Geosites, Journal of Financial Services Marketing, and Current Journal of Applied Science and Technology.

Sharifah Nurafizah Syed Annuar is a Senior Lecturer in Universiti Teknologi MARA Sabah Branch, Malaysia, and currently holding a post as a Deputy Rector in Research, Industry Linkages, Community and Alumni Network in Universiti Teknologi MARA Sabah Branch, Malaysia. Annuar received her Doctoral degree in

Marketing from the Universiti Malaysia Sabah in 2017. Her research interests are primarily on marketing communications, social and health marketing, consumer behavior, and entrepreneurship. She has broad experience in facilitating engagements between governmental bodies and corporations across Southeast Asia and has worked with the community by providing consultancy services to local start-ups and small-medium enterprises (SMEs) in Sabah. She is also, at the moment, appointed as a Senior Associate and Consultant to a local company in Brunei Darussalam and a core member of Southeast Asia Research Academy (SEARA).

Rashmi Ranjan Choudhury is enthusiastic about nature, a traveler by passion, and a researcher by profession, and is presently pursuing his Doctoral program at North-Eastern Hill University, Shillong, India. He has completed his Masters in Tourism Management and is a Junior Research Fellow awardee from the University Grant Commission, Govt. of India. His research interests include community-based tourism, technology in tourism, sustainability, and smart tourism. He has presented several research papers at various conferences and published articles in refereed research journals, including two book chapters to his credit.

Nimit Chowdhary is an engineer, MBA, and PhD in Management. He has more than 26 years of postgraduate teaching and research experience. He has been a Full Professor for close to 14 years, serving at Mizoram University, IITTM (Gwalior, Noida, and Nellore), and currently serving as the Head of the Department of Tourism and Hospitality Management at Jamia Millia Islamia, New Delhi. Recently, Arizona State University accorded him the honor of Adjunct Professorship. He is a recipient of AICTE Career Award for Young Teachers; SIDA Fellowship, Sweden; Guest Scholarship, Sweden; Linnaeus Palme Exchange Programme Grants, Sweden; PIMG Research Excellence Award, Gwalior; Scholars' Grant (EMTM), Erasmus Mundus, Europe, among others. Recently he was chosen for the prestigious LEAP program at Oxford University. His research has focused on tourism, travel, service experiences, and now transformation. He is a referred researcher in many international journals. He has researched for UGC, ICSSR, AICTE, and Ministry of Tourism, Govt. of India worth around US$ 5 million. He has supervised 15 PhDs, authored eight books, edited two books, and contributed more than 116 papers.

Cynthia Robert Dawayan is currently a Senior Lecturer in the Faculty of Business and Management Universiti Teknologi MARA (UiTM), Sabah Branch, Malaysia. Her passion in the tourism industry has led her to take up a Diploma in Tourism Management right after completing high school, followed by Bachelor of Science (Hons) Tourism Management, after which she decided to attach herself to the industry. Not long after, she decided to further her studies and expand her field into marketing. She completed her MBA (Marketing) in Universiti Malaysia Sabah and finally landed herself a lecturing job with UiTM Sabah. Currently, she is in the midst of completing her PhD in Marketing, where she focuses on the area of homestay marketing. Other than tourism and marketing, Cynthia is also interested in the areas of entrepreneurship and community based tourism topics.

Chinthana Duminduhewa is currently a Tourism Consultant and Visiting Lecturer University of Colombo and Open University. He is reading to PhD Tourism Economics and Having Master Degree and Postgraduate Diploma in Tourism Economics and Hospitality Management. His core subjects are Tourism, Sustainable Tourism, Sustainable Development, Responsible Tourism, Community Tourism, Nature Tourism, and Wildlife Tourism.

Sonia Ferrari is an Associate Professor of Tourism Marketing and Place Marketing at the Department of Business and Law Science, University of Calabria, Italy. She has been serving as the President and Commissioner of Sila National Park for nine years. She is a Member of the World Commission on Protected Areas WCPA of IUCN since 2016. Her main research interests are tourism and place marketing, tourism and management in natural protected areas, service management, event marketing, roots tourism.

Anurodh Godha is working as Assistant Professor at Vardhman Mahaveer Open University, Kota, Rajasthan, India, since August 2009. He is also looking after the affairs of Centre for Entrepreneurship and Skill Development (CESD) as Dy. Director at VM Open University, Kota. Previously, Dr Godha was associated with Centre for Entrepreneurship and Small Business Management, M. D. S. University, Ajmer as Post-Doctoral Fellow. His field of specialization is finance, tourism & entrepreneurship. He is also working on various perspectives and dimensions of the Management of Distance Education Institutes. As an editor, he has many books in his account and also contributed to various edited books at the national level. He has also published many research papers in various journals. In an administrative capacity, he is working as Dy. Director, School of Commerce & Management at VM Open University, Kota.

Sachin Gupta Assistant Professor, Department of Business Administration, Mohan lal Sukhadia University, Rajasthan, India. Dr Gupta has a brilliant academic record. He has earned his Doctorate (PhD) from the Department of Business Administration, University of Rajasthan, Jaipur. His Academic records include MPhil, MCom, and MBA in Finance. Dr Gupta cracked UGC NET Exam in Management twice (June-2012 & December 2012). His professional activities include roles as Associate Editor, Editorial Board Member, Advisory Board Member, and Reviewer for many Journals and Conferences. He has written more than 50 Research Papers in well-known journals and has presented papers in more than 25 National and International Conferences and Seminars. His core subjects are Finance, Entrepreneurship & Innovation, Business & Corporate Laws, Industrial Relations Laws, Business Ethics and Corporate Social Responsibility, Corporate Governance, Corporate Strategy, Marketing Research & Analytical Marketing, etc. Five PhD Research Scholars have been conferred PhD Doctoral degree under Dr Gupta's guidance and have guided one MPhil dissertation. He is so thankful to the education fraternity for the opportunities in the last ten years and really looks forward to learning much more and achieving new horizons.

Manori L. Guruge is currently following a Master in Tourism Economics and Hotel Management, Department of Economics, University of Colombo, Sri Lanka, and Director of Aquila Research and Consultant Pvt. (Ltd.). She has done her Bachelor in Economics (Special) and Master of Economics, Department of Economics, University of Colombo, Sri Lanka. Her core subject areas are Research Methodology, Statistics, Econometrics, Economics, and Tourism. She has published more than 20 research articles nationally and internationally.

Azizul Hassan is a member of the Tourism Consultants Network of the UK Tourism Society. Dr Hassan has been working for the tourism industry as a consultant, academic, and researcher for over 20 years. His research interest areas are technology-supported marketing for tourism and hospitality, immersive technology applications in the tourism and hospitality industry, and technology-influenced marketing suggestions for sustainable tourism and hospitality industry in developing countries. Dr Hassan has authored over 150 articles and book chapters in leading tourism outlets. He is also part of the editorial team of 25 book projects from Routledge, Springer, CAB International, and Emerald Group Publishing Limited. He is a regular reviewer of a number of international journals.

Nicolaia Iaffaldano is a Researcher of Economics and Business Management at the Ionian Department of Law, Economics and Environment, University of Bari, Italy. She has taught Marketing, Economics, and Management of Business Networks in the course of studies in Economics and Business Administration and Cultural Heritage Sciences for Tourism and Environment at the University of Bari. Her main research interests are management and marketing of tourist destinations, service management, and marketing, green tourism, sustainable tourism, the relationship between business strategy and local development, with particular attention to sustainability.

Aysen Ercan İştin is an Assistant Professor. İştin completed her undergraduate education in 2010 at Mustafa Kemal Üniversity, Department of Tourism and Hotel Management. Also, she completed her Master's degree in 2014 at Akdeniz University, Department of Tourism and Hotel Management, and her PhD degree in 2018 at Mersin University, Department of Tourism Management. She, who started to work as a Research Assistant in 2017, was appointed as an Assistant Professor in 2018 at the Department of Tourism and Hotel Management at Şırnak University. She carries on her career as the Head of Gastronomy and Culinary Arts Department in the same university. She instructs undergraduate and graduate levels in the fields of management and organization and issues related to tourism businesses. She has articles, book chapters, and papers in the same fields and carries on her studies.

Neşe Kafa is an Assistant Professor in Çanakkale Onsekiz Mart University Gökçeada School of Applied Sciences in Turkey. She completed her undergraduate education at Balikesir University in the field of Tourism and Hotel Management in

2007. She received her Master's degree in Tourism and Hotel Management from Balıkesir University in 2010 and PhD degree from Balikesir University in 2014. She has worked as an academician since 2009. Her main research interests are rural development and tourism, pro-poor tourism, poverty elimination in tourism, tourism marketing. She teaches courses types of tourism, tourist behavior, and entrepreneurship.

Snigdha Kainthola is a Research Scholar in the Department of Tourism and Hospitality Management, Jamia Millia Islamia, a top-ranked university in India. She is a graduate in history from Delhi University and has completed her MBA in Tourism with a specialization in tourism leisure. She has worked in the tourism sector for two years as a tour leader. She is an avid traveler and keen learner. She decided to pursue academics and make a career in the tourism education sector. Her research inclination is in tourism marketing, spiritual tourism, and destination management. She has contributed chapters in two different books on overtourism under well-known publications.

Mubarak Kaldeen is currently a Senior Lecturer at the Department of Marketing Management, Faculty of Management and Commerce, South Eastern University of Sri Lanka. He is a Commonwealth fellow (UK) and has earned his PhD in Tourism Marketing from Malaysia. His core subjects are Tourism Marketing and Marketing Communications. He has edited three books and published 45 research papers in national and international Journals. He is an editor of SEUSL-Journal of Marketing and co-editor of Journal of Tourism Economic and Applied Research.

Yesim Koba is an Associate Professor at the University of Dokuz Eylül, Turkey, since 2019. Koba graduated from Mersin University, School of Tourism and Hotel Management in 1996. She completed her BA in Business Administration at Anadolu University (2003) and respectively got her MA (2006) and PhD (2013) Degrees in Tourism Management from Dokuz Eylül University. She worked at Muğla University between 2007 and 2012. She has been working as a Lecturer at Dokuz Eylül University, School of Applied Science, since 2012.

Halil Korkmaz is an Assistant Professor in the Faculty of Tourism, Çanakkale Onsekiz Mart University in Turkey. He completed his undergraduate education at Bilkent University in the field of Tourism and Hotel Management in 2007. He received his Master's degree in Tourism and Hotel Management from Balıkesir University in 2013 and PhD degree from Çanakkale Onsekiz Mart University in 2016. He has work experience in various tourism businesses until 2011. He has worked as an academician since 2011. His main research interests are tourism marketing, environmental awareness in tourism businesses, green marketing, and service quality. He teaches courses in marketing, behavioral sciences, and entrepreneurship.

Maximiliano E. Korstanje is a leading global cultural theorist specializing in terrorism, mobilities, and tourism. Dr Korstanje serves as Senior Researcher at the University of Palermo, Buenos Aires, Argentina, and Editor in Chief of the International Journal of Safety and Security in Tourism and Hospitality. In 2015, he was Visiting Professor at the Centre for Ethnicity and Racism Studies (CERS) at the University of Leeds, United Kingdom; TIDES at the University of Las Palmas de Gran Canaria, Spain in 2017; and the University of La Habana, Cuba, in 2018. Korstanje is Book Series Editor of Advances in Hospitality, Tourism and Service Industries (IGI Global, USA), and Tourism Security-Safety and Post Conflict Destinations (Emerald Publishing, UK). He was elected as a foreign faculty member of the Mexican Academy of Tourism Research (Mexico) as well as a foreign member of The Tourism Crisis Management Institute (University of Florida, USA). In 2018, his biography was selected to be part of the roster of Alfred Nelson Marquis Lifetime Achievement Award (Marquis Who's Who).

N. H. D. Kumara is a Master's student in Tourism Economics and Hotel Management, Department of Economics, University of Colombo, Sri Lanka.

Priyakrushna Mohanty is an Assistant Professor in the Department of Tourism & Travel Management at Jyoti Nivas Autonomous College, Bengaluru, India. He is also a UGC Senior Research Fellow under the guidance of Dr Anu Chandran at the Department of Tourism Studies, Pondicherry University, India. He is an awardee of the prestigious Travel Corporation (India) Gold Medal for his outstanding performance in Master's Degree in Tourism Studies from Pondicherry University, India. He also holds a Master's Degree in Commerce along with three PG Diploma Degrees in Rural Development, Research Methodology, and Teaching Skills. Mohanty has served the Indian Railway Catering and Tourism Corporation Ltd. for two years, following which he was recruited as a Guest Faculty in the Department of Tourism Studies, Pondicherry University. He has published more than fifteen articles and chapters in both international and national journals and edited books. Mohanty is passionate about academic areas of Tourism Sustainability, Sustainable Livelihood, Technology, and Tourism, along with Gender issues in Tourism Development.

Aslı Özge Özgen Çiğdemli is an Assistant Professor at the Recreation Management Department, Osmaniye Korkut Ata University, from March 2020. She graduated from the Department of International Relations, Faculty of Political Sciences of Ankara University in 2006. She completed a Master degree in 2010 in International Relations from the Süleyman Demirel University, and her thesis was titled "European Union Immigration Policies: Historical Background and Current Situation". She completed PhD in 2016 from the Hospitality and Tourism Management Department of Mediterranean University. The thesis was titled "The Comparative Analysis of Mediterranean Cruise Destinations: The Position of Turkey". ÇİĞDEMLİ worked at the Republic of Turkey's Ministry of Trade between 2009-2017. She worked as a Faculty member at the School of Tourism and Hotel Management of Şırnak

University between 2017-2020. ÇİĞDEMLİ's main areas of interest are tourism marketing, tourism policy, cruise tourism, consumer behavior in tourism, and strategic marketing management.

Samik Ray is an ex-faculty of Department of Folklore, University of Kalyani; Travel and Tourism Management in MPTI (Kolkata); and WTCC School of Trade and Commerce (Kolkata); Ex-trainer and Faculty of Regional Level Guide (RLG) Training (Govt. of India, Department of Tourism), presently working as RLG (Govt. of India, Department of Tourism); Visiting Faculty in the Department of Tourism Management, Rãmãkrishna Mission Vidyãmandira autonomous college with potentials for excellence, UGC; Editor of "Tourism Theory and Practice", the author of several essays on tourism studies and management, social science, and literary criticism, and received National Tourism Award in the category of "The Best Tourist Guide".

Sehar Saeed is a PhD student in the Faculty of Business and Management in Universiti Teknologi MARA (UiTM) in the field of Consumer Marketing. Her concerned areas are Consumer behavior, Green marketing, E-Marketing, and Tourism. Her study interest in Green marketing has made her look at and choose electronic marketing dynamics and normative behavior for her significant wasted contributions in the future. She works as a Lecturer in Hamdard Institute of Management Sciences (HIMS), Hamdard University (HU), Karachi, Pakistan. She has experience teaching Business and Marketing research, Consumer Behavior, strategic marketing, Sociology, and Research strategy. She also served her expertise in the institute's quality enhancement and remained a member of the Board of Research and Board of Studies at her Faculty.

Işıl Arikan Saltik is currently Associate Professor of Tourism Management, Fethiye Faculty of Business Administration at Mugla Sitki Kocman University, Tourism where she teaches and does research on hospitality management and organization, destination management, human resource management, organizational behavior, and tourism guidance since 2020. She joined the Muğla Sıtkı Koçman University, Turkey, as a Research Assistant in February 2009, after having worked as a Professional Tourist Guide in Spanish and English languages. Dr Arikan-Saltik has presented the result of her research in several international conferences, and she has published in several refereed international and national journals, books, and chapters.

Hugues Séraphin is a Senior Lecturer in Event and Tourism Management Studies. Hugues Seraphin holds a PhD from the Université de Perpignan Via Domitia (France) and joined the University of Winchester Business School in 2012. He was the Programme Leader of the Event Management program between 2015 and 2018. Dr Séraphin has widely published in the area of tourism management.

Sofia Devi Shamurailatpam holds a PhD in Economics with a specialization in the area of Banking and Financial Economics. Currently, she is serving as an Assistant Professor in the Department of Banking and Insurance, Faculty of Commerce, The Maharaja Sayajirao University of Baroda. She has published several research papers in her credit and authored a book entitled "Banking Reforms in India: Consolidation, Restructuring, and Performance, published by Palgrave Macmillan, UK (2017). Her major research area of interest includes Economics of Banking, Financial Economics, Economics of Gender, Agricultural Economics, and Development Economics, particularly Contemporary issues on sustainability.

Anukrati Sharma is currently an Associate Professor and Head of the Department in the Department of Commerce and Management, University of Kota, Kota, Rajasthan, India. She has edited and authored several books and has attended a number of national and international conferences, presenting over 45 papers. She has been invited to talks, lectures, and panel discussions by different universities. Dr Sharma handles training sessions at the Rajasthan Police Academy, Jaipur, India. Her special interest areas are tourism, tourism marketing, strategic management and international business management.

V. S. S. Shreya is currently working as a Research Scholar at the Department of Tourism Studies, Pondicherry University. She possesses a Master's degree in Tourism and Hospitality Management and a PG Diploma in Statistics and Research Methods. Her research interests are sustainable management and the development of destinations. Her doctoral research focuses on Overtourism and residents' perceptions.

Parag Shukla holds a PhD Degree in Commerce and Business Management with a focus on Strategic Marketing in the area of "Retailing". He has been working as an Assistant Professor at The Maharaja Sayajirao University of Baroda since 2009. He has presented and published many research papers in contemporary areas of Marketing in National and International Journals. He is also an author in a book entitled "Retail Shoppers' Behavior in Brick and Mortar Stores -A Strategic Marketing Approach" published by a reputed publisher. His major research area of interests includes Retailing, Services Marketing, and Consumer Behavior to name a few.

D. A. C. Suranga Silva is the Founder of the Tourism study program at the University of Colombo, Sri Lanka, and a Professor in Tourism Economics. He has a wide experience related to tourism and headed various positions in National Institutes in Sri Lanka. He is a tourism strategist and contributes as a policymaker and consultant for various leading projects and programs related to Tourism in Sri Lanka.

Lakhvinder Singh is currently working as Assistant Professor of Tourism in Government College, Kaithal affiliated to Kurukshetra University, Kurukshetra (Haryana), India. He holds PhD Degree with UGC-NET qualification in Tourism Administration & Management. He has more than four years of teaching & research experience and has more than 30 research publications on his credit. He is in the Editorial/Advisory Board of different Journals of International and National repute. His areas of interest include tourist behavior, tourism marketing, tourism impact, culture & religious tourism, e-tourism, and ICT in tourism.

Manisha Solanky is a Research Scholar pursuing PhD from Mohan Lal Sukhadia University Rajasthan, India. She has cleared UGC NET twice (Dec 2015 and Dec 2019). Her academic records include MCom in Business Administration and BBM. She has a research experience of 3 years and is currently serving as a guest faculty in the Department of Business Administration and holds five years' teaching experience. Her major area of interest revolves around Organisational Behavior, Marketing, Tourism, Human Resource Management, and Entrepreneurship. She has written a few research papers and chapters for well-known journals and books and also presented research papers at National and International conferences. She continues to remain dedicated to her research and academic works until she achieves excellence.

Pinaz Tiwari is a Research Scholar in the Department of Tourism and Hospitality Management in Jamia Millia Islamia, India. She has worked in the tourism sector for two years. She has expertise in customer management and has been involved in travel agents' capacity development projects in the organization. She has done her graduation in Commerce from Delhi University and completed her MBA in Tourism with a specialization in International Tourism Business. She cleared the NET-JRF exam conducted by UGC in 2018 and looks forward to making a career in teaching and creating a difference in the tourism industry with her research works. Her interest areas are in tourism marketing, destination management, tourism education, stakeholders' capacity development, and responsible tourism. She has contributed different chapters related to overtourism in different books under well-known publications.

Uygar Turgut is an English Teacher at Koycegiz District Directorate of National Education. During his sixteen-year career teaching at state schools, he has coordinated different European School Projects such as Commenius and Erasmus+. He completed a Masters of Science degree in Tourism Management at Muğla Sıtkı Koçman University. He is now doing his Doctorate in Tourism Management at the same university.

Eray Turpcu is an Assistant Professor. Turpcu completed his undergraduate education in 2010 at Mustafa Kemal University, Department of Tourism and Hotel Management. He also completed his Master's degree in 2012 at Mustafa Kemal University, Department of Tourism and Hotel Management, and his PhD Degree in

2017 at Atatürk University, Department of Tourism Management. He started his academic life as a Lecturer and worked from 2012 to 2017. He was appointed as an Assistant Professor in 2018 at the Department of Tourism Management in Bulancak Kadir Karabaş School of Applied Sciences at Giresun University. His study fields include tourism marketing, destination management, and current issues related to the tourism industry.

Mehreen Waheed is a PhD Scholar of Management Sciences in the School of Business & Economics (SBE) at the University of Management & Technology (UMT) Lahore, Pakistan. Her areas of interest reside in Knowledge Management, Service System Management, Business Research, Organizational Behavior, Management & Marketing, and Tourism, etc. She has published one Research Paper in the International Journal of Business Competition & Growth Vol. 7, No. 1, 2020, title "Exploring divergence and convergence of training and development using service-dominant logic". She is serving as Lecturer in Management Sciences Department at NUML (National University of Modern Languages) Lahore, Pakistan. She has Experience Teaching in different universities. Her Major Subjects are Business Research Methods, Qualitative Research, Knowledge Management, Learning Organization, Organizational Behavior, Marketing Management, Business, and Human Resource Management.

Venkata Rao Yedla is a Full Professor of Tourism and Travel Management at Pondicherry Central University of Puducherry Union Territory at Puducherry. He received MHRM from Andhra University and MBA from the University of Madras and completed FDP from Indian Institute of Management, Ahmedabad. He obtained PhD in Management from Tezpur Central University, Assam. Dr. Rao has published seven books in the area of Management, Tourism Business, Tourism Entrepreneurship, Travel Agency and Tour Operations, Compensation Management, Destination Planning, and Development. Eighteen of his research papers were published in edited books, and about twenty seven papers were published in reputed national and international journals. He organized national and international conferences and published conference proceedings as edited books.

Part I
Overtourism and Technology

Chapter 1
Theorizing New Technologies as Potential Incremental Management Tools to Tackle Overtourism

Hugues Séraphin

Abstract As a global issue, Overtourism has highlighted three main things: (1) Tourism is not the cure for all socio-economic issues faced by destinations; (2) sustainability is still an issue to be overcome by the tourism industry; and (3) incremental strategies (as opposed to radical strategies) are not the way forward for the sustainability of the industry. As a result, this chapter is suggesting that new technologies (incremental strategies) could contribute to tackling overtourism, but only if substantiated with a driver of deep and meaningful changes (radical changes) such as an external shock (COVID-19, for instance). In that case, new technologies could be assimilated to an amplifier and the external shock as a driver. This conceptual chapter, based on secondary data (literature review), provides an overview of research on overtourism, and new technologies in tourism. The chapter then provides evidence that new technologies only work as enablers but not as drivers of changes. Finally, taking wedding tourism as an example, this chapter shows that as opposed to incremental strategies (new technologies), COVID-19, as an external shock, is pushing the industry towards radical changes in which technology could potentially play a significant role. This study is offering a strategy (theory) to tackle overtourism, which has not been suggested before, namely combining incremental (enablers) and radical (drivers) strategies. This approach falls under the ambidextrous management approach.

Keywords Overtourism · COVID-19 · New technologies · Innovations · Wedding tourism

---

H. Séraphin (✉)
Department of Marketing, Event Management and Project Management, University of Winchester, Winchester, UK
e-mail: Hugues.Seraphin@winchester.ac.uk

A. Hassan, A. Sharma (eds.), *Overtourism, Technology Solutions and Decimated Destinations*, https://doi.org/10.1007/978-981-16-2474-2_1

Introduction

The term innovation, which is to be associated with the terms "renewed"; "changed"; and "growth", is considered as important for the sustainability of the tourism industry and can either be incremental or radical (Brooker & Joppe, 2014). Incremental innovations are changes that are reactionary, as opposed to proactive, with the objective of maintaining a certain status-quo, rather than triggering a proper change of direction. The objective is, in general short-term (Brooker & Joppe, 2014). As for radical innovations, they are proactive, and disrupt current conventions and are based on a constant reinvention of what is currently done, and are also being unconcerned with previous practices (Brooker & Joppe, 2014). At the moment, 'incremental improvements are the norm in the tourism sector, and radical innovation is an abnormality' (Brooker & Joppe, 2014: 500). The use of technology to deal with issues faced by destinations are to be considered as incremental innovations or changes, as has happened as a result of a context. Indeed, 'the rapid and dynamic rate of urbanization has resulted in a need to find sustainable ways of dealing with the excessive strains and pressures that come to bear on existing infrastructures and relationships. Increasingly during the twenty-first-century, policymakers have turned to technological solutions to deal with these challenges' (Stokes et al., 2015: 3). Smart cities, which are cities incorporating technology-based solutions in their planning development and operation (Stokes et al., 2015), are examples (among others) of this contemporary use of technology.

The development of technologies has led to the emergence of the concept of smart cities, which in turn has led to the concept of Smart Tourism Destinations (Buhalis & Amaranggana, 2014), and also mobile living, facilitated with the evolution of the mobile web (Yeoman, 2013). Boes et al. (2016) are arguing that technology is an enabler for smart tourism destinations. Along this line of thoughts, Jovicic (2019), is arguing that technology contributes to achieving successful market valorization of destinations, as technology can help a destination to address both the needs of customers and subsequently increase their competitiveness (Buhalis & Amaranggana, 2014), but also the needs of locals in terms of quality of life (Boes et al., 2016). Technology can also contribute to educating visitors about a destination. This is all the more important for destinations with a negative image (Séraphin et al., 2016). However, despite the reliance on destinations such as China (Li et al., 2017) on technology (Gretzel et al., 2015), Boes et al. (2016) are also highlighting the fact that alone technology, which is an enabler, is insufficient to introduce smartness, as other elements such as leadership, social capital, and human capital are necessary (Fig. 1.1).

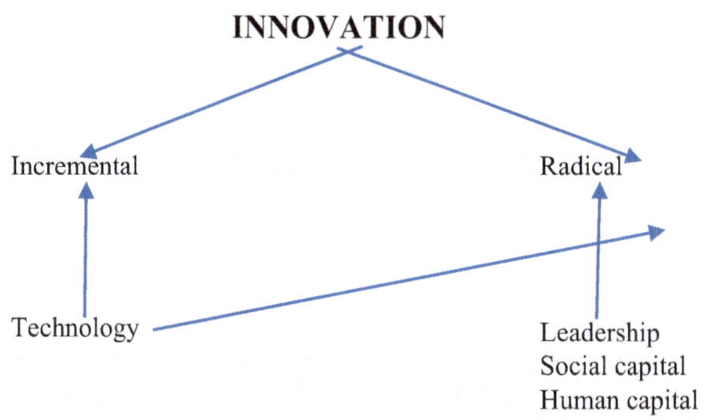

Fig. 1.1 The role of technology in tourism innovations. (Source: The author, 2020)

Overtourism Versus COVID-19

Internal Shock (Overtourism)

Overtourism, which is the concentration of visitors in some of the most popular destinations (Milano et al., 2019), was a worldwide issue, and not just the issue of major tourist destinations in the world. Dodds and Butler (2020) compared overtourism to a volcano that has been threatening to erupt for years, highlighting the fact that sustainability still has not been achieved in the tourism industry (Dodds & Butler, 2020). Even unexpected places such as the Himalayas; Rio de Janeiro favelas, etc. have become overvisited tourist destinations (Milano et al., 2019). As a destination, Venice is epitomizing the negative impacts of overtourism (Séraphin et al., 2018). A destination such as Chin has gone from nowhere to being one of the most visited destination in the world (Milano et al., 2019). Among the reasons at the origin of overtourism is the increase of population mobility facilitated with the development of transport, and more specifically, low-cost carriers; the lack of proactive management of destinations by governments; etc. (Milano et al., 2019). The consequences of overtourism can be mentioned, water shortages, traffic congestion, sewage problem, decrease of quality of life of local residents, tourismphobia, etc. (Dodds & Butler, 2020; Milano et al., 2019). That said, it is also important to mention the fact that there is a discrepancy among visitors in terms of popularity with visitors (Séraphin et al., 2020). Indeed, Milano et al. (2019) explained that if overtourism is very often pointed as an issue, under-visited destinations would love to be suffering from overtourism. Among the strategies suggested to tackle overtourism could be mentioned the use of technology (social media); the closure of popular sites; the development of policies by governments; tourism education of children who will be future leaders and tourists; etc. (Séraphin et al., 2020).

External Shock (COVID-19)

With the breakout of the 'first real global health emergency experienced by many members of today's society' (Jamal & Budke, 2020: 2), traveling and gathering have been banned (Jamal & Budke, 2020), causing all destinations in the world to suffer financially (Chaney & Séraphin, 2021). Despite the negative impacts of COVID-19 on destinations, some positive aspects have been identified, among these are the empowerment of children (Séraphin, 2021a); the opportunity to review some practices in the tourism industry, and to adopt new ones (Prideaux et al., 2020); etc. Along these lines, Chaney and Séraphin (2021) even argued that COVID-19 is offering the world an opportunity to adopt radical changes to tackle overtourism. Chaney and Séraphin (2021) even added that these changes (radical), if well-executed, can lead to the important and long-term transformation of the industry. As an example of radical change in the industry, could be mentioned the fact that Lourdes Pilgrimages was for the first time delivered online (Chaney & Séraphin, 2021). These radical changes are needed because Séraphin and Dosquet (2021) are arguing that COVID-19 has not led to the end of overtourism but to its mutation into something different, namely second home tourism. That said, despite some limitations highlighted by Séraphin et Dosquet (2021), COVID-19 has pretty much put an end to (over)tourism everywhere in the world (Jamal & Budke, 2020). Wedding tourism, which could to some extent be connected to overtourism (Del Chiappa & Fortezza, 2015; Ritter et al., 2020), has pretty much reached an end with the breakout of COVID-19. This point will be discussed more in-depth in the following section.

Marriage A-la-Mode COVID-19: From Wedding Tourism to "Wedding Staycation"

Overview

In *Marriage-A-la-Mode*, Hogarth painted the story of a marriage between the daughter of a wealthy bourgeois with the son of a poor nobleman. The purpose of this alliance was to ennoble the bourgeois and consolidate the social status of the nobleman. The painting also provides an overview of the reception room; attendees; and equally important, depicts a marriage party in accordance with the customs of the higher nobility (Bush, 2001). Hogarth's work was connecting contemporary life and events. As a result, he was considered an illustrator of contemporary life and morality (Bush, 2001).

This section of the chapter is investigating *Marriage-A-la-Mode* COVID-19, as the breakout of the virus has disrupted lives, habits, rituals, economies, etc. everywhere in the world (Gossling et al., 2020; Hall et al., 2020; Higgins-Desbiolles, 2020; Jamal & Budke, 2020; Yang et al., 2020). As far as weddings are concerned, in the UK, for instance, nearly three-quarters have been rescheduled; 44% of hen and stag dos have been canceled; etc. (Pye, 2020 [Online]). It is all the more relevant to consider weddings in the context of COVID-19, as this ritual has often been associated with a breakout of illness (Lappi et al., 2013). More specifically, this section is investigating how weddings are going to adapt to the COVID-19 context, as Getz (2012: 383) explained that: 'because events have met essentials needs throughout history, in all civilizations, it is safe to conclude that they will endure and adapt'. As a result, the focus of the study is on the typology of attendees and the potential impacts on wedding tourism, a niche form of tourism that has been booming over the past two decades (Del Chiappa & Fortezza, 2015), and which has also played a significant role in the hospitality sector (Ritter et al., 2020). This choice of focus is based on the fact that it appears that after a global pandemic, the typology of attendees at events are amongst the most affected elements (Getz, 2012), which is going to naturally impact wedding tourism, an affection based motivation to travel (Chen et al., 2020), largely influenced by family, friends, and experience of traveling (Ritter et al., 2020).

Weddings and Tourism

Weddings, which are private events or functions, but also the celebration of consumerism, romance (Castren & Maillochon, 2009), are celebrated in all communities in the world (Dawson & Bassett, 2018; Duncan, 2016; Getz, 2012). Despite the evolution of society, weddings have maintained some traditional aspects (Currie, 2019). There is such an expectation in society for individuals to get married that products and services to match individuals who are single are proliferating (Ahuvia & Adelman, 1992). It is also worth highlighting the fact that wedding is also a big business, and as such represent a significant part of the event industry. In the UK, for instance, the industry generates each year roughly £10 billion (Dawson & Bassett, 2018). Wedding tourism indicates the flow of individuals generated by weddings to a destination where neither the bride nor the groom lives in order to fulfill some family obligations; reproduce social networks, and display reciprocal love (Bertella, 2015). Also called an overseas wedding, wedding tourism, which combines wedding, honeymoon, and holidays are for couples an alternative to expensive traditional weddings (Bertella, 2015; Seebaluck et al., 2015). Many products and services are developing around this new trend. Some destinations are also specializing themselves in this emerging form of tourism. Apart from the reasons previously mentioned, wedding tourism is getting popular because of the ease of formalities in some destinations; the variety of products and services available; and the uniqueness of the experience (Seebaluck et al., 2015).

COVID-19, Weddings, and Wedding Tourism

COVID-19 could be said to have impacted the happenings of some rites and rituals (Séraphin, 2021b), defined as patterned forms of behaviors within the occurrence of events (Getz, 2012). COVID-19 is also limiting the development of social capital, which includes a sense of belonging, the development of loyalty, networking; (Huang et al., 2016; Raj et al., 2017; Séraphin, 2021b). As a wedding, and wedding tourism are associated with terms such as holidays; networking; the flow of individuals (Bertella, 2015; Seebaluck et al., 2015); and COVID-19 associated with terms such as limitation; social distancing; disruption of habits; rituals; etc. (Gossling et al., 2020; Hall et al., 2020; Higgins-Desbiolles, 2020; Jamal & Budke, 2020; Séraphin, 2021b; Yang et al., 2020), this study is arguing that the breakout of the virus is going to put an end to the growing trend of wedding tourism (Del Chiappa & Fortezza, 2015), and is going to replace it with what this study is self-acclaiming: 'wedding staycation'. This hypothesis is based on the fact that: First, the wedding is not only an important ritual in society but an expectation (Ahuvia & Adelman, 1992; Getz, 2012); second; small-at-home ceremonies are often considered as the most appropriate to avoid any incident (Bruce, 2012; Lewin, 1998); and finally, staycation is an alternative to travel in time of crisis (James et al., 2016). This study is also arguing that the incomes generated by the now bespoken 'wedding staycation' are going to be small, compared to what wedding tourism used to be. This second hypothesis is based on the Theory of Planned Behaviour (TPB), which argues that individuals make decisions to engage in specific behaviors by evaluating the information available to them (Ajzen & Fishbein, 1980).

Guest list: Pre-COVID-19 – Weddings are also the encounter between two segments: the family segment and the friend segment. However, the main difference between both universes is that at weddings, family segments are invited because it is perceived almost as a social obligation, whereas friends are invited because their presence is wanted by both brides and grooms (Maillochon, 2012). Other people are also invited, not because their presence is particularly wanted, but because brides and grooms feel obliged to invite them because those people either: invited them at their wedding; they are close friends of parents; they belong to their social and professional circle; they are former friends who are now professionally successful, and worth having back in their lives; to rebuilt relationships among members of the family (Maillochon, 2012). The typology (determined by Maillochon, 2012) of guests at weddings (in a non-crisis context) is as follows: (1) blood related (2) legally related (3) have strong positive feelings (4) share positive experiences and memories. By identifying criteria to select individuals to invite at a wedding, Maillochon (2012), used without naming it that way, a multi-criteria decision analysis (MCDA). It is a method to support decision making in problems involving several criteria and choices (Botti & Peypoch, 2013).

Guest list: COVID-19 context – Criteria established by Maillochon (2012) have to be reviewed in light of the current context. Subsequently, this study is acknowledging the criteria identified by Maillochon (2012), which are considered as core criteria, while adding to them, other criteria (contextual) dictated by the current crisis context. The approach adopted in this study is supported by Haywood (2020), who argues that in this crisis context, some things have to change, and others have to remain constant. Those contextual criteria are as follows: (1) Mobility – With the outbreak of the virus, mobility has been reconsidered (Renaud, 2020), with a greater tendency for individuals to think more in terms of community and to stay local (Lapointe, 2020). This trend was already predicted by Getz (2012). Instead of organizing their wedding abroad as it was more and more the case (Bertella, 2015; Bertella, 2016), couples are likely to go for more local options. As a result, the geographical location of the place of residence of guests will now be taken into consideration when 'short listing' them. This is for health and safety reasons. Indeed, Séraphin and Dosquet (2020) have evidenced that with the outbreak of COVID-19, individuals living in major cities when moving to their second home in the countryside or seaside areas have contributed to the spread of the virus. Hence the reasons borders have been closed (Mostafanezhad, 2020). This is also for environmental reasons. The COVID-19 outbreak offers an opportunity to rethink travel and strategies to combat climate change (Prideaux et al., 2020). (2) Health condition – The outbreak of the virus has highlighted how important health is (Cooper & Alterman, 2020). Individuals with pre-existing illness (cardiovascular disease, diabetes, cancer, hypertension, etc.) are at a higher risk of dying from a COVID-19 infection (Worldometer, 2020 [Online]). This is to be added to the potential breakout of illness related to food at weddings (Lappi et al., 2013). (3) Age – Data shows that the older individuals who have contracted the virus, the more likely they are to die. The 70+ years are the most affected (Statista, 2020 [Online]). The updated multi-criteria decision framework developed by this study is as follows (Fig. 1.2).

The potential shift from wedding tourism to wedding staycation is another example of the transformational impact of COVID-19 on the industry (Sigala, 2020), which does not seem very beneficial as some services such as wedding planning; or hotels reception rooms; etc. might not be needed to the extent they were needed before. What is suggested here is in line with the substantial fall in valuation in many sectors of the tourism industry (Sharma & Nicolau, 2020). Additionally, as the distinction between what is traditional and modern in weddings is blurred, and also because the tradition in the wedding is constantly being re-invented (Carter & Duncan, 2018), COVID-19, is potentially going to play a role in this constant reinvention on weddings, and potentially enforce the transition to 'wedding staycation'. The same way there are determinants (educational; race; age; money; etc.) for who marries whom (Choi, 2019; Sassler & Lichter, 2020; Smock et al., 2005), this study has also presented one of the potential determinants of wedding staycation, basically, whom are likely to be invited (or not) at weddings. Existing literature does not really cover this aspect of weddings, despite the importance of guests in making the day successful (Castren & Maillochon, 2009), but also in terms of flow to the wedding destination, which equates to more business to the stakeholders in the sector. One of the options for wedding tourism to overcome the negative impacts of the crisis is for all stakeholders in the sector to rebuild the connection with their target through shared emotions, which contributes to establishing emotional attachment

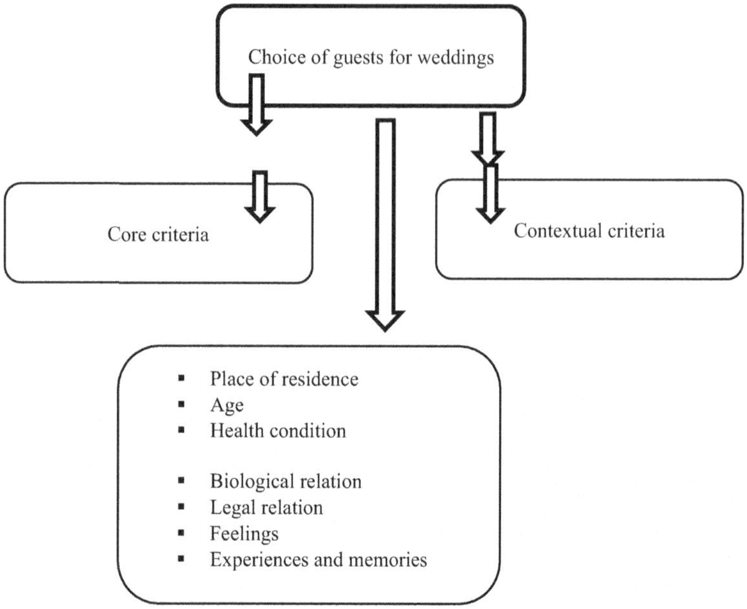

Fig. 1.2 Wedding guests list multi-criteria decision framework. (Source: The author, 2020)

(Hang et al., 2020). For this recovery strategy to work, team effort (all stakeholders in the wedding sector) could help to understand (better and quicker): the expectations of customers; their perception of the context; and equally important to innovate, and as a result (re)boost their competitiveness (Amaro & Duarte, 2015; Jackson, 2006; Kibicho, 2008; Novelli et al., 2006).

Conclusion

Incremental and radical innovations are the main types of changes. In the tourism industry, incremental innovation is the most common approach adopted. Despite the fact, technology is a key component of innovation in tourism, tourism, and non-tourism literature, both agree that it is just an enabler. In other words, technology is to be considered as a form of incremental innovation or change, as opposed to radical (Fig. 1.1). Subsequently, the first proposition (P1) of this chapter is that internal shocks to the tourism industry, such as overtourism, are not enough on their own to trigger radical changes. The second proposition (P2) of this chapter is that external shocks such as COVID-19, are more likely to trigger radical changes towards sustainability in the industry. Indeed, overtourism (internal shock) has been a long-lasting issue which only stopped immediately with the breakout of COVID-19 (external shock). Based on this, it could therefore be concluded that internal shocks may lead to changes, but they take a long time to be visible and/or be implemented.

As for external shocks, they lead to immediate changes. The third and final proposition of this chapter (P3) is that the most suitable innovation should be a combine a radical and incremental, as a result of shocks that are mild (a mixture of internal and external shocks). This proposition is a reminder of the importance of ambidextrous management in tourism, which is an approach closely related to sustainability and improvement of performance in the tourism industry (Vo-Thanh et al., 2020). Beyond the topic of wedding tourism, this chapter is also on evolutionary psychology, as it discusses to some extent, tourism customers' psyche (Kock et al., 2020). The conclusions of this study are in line with Kock et al. (2020: 10), who argue that "the Coronavirus pandemic has revived people's existential anxieties by reminding them of their physical and economic vulnerability. It is during these times that our deeply-rooted evolutionary protection mechanisms are activated".

References

Ahuvia, A. C., & Adelman, M. B. (1992). Formal intermediaries in the marriage market: A typology and review. *Journal of Marriage and the Family, 54*(2), 452–463.

Ajzen, I., & Fishbein, M. (1980). *Understanding attitudes and predicting social behaviour.* Prentice-Hall.

Amaro, S., & Duarte, P. (2015). An integrative model of consumers' intentions to purchase travel online. *Tourism Management, 46*, 64–79.

Bertella, G. (2016). Celebrating the family abroad: The wedding tourism experience. *Annals of Leisure Research, 18*(3), 397–413.

Bertella, G. (2015). The emergence of Tuscany as a wedding destination: The role of local wedding planners. *Tourism Planning & Development, 14*(1), 1–14.

Boes, K., Buhalis, D., & Inversini, A. (2016). Smart tourism destinations: Ecosystems for tourism destination competitiveness. *International Journal of Tourism Cities, 2*(2), 108–124.

Botti, L., & Peypoch, N. (2013). Multi-criteria ELECTRE method and destination competitiveness. *Tourism Management Perspectives, 6*, 108–113.

Brooker, E., & Joppe, M. (2014). Developing a tourism innovation typology: Leveraging liminal insights. *Journal of Travel Research, 53*(4), 500–508.

Buhalis, D., & Amaranggana, A. (2014). Smart tourism destinations. In Z. Xiang & I. Tussyadiah (Eds.), *Information and communication technologies in tourism* (pp. 553–564). Springer.

Bush, W. (2001). Hogarth's marriage-A-la-mode: The dialectic between precision and ambiguity. In D. Bindman, F. Ogee, & W. Peter (Eds.), *Hogarth: Representing nature's machines* (pp. 195–218). Manchester University Press.

Bruce, K.R.E. (2012). *Doing coupledom: Imagining, managing and performing relationality in contemporary wedding and civil partnership rituals*. PhD thesis, University of Southampton.

Carter, J., & Duncan, S. (2018). Individualised conformity: Creating a wedding. In *Reinventing couples* (Palgrave Macmillan Studies in Family and Intimate Life). Palgrave Macmillan.

Castren, A. M., & Maillochon, F. (2009). Who chooses the wedding guests, the couple or the family? *European Societies, 11*(3), 369–389.

Chaney, D., & Séraphin, H. (2021). Covid-19 crisis as an unexpected opportunity to adopt radical changes to tackle overtourism. *Anatolia*. https://doi.org/10.1080/13032917.2020.1857804

Chen, G., So, K. K., Poomchaisuwan, M., & Hu, X. (2020). Examining affection-based travel: Development and validation of a measurement scale for honeymooners' motivation. *Journal of Destination Marketing & Management, 17*, 100452.

Choi, K. H. (2019). Racial diversity in the marital assimilation of Hispanics. *Journal of Marriage and the Family, 82*(2), 675–690.
Cooper, J. A., & Alterman, D. H. (2020). Cancelling March Madness exposes opportunities for a more sustainable sports tourism economy. *Tourism Geographies, 22*(3), 525–535.
Currie, D. H. (2019). Here comes the bride: The making of a modern traditional wedding in western culture. *Journal of Comparative Family Studies, 24*(3), 403–421.
Dawson, R., & Bassett, D. (2018). *Event planning and management: Principles, planning and practice*. Kogan Page.
Del Chiappa, G., & Fortezza, F. (2015). Wedding-based tourism development: Insights from an Italian context. In *Marketing places and spaces* (Advances in Culture, Tourism and Hospitality Research, Vol. 10) (pp. 61–74). Emerald Group Publishing Limited.
Dodds, R., & Butler, R. (Eds.). (2020). *Overtourism*. De Gruyter.
Duncan, M. L. (2016). Wedding industry. *Encyclopedia of Family Studies*. https://doi.org/10.1002/9781119085621.wbefs548
Getz, D. (2012). *Event studies, theory, research, policies for planned events*. Routledge.
Gossling, S., Scott, D., & Hall, C. M. (2020). Pandemics, tourism and global change: A rapid assessment of COVID-19. *Journal of Sustainable Tourism, 29*(1), 1–20.
Gretzel, U., Sigala, M., Xiang, Z., & Koo, C. (2015). Smart tourism: Foundations and developments. *Electron Markets, 25*, 179–188.
Hall, C. M., Scott, D., & Gossling, S. (2020). Pandemics, transformations and tourism: Be careful what you wish for. *Tourism Geographies, 22*(3), 577–598.
Hang, H., Aroean, L., & Chen, Z. (2020). Building emotional attachment during COVID-19. *Annals of Tourism Research, 83*, 103006.
Haywood, K. M. (2020). A post-COVID future: tourism community re-imagined and enabled. *Tourism Geographies, 22*(3), 599–609.
Higgins-Desbiolles, F. (2020). Socialising tourism for social and ecological justice after COVID-19. *Tourism Geographies, 22*(3), 610–623.
Huang, W. J., Ramshaw, G., & Norman, W. C. (2016). Homecoming or tourism? Diaspora tourism experience of second-generation immigrants. *Tourism Geographies, 18*(1), 59–79.
Jackson, J. (2006). Developing regional tourism in China: The potential for activating business clusters in a socialist market economy. *Tourism Management, 27*(4), 695–706.
Jamal, T., & Budke, C. (2020). Tourism in a world with pandemics: Local-global responsibility and action. *Journal of Tourism Futures, 6*(2), 181–188.
James, A., Ravichandran, S., Chuang, N. K., & Bolden, E. (2016). Using lifestyle analysis to develop lodging packages for staycation travelers: An exploratory study. *Journal of Quality Assurance in Hospitality & Tourism, 18*(4), 387–415.
Jovicic, D. Z. (2019). From the traditional understanding of tourism destination to the smart tourism destination. *Current Issues in Tourism, 22*(3), 276–282.
Kibicho, W. (2008). Community-based tourism: A factor-cluster segmentation approach. *Journal of SustainableTourism, 16*(2), 211–231.
Kock, F., Norfelt, A., Josiassen, A., Assaf, A. G., & Tsionas, M. G. (2020). Understanding the COVID-19 tourist psyche: The evolutionary tourism paradigm. *Annals of Tourism Research, 85*, 103053.
Lapointe, D. (2020). Reconnecting tourism after COVID-19: The paradox of alterity in tourism areas. *Tourism Geographies, 22*(3), 633–638.
Lappi, V., Archer, J. R., Cebelinski, E., Leano, F., Besser, J. M., Klos, R. F., Medus, C., Smith, K. E., Fitzgerald, C., & Davis, J. P. (2013). An outbreak of foodborne illness among attendees of a wedding reception in Wisconsin likely caused by Arcobacter butzleri. *Foodborne Pathogens and Disease, 10*(3), 250–255.
Lewin, E. (1998). *Recognising ourselves*. Columbia University Press.
Li, Y., Hu, C., Huang, C., & Duan, L. (2017). The concept of smart tourism in the context of tourism information services. *Tourism Management, 58*, 293–300.

Maillochon, F. (2012). Le Coeur et la raison. Amis et parents invites au marriage. *Geneses, 83*, 93–117.

Milano, C., Cheer, J. M., & Novelli, M. (2019). *Overtourism*. CABI.

Mostafanezhad, M. (2020). COVID-19 is an unnatural disaster: Hope in revelatory moments of crisis. *Tourism Geographies*, *22*(3), 639–645.

Novelli, M., Schmitz, B., & Spencer, T. (2006). Networks, clusters and innovation in tourism: A UK experience. *Tourism Management, 27*(6), 1141–1152.

Prideaux, B., Thompson, M., & Pabel, A. (2020). Lessons from COVID-19 can prepare global tourism for the economic transformation needed to combat climate change. *Tourism Geographies*, *22*(3), 667–678.

Pye, H. (2020). *Hitched survey*. Retrieved from: https://www.hitched.co.uk/wedding-planning/news/covid-19-wedding-couple-survey/. Accessed 18 Sept 2020.

Raj, R., Walters, P., & Rashid, T. (2017). *Events management*. Sage.

Renaud, L. (2020). Reconsidering global mobility – distancing from mass cruise tourism in the aftermath of COVID-19. *Tourism Geographies*, *22*(3), 679–689.

Ritter, M., Boger, C. A., & Draper, J. (2020). The effects of time living away from home and family support on intention to host a destination wedding. *Journal of Destination Marketing & Management*, *17*, 100461.

Sassler, S., & Lichter, D. T. (2020). Cohabitation and marriage: Complexity and diversity in union-formation patterns. *Journal of Marriage and Family, 82*(1), 35–61.

Seebaluck, N. V., Munhurrun, P. R., Naidoo, P., & Rughoonauth, P. (2015). An analysis of the push and pull motives for choosing Mauritius as "the" wedding destination. *Social and Behavioural Sciences, 175*, 201–209.

Séraphin, H., Sheeran, P., & Pilato, M. (2018). Over-tourism and the fall of Venice as a destination. *Journal of Destination Marketing & Management, 9*, 374–376.

Séraphin, H., & Dosquet, F. (2020). Mountain tourism and second home tourism as post COVID-19 lockdown placebo? *Worldwide Hospitality and Tourism Themes*, *12*(4), 485–500.

Séraphin, H. (2021a). COVID-19 and the acknowledgement of children as stakeholders of the tourism industry. *Anatolia*, *32*(1), 152–156.

Séraphin, H. (2021b). COVID-19: An opportunity to review existing grounded theories in event studies. *Journal of Convention and Event Tourism, 22*(1), 3–35.

Séraphin, H., Butcher, J., & Korstanje, M. (2016). Challenging the negative images of Haiti at a pre-visit stage using visual online learning materials. *Journal of Policy Research in Tourism, Leisure and Events, 9*(2), 169–181.

Séraphin, H., & Dosquet, F. (2021). COVID-19 and the mutation of overtourism. The case of France. In H. Séraphin & A. Yallop (Eds.), *Overtourism and tourism education. A strategy for sustainable tourism futures*. Routledge.

Séraphin, H., Gladkikh, T., & Vothanh, T. (Eds.). (2020). *Overtourism: Causes, implications and solutions*. Palgrave Macmillan.

Sharma, A., & Nicolau, J. L. (2020). An open market valuation of the effects of COVID-19 on the travel and tourism industry. *Annals of Tourism Research, 83*, 102990.

Sigala, M. (2020). Tourism and COVID-19: Impacts and implications for advancing and resetting industry and research. *Journal of Business Research, 17*, 312–321.

Smock, P., Manning, W. D., & Porter, M. (2005). Everything's there except money: How money shapes decisions to marry among cohabitors. *Journal of Marriage and Family, 67*(3), 680–696.

Statista. (2020). *Study: Elderly most at risk from the Coronavirus*. Retrieved from: https://www.statista.com/chart/20860/coronavirus-fatality-rate-by-age/. Accessed 18 Sept 2020.

Stokes, P., Larson, M., Russell, N., Adderley, S., Moore, N., Mathews, M., Smith, S., Lichy, J., Scott, P., Ward, T., & Brindley, C. (2015). Smart cities. Dynamic sustainability issues and challenges for 'old world' economies: A case from the United Kingdom. *Dynamic Relationships Management Journal, 4*(2), 3–22.

Vo-Thanh, T., Séraphin, H, Okumus, F., & Koseoglu, M.A. (2020) Organizational Ambidexterity in Tourism Research: A Systematic Review.*Tourism Analysis, 25*(1), 137–152.

Worldometer. (2020). *Age, sex, existing conditions of COVID-19 cases and deaths*. Retrieved from: https://www.worldometers.info/coronavirus/coronavirus-age-sex-demographics/. Accessed: 18 Sept 2020.

Yang, Y., Zhang, H., & Chen, X. (2020). Coronavirus pandemic and tourism: Dynamic stochastic general equilibrium modelling of infectious disease outbreak. *Annals of Tourism Research, 83*, 102913.

Yeoman, I. (2013). A futurist's thoughts on consumer trend shaping future festivals and events. *International Journal of Events and Festivals Management, 4*(3), 249–260.

Chapter 2
Can Technology-Tourism Interaction Promote Responsible Practice Preventing Overtourism!

Samik Ray

Abstract Technology and tourism are intertwined. The interaction between the two then remains instrumental to the tourism growth-mobility making changes in its production, consumption behavior, and business ecosystem. Primarily interaction transformed scattered and unorganized travel into orderly and organized tourism. It also assisted organized tourism to spread amid mass milieu beyond the elitist edge in the post-1950s. Further, it played a crucial role in a qualitative and quantitative transformation of tourism from mass to conscious and sustainable and eventually to responsible practices. Qualitative and quantitative changes from scattered to organized mobility, then mass-mobility beyond carrying capacity, controlled or sustainable-mobility, and finally to the responsible-mobility are technology-tourism interaction driven. The post-1950 increasing mobility in the tourism sector became the indicator of the economic success evident in socio-economic makeover at many places overtly. The overtourism phenomenon emerged as the logical extension of tourism's rapid increment, then. With the growth of information and communication technology (ICT) dependence within civil society and tourism, this phenomenon and its countercultural reaction turned global by the early twenty-first century. Consequently, a responsible practice emerged comprehending neither limitless growth nor controlled mobility can make a place better to live and better for visit and experience. The increasing application of Virtual Reality (VR) technology during the 2020 pandemic also makes the World perceive this new understanding. The present chapter will focus on how technology-tourism interaction could translate the new understanding into practice resisting the overtourism phenomenon.

Keywords Contradictions · Growth · Interaction · Overtourism · Responsible practice · Technology

---

S. Ray (✉)
Regional Level Guide, Department of Tourism, The Government of India, Kolkata, India

A. Hassan, A. Sharma (eds.), *Overtourism, Technology Solutions and Decimated Destinations*, https://doi.org/10.1007/978-981-16-2474-2_2

Introduction

Technology advances in civil society contribute to growing affluence (World Travel & Tourism Council, 2017) and better quality living facilitating decrement in work-hour; work hazards; physical strain and increment of disposable income; leisure-time and recreation pursuits; comfortable life. The surplus leisure-time sanction created enormous potentials of leisure and recreation linked tourism growth in the post-1950 world. Applying technological innovations tourism entrepreneurs attempted to translate potential into an opportunity to develop more attractive, comfortable, efficient, differentiated, competent, and competitive travel-tourism service offers. It led tourism to rise vertically and expand horizontally at a faster rate. Hence interaction between tourism and technology innovations contributes to the high growth of tourism. The vertical escalation in tourist footfall and the horizontal expansion of the tourism business multiplying the tourism growth at the outset and then overgrowth compose a new narrative of destinations'economic makeover. The horizontal and vertical mobility en masse sparks off the overtourism issues. By the end of the 2010s decade, overtourism becomes more of a perceived sensibility response of the affected community towards tourist crowding and over-expansion of tourism business than mere data. The contribution of interaction between technology advances and tourism towards the growth and then overgrowth of tourism becomes crucial.

The outbreak of the 2020 pandemic was a shock to the tourism economy as tourism incurred a loss of trillion US$s and million jobs. Stakeholders become puzzled and bewildered. There is hardly any clue for concrete and comprehensive revival plan. Proposals for post-pandemic tourism revival come with a new format prioritizing the notions of distancing, health safety protocols, wellbeing, mindfulness, home proximity, and quality experience. Thus most people perceive that issues of overtourism will automatically be extraneous. Between April and October 2020, the US travel association and MMGY travel intelligence team survey the impact of COVID-19 on US travel behavior (MMGY, 2020) and find that the willingness to travel among respondents fluctuates with the growth and decline of COVID-19 positive cases. Leisure travel expectation oscillates between 36% and 66% and business travel intentions between more than 30% and less than 40%. Global reactions to the ease of lockdown or stay-home restriction are mixed and sometimes contradictory. Some support the easing decisions while some oppose. Some fear a further surge. Some are delighted, thus people from Athens to Amsterdam flock to museums, bars, and beaches. Californians flock to the beaches along 75 miles long coastline. Istanbul's tourist places return to business. Many Germans gather at local beaches for Sunbathe. Indeed, those phenomena and reports indicate that the relevance of overtourism will not automatically be off the point.

Early Approaches

Technology-tourism interaction does not receive much attention except a few recent research, although it is a diachronic phenomenon with multifaceted social implications. Indeed, the increased use of information and communication technology (ICT) in tourism prompts academic interests to study the issues and impacts of ICT applications in the tourism business and management. Among the general discussions, researches on scopes and prospects of ICT applications in tourism-hospitality (Ansel & Dyer, 1999; Griffin, 1998) are crucial. The majority of researchers deal with the issues and effects of ICT applications in promotion and marketing strategies (Choi et al., 2017; He et al., 2018; Kim et al., 2014; Liu & Attila, 2016; Ray, 2019), customer and operation management (Cho et al., 1996; Chathoth, 2007; Breukel & Go, 2009; Law et al., 2012; Leung & Law, 2012; Noone & Coulter, 2012; Vogt, 2010), demand-supply networking (Jadhav & Mundhe, 2011), destination imaging (Ray, 2018), service innovation and smart or e-business development (Arana & Leon, 2013; D'Ambra & Mistilis, 2009; Morosan, 2011; Tribe & Mkono, 2017; Tussyadiah & Zach, 2012; Mackay & Vogt, 2012; Jung et al., 2015), tourists behavior and organizational performance (Cheng & Cho, 2010; Choi et al., 2017; Vu et al., 2018; Vu et al., 2017), and value and competitive advantage creation (Kim et al., 2009; Lin, 2016; Minghetti & Buhalis, 2009; Straker & Wrigley, 2018; Yepes, 2014). Recent studies inquire about the role and impacts of ICT towards travel behavior changes and overtourism control (Condé Nast Traveller, 2019). Works on technology implications on tourism and hospitality (Hassan & Sharma, 2021), shifts and impacts of technology and tourism interaction over time (Ray, 2021a), social media role in the transformation of tourist behavior (İştin, 2021), the contribution of ICT in emotional branding (Moturu et al., 2021), impacts of online marketing (Albattat, 2021) and website designing (Türker, 2021), ICT based marketing (Singh & Dhankhar, 2021), and e-HRM applications (Said & Umachardran, 2021) also make crucial contributions. Hence studies on technology-tourism or technology-overtourism interaction haven't received much attention yet.

Technology Innovations – Overtourism Interaction

Overview

Overtourism or 'turismofobia' (Martin et al., 2018; Martins, 2018; Milano, 2017) is a complex phenomenon, began to grow in the late twentieth century and turned global by the present century, though destinations and society perceive its consequences of late. The conflict between the limitless tourism growth approach and its consequentialexperiences made the overtourism notion and perception emerged. Common mass senses its appearance in resource degradation and severe experience-deterioration (World Tourism Organization, 2018) to dwell in and visit (Goodwin,

2017). Thus a repulse or hostile feeling or counter-response of tourists and host communities to the tourism overgrowth exceeding carrying capacity limit is quite logical and natural. Interaction between tourism and technology innovations is a diachronic evolvement thus over time phenomenon. Since overtourism is the most critical phenomenon of post-1950 tourism, its interaction with technology also has direct bearings on its growth. Technology innovations and advancements turn crucial in overtourism-growth and its effect-experiences making qualitative and quantitative changes in tourism business ecology; travel phenomenon; tourist mobility (Ray, 2021a, 2021b).

Generally, interaction occurs with two-way effects upon participants with each other. Technology and tourism also affect each other while interacting. It is a diachronically proven fact. The seventeenth century Grand tour concept turned to be an organized holidaying or vacationing by the mid-nineteenth century when the inclusion of rail-cruise transit made qualitative changes in tour-offers and the printing-technology or print media advances set to develop better publicity; promotion; operation techniques. Adoption of technology innovations in tourism in the early and mid-twentieth century was evident in the better motel construction, application of multi-use architectural advances, the introduction of the luxury road; water; rail transports, initiation of luxurious cruise lines with stateroom/ cabin; a private lounge; spacious decks facilities, and the first passenger aircraft take-off etc. Conversely, the interaction with tourism led technology innovations to be equipped to acclimatize better travel-leisure logistics or consumption experience requirements and serve the cause of tourism's qualitative development and quantitative progression.

The post-1950 tourism too pushes technology innovations to become pertinent to its growth-mobility customizing advances to suit the tourist-friendly facilities and comfort as is evident in changes of surface and air transport technology, reservation system and many more. And technology innovations facilitate tourism's vertical growth and overgrowth alongside horizontal expanse and over-expanse advocating qualitative transformations in leisure-pleasure facilities; travel-logistics; business ecosystem. Effects of interaction between tourism and ICT innovations disrupt the existing tourism business ecosystem and lead ICT application to be more innovative and proactive towards the cause of rapid increment of tourism growth. Thus, post-1950-interaction features tourism-technology mutual dependency; tourism growth and overgrowth.

Interaction of any sort perceived best in the phenomena generated out of it. The rapid escalation in tourist mobility and fast expansion of the tourism business is the most crucial phenomenon, developed due to the post-1950 interaction between tourism and technology. It gradually leads the overtourism phenomenon to be emerged by the end of the twentieth century, although its impacts and reactions to counter the effects surface only in the twenty-first century. Interaction between tourism and most recent ICT innovations accelerates the overgrowth thus overtourism, making changes in the production-consumption system or business ecology (Ray, 2021a, 2021b) introducing co-evolve web relationship or value network (Allee, 2003) between demand and supply-side. Interaction between technology innovations and overtourism is thereby not independent of its generated outcome. Changes in the growth-mobility paradigm, transformation in the production and consumption-behavior pattern, and transmutations of the business ecosystem are the most visible outcome in this regard. Thus the interaction in progress between overtourism and technology innovations; adoptions; applications become evident in those changes, transformations, and transmutation.

Changes in Growth

The pre-1950 interaction between technology innovations and tourism crafted the vertical demand-growth of organized holidaying visible by quantity. Once quantitative growth was noticeable new entrepreneurs involved in the organized tourism venture. Horizontal expansion of tourism business set in progress as a logical and subsequent extension of the rise-phenomenon in the organized holidaying, then. Tourism thereby entered into the growth era. Applications of technological advances in transportation, communication, home appliances, and civic amenities restructured the post-world war civic lifestyle making civil society experienced better comfort in life, living, and leisure-recreation pursuit. Better comfort experience turned out to be the demand-stimulator for leisure-recreation-tourism in contemporary society. And demand-rise attracted more investments and new capital triggering better tourism growth potentials. Interaction outcome thereby contributes to the qualitative changes which stimulate vertical demand-rise. The rise facilitates vertical growth as evident in the increment of tourist footfall and horizontal expansion involving more entrepreneurs who in turn become interested to use technological advances for the cause of tourism. Hence tourism-technology interaction outcome-driven tourism growth is a never-ending cyclic motion (see Fig. 2.1).

Pre-1950 interaction outcomes were apparent in the qualitative transformations of tourism and served to be the precondition of post-1950 tourism expansion at

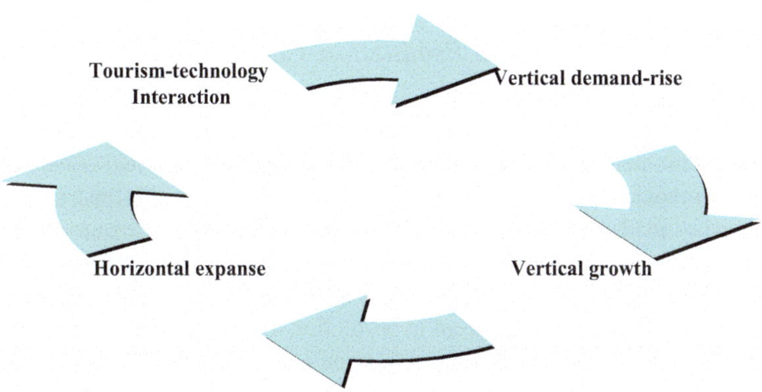

Fig. 2.1 Interaction outcome effect circulation. (Source: developed by the author)

mass magnitude featured by leisure-recreation-tourism demand escalation and fast entrepreneurial growth. Tourism entrepreneurs set to translate interaction outcome to the horizontal expanse and vertical growth opportunities by volume, then. The world economy registers faster qualitative transformations and quantitative growth of tourism-hospitality. Consequently, tourism's growth entered into the mass-mobility era marked by the vertical escalation of tourist footfalls at mass scale and horizontal expanse of tourism-entrepreneurial activity at a mass magnitude. And mass-mobility turns to be an economic success indicator in many destinations. Interaction, thereby, establishes a strong link to mass-scale quantitative growth.

The fast adoption of ICT innovations by civil society motivates tourism to interact with it since the late twentieth century. The overt outcome of the interaction is apparent in the rise of the overgrowth phenomenon. Indeed, the interaction outcome primarily contributes to the qualitative transformations or changes in the production-consumption system or business ecology (Ray, 2021a, 2021b) facilitating faster exchanges of static and dynamic tourism information and knowledge about destinations; tourism offers; services; competitive issues in real time scale between demand and supply-side through co-evolve web relationship or value network (Allee, 2003) among those (Ray, 2021a). Thus, through baring conventional spatial and temporal constraints of communication, tourists and entrepreneurs prepare themselves to gather, store, and disseminate information and knowledge. Traditional time-space concept independent communication between demand and supply-side contributes to the overgrowth of tourism and its vertical and horizontal mobility, the most dramatic and crucial outcome of the interaction between ICT and tourism.

Tourism entered into the overgrowth or high mobility era. Overgrowth in tourism referred to as overtourism with the rise of contradiction and conflict between the overgrowth phenomenon and counter feeling to its consequences marked by qualitative deterioration in experience to live in and visit, a logical extension of disruptions that occurred in the pre-overgrowth bio-socio-economic ecosystem or state of tourism. Hence the contribution of tourism-technology interaction outcomes to the changes in the tourism growth paradigm is a proven fact.

Changes in Promotion and Production-Consumption

Outcome-effects of the late nineteenth and early twentieth century better tourism logistics requirement driven interaction between travel-tourism and contemporary technology innovations in communication, transport and related infrastructure, and other logistics made simultaneous and remarkable changes in the tourism promotion; production; consumption, the closely associated chain of actions. Both the qualitative and quantitative transformations and paradigm shifts from existing to a new state (Ray, 2021a) were apparent in promotion techniques and process, tourism service and offers, and production-consumption pattern and behaviour.

The most crucial outcome effect in the travel-tourism pattern was the emergence of the organized tourism product by the end of the nineteenth century. The concept of modern tourism emerged with a shift from unorganized; scattered; individually planned and arranged travelling tradition to hazard-free, orderly, well-framed, and organized practice, then. Thus designing and making arrangements of travels became a specialized work sphere which gradually led to developing organized entrepreneurial activities. The sense of tourism product in the form of an organized integrated group package holiday grew in the field of travel and tour. Interaction with early twentieth-century technological innovations linked to travel logistics widened the scope of vertical and horizontal growth of diversified; segmented; more flexible group package holiday products, thereby contributed to the changes in the production pattern and product nature. The post-1950 qualitative advancement and quantitative growth in air and surface transport, accommodation, and other leisure logistics contributed to the quantitative growth of group packages and the emergence of organized, integrated, individual package holidays at a mass scale. It fuelled Mass tourism's appearance and growth in the post-1950 era. Rapid increment of the tourist footfalls and fast expansion and growth of tourism entrepreneurship together with ICT and tourism interaction gradually turned mass tourism into overtourism phenomena.

The world experienced changes in tourism service and offers with the emergence of integrated holiday package which includes travel; stay; recreation; guide or tutor service arrangements, availability of printed standard travel itineraries; vouchers; brochures; travellers-cheque, and planning of rail-road-cruise transit-tour. Consumers could link destination with accommodation, transportation and recreation in an integrated way since then. Travel-tourism services became specialized and touristized substituting old fashioned disorganized general services. The growth of holiday travel packages transformed consumers from conventional traveller to modern tourist and extended the opportunity of travelling for leisure and recreation outside aristocrat and elitist edge. As a result, class-segregated travelling transformed to class-desegregated touring habit and the nature of consumption changed from the tradition of intemperance leisure-recreation to wanderlust. The introduction of diversified; segmented; more flexible group package holiday products widened the consumption choices. With the paradigm shift in promotion technique from mouth publicity to printed publicity tour companies of organized holiday package attempted to set societal standards, trends of purchase by bringing the messages to the masses or into the personal domain of travel-buyers (Ray, 2021b). Hence revolutionary changes in buyer-behaviour, culture, values, and expectations (Dewar, 1998; Eisenstein, 1979) set in progress. All the pre-1950 qualitative changes translated into quantitative growth at entrepreneurs' behest in the post-1950. Outcome-effects of interaction between tourism and technology in the post-1900 like the introduction of the technology-fassisted advanced and comfortable modern cruise; hotel; car rental services; tourist charter flights; automated reservation systems like Computerized Reservation System (CRS), Global Distribution System (GDS), and Mach Associates Reservations Systems (MARS); Euro-rail and Indrail pass contributed to the growth making travel-tour more comfortable and easy, thus facilitated gradual increase of tourist mobility and eventually led to mass-scale mobility.

The post-world war improvement of the camera; audio-visual; colour technology facilitating the use of creative illustrations in promotional advertisements assisted the promoters to develop an alive and more self-interpretative image near to the reality that a destination could offer and possess. The shift from monotonous; toneless; blunt; flat; wordy printed promotional texts to self-expressive and interpretative, dynamic, and illustrative audio-visual and printed promotional advertisements together with the increase in spending power and internationalization in modern communities by the end of the twentieth century fuelled the growth of urge for travelling or touring within a wide cross-section of the society. This trend increased tourist mobility and expanded the tourism business at a faster rate. Thus tourism-technology interaction output-led qualitative changes in promotional technique triggered the quantitative growth of leisure-recreation linked travel-tourism demands directing production and consumption increase at mass scale. The post-1950 trend of tourist consumption features affluence, surplus leisure time, upward social mobility, leisure-recreation, wanderlust, conspicuous and vicarious consumption and spending beyond mere subsistence. By the end of the twentieth century, this trend has become one of the dominant global social phenomena, cutting across various social differences like religion, class, gender, ethnicity, and nationality etc. The need for relief from work pressure; increased desire for recreation-leisure; demand growth for business meets; urge to display and attain social status advocated vacationing to familiar or exotic; nearby or distant places and bringing souvenirs; memories; experiences back at an increased rate.

Interaction between ICT, civil society and tourism in the late twentieth and early twenty-first century further accelerated the tourism growth rate and made changes in consumption and production pattern. Thus the increase of tailor-made group and individual holidaying against organized group holiday package with the fixed departure date, shift from all-inclusive to semi or non-inclusive packages, the gradual substitution of entrepreneur chosen fixed packages by the customized packages, the emergence of a wide range of customization and likings; purpose; motivation; budget; meal/ accommodation/ attraction/ activity/ amenity/ travel-route choice based tailoring is evident. Consequently, the concept of a varied meal plan, tour plan, sightseeing plan, and activity plan emerged. Widening of consumer choices and substitution of consumerism by mass consumerism are apparent by then. A rise in Gross Domestic Product (GDP) growth at many places and destination's post-world war economic makeover occurred as a logical extension of the rapid increment of the mass consumerism or excess spending trend in tourism.

Changes in the production of tourism services and offers and tourist spending were simultaneous. In some place increasing demand or escalating tourist mobility made the changes in production and supply while someplace else experienced conversely. Simultaneous changes in the production of tourism services and offers were mainly the outcome of tourism and technological innovations interaction and the increased demand-supply match requirement driven. Contemporary technology innovation thereby abetted communication and media system; construction of tourist accommodation and manmade attractions; renovation, restoration, and reproduction of cultural capitals; conversion of traditional and inherited resources of living into tourism resource. Conversion of space; place; landscape; natural habitat; tangible and intangible ethno cultural capitals of host destination into commodities at increased scale ignoring hosts' traditional interests, priorities emerged as a common phenomenon across the world. The commodification approach ensures the entrepreneur's hegemony over the resources, market; product; tourist choice; profit share undisputedly, thus over the growth mobility. Hence norms of the vacation experience and tourism-leisure-recreation business happened to be producers or entrepreneurs' vested interest and expectation-driven.

The overt outcome of ICT innovations and civil society interaction brought user-created varied static and dynamic tourism information and knowledge regarding destination's tourism infrastructure, attractions, amenities, scopes of activities, offers, services and any other travel-related content to the masses or into the personal domain of travel enthusiasts or future tourists in the quickest possible way. Fast exchanges of those real-time contents through virtual media increase the travel demand and eventually contribute to the overgrowth or high mobility phenomenon thus to overtourism. ICT application-led changes in consumption and production pattern accelerated the overgrowth phenomenon. Increased usage of email, internet, and World Wide Web technology, near-instant communication, instant messaging, Voice over Internet Protocol (VoIP) telephone calls, two-way interactive video calls, discussion forums, blogs, YouTube, and other social networking systems empower the tourists to browse static and dynamic information related to travel and visit a destination at their convenience, choose a destination; service offers; service providers, online booking and transaction at their choice directly with the ground service providers in a spatially and temporally flexible environment. The growth of direct online communication between consumers and service providers made travel cost-effective and easy which ultimately contribute to the overgrowth of the footfall of tourists at the destination. After easing lockdown or stay-home restriction, ICT assisted direct online business between consumers and local or ground service providers become quite effective since all the rooms of tourist accommodations at the nearby destinations during the weekend are being sold-out marginalizing the role of mediating or sales and marketing agents of the source market. The digital native and digital resident also contribute to the overgrowth while spreading the travel-related contents of any destination widely. Web abetted transaction networking also contributed to the overgrowth making the flow of tangible value like order, payment, contracts, invoices, confirmations, receipts and payments easy and cost-effective. Internets, extranets, intranets and other ICT applications building efficient reservation and payment mechanism, companionable entrepreneurs' consortia and demand-supply interconnectivity stimulate the urge of frequent travelling which also eventually leads to overgrowth. Further technology innovations made the growth of low-cost carriers or LCC in air travel which promotes the overgrowth of tourist mobility making the travel easy, cost-effective and pocket friendly to a wider section of the consumers.

Changes in production-consumption become apparent with the growth of the overtourism phenomenon. Indeed, the over-mobility of tourists causes the consumption experience-deterioration and the qualitative deterioration in production.

Deterioration in consumption experience becomes apparent when over-congestion at the destination bars the visitors to perceive the authentic meaning of host culture, gaze the monuments or landmarks peacefully or well, experience litter and pollution-free environment, receive expected efficiency in services, get the expected value of the money paid for the holiday and expected benefits which they foresee, spend quality time, connect hosts, access to amenities easily, find privacy and experience friendly approach from locals who feel culturally invaded by the tourists, then. Deterioration in production becomes evident when services marked by the inefficiency, authenticity transformed to staged authenticity, local intangible attraction restructured to suit the touristic purpose, the meaning of host culture and host ethnic identity misinterpreted deliberately (Ray, 2012), tourism offers created based on preconceived notions of what the cultural experiences of the tourists should be (Clarke, 2000), and traditionally inherited cultural forms encountered its touristification. Indeed, audio-visual and other contemporary technological innovations contributed to recreate cultural hoax or professional fantasy, and myths; fake images or pseudo to suit the purpose of the tourists or sometimes to make it easily digestible for the visitors or strangers.

Contribution of Technology-Tourism Interaction Towards a Shift from Overtourism Phenomenon to Responsible Practice

The growth and then the overgrowth of tourism in the post-1950 is predominantly technology-tourism interaction driven. The application of technology in tourism is always strategic. Primarily, in the last century tourism sector used technology innovations to provide better logistics to the consumers who already got the taste of technology abetted better and comfortable civil life. It contributed to growing urges to travel at an increased scale across the world which in turn triggered the demand growth for leisure-recreation-tourism in the post-1950 era. Entrepreneurs, then, began to translate the technology-tourism interaction outcome into optimum growth opportunity and perceive the outcome benefits by increasing growth value. Initiatives to interpret tourism as the most critical driving force of smoke-free economic development and growth were apparent (Ray, 2021b). Increasing vertical and horizontal mobility eventually led to the overtourism phenomenon by the end of the twentieth century when growth culminates in the overgrowth phenomenon. Indeed, the notion of overtourism developed in the perspective of contradiction between the overgrowth phenomenon and its consequences felt in the decline of the destination's lifestyle and qualitative deterioration in the visitation experience (Ray, 2021b).

A dialogue opens over the premises to control the overtourism phenomenon with an effort to minimize the negative consequential impacts on residents and visitors. It leads the industry to interact with sustainable technological innovations. The outcome of this interaction contributes to the emergence of sustainable-technology supported and sustainable goal-oriented tourism practice. At the outset, it yields apparent success in controlling deterioration of natural resources within the sustainable limit and limiting the growth within the destination's carrying capacity limit establishing control over across-the-board tourist-mobility and control free socialization between guests and host overtly. Conversely, this approach exposes much truth like denials of tourism right for all, the right to access the tourist enclaves for locals, the demand for authentic experiences in contact and connection with hosts and many more. Questions arise about the visitor-experience occurs within a counterfeit fortress made for touristic purpose only and regarding arbitrary sanctions over control-free socialization between the guests and locals. Thus consumption experience does not end up differently.

The growth of ICT applications across various social spheres makes the greater society understand the growth-success myth-based post-1950 overtourism phenomenon and also the counter-myth developed to control the overtourism. The myth and counter-myth come into conflict with the truth of shared individual- travel experiences of holidaymaker and vacationers in social media, authenticated through uploaded contents of live realities. It opens another dialogue in the form of a countercultural ideological rejection of all kinds of post-1950 tourism agendas and propositions. Increased use of ICT across the globe among people from cross-sections of the society where vacationers and holidaymakers also reside triggers contradictions between the approaches of restricted and control-free socialization, neo-capital based economy and local economy, tourism rights for all and rights restricted to elites, growth and its consequential experiences and authentic and pseudo consumption experience. ICT and tourism interaction turns crucial in the growth of a countercultural ideological development against overtourism and its counter premises in the form of responsible tourism practice aiming to develop a control independent sustainable place for better living and better visitation. Sharing habit of travel and tourism information-impression using ICT facilities abets the vacationers, destination marketers and developers and other stakeholders understand the significance of the truth of holiday experiences, makes the world of tourism perceive the importance of a destination which offers a better opportunity to live and visit, increases longing for reality experience at the visitation and authentic imaging at post-visit state by the tourist (Munar & Jacobsen, 2014; Ray, 2018). Further with the increase of ICT application-aided value networking, authority and responsibility of imaging destination, creating and formulating offers and making a tourist destination better to live and experiencing at the visitation begin to be shared among tourists, market generating entrepreneurs and service providers. Eventually, the ICT aided online interactions empower travelers to share information about their responsible visits without disrupting the local lifestyle, thus facilitates the faster awareness campaign for responsible practice and creates an entirely new category of tourism stakeholders. Tourism then resurged with new dynamics blurring the myth of irreplaceable divides between entrepreneurs and tourists since both participate in responsible imaging, promotion, marketing, and developing new products collectively while interacting through ICT-based media. Thus collective gaze transforms into an individual gaze, standardized package changes to tailor-made responsible tour program, and bubble gazing shifts to authentic experiencing connecting the hosts by heart. Realizing the neo reality tourism stakeholders encourage co-evolves web relationship or value networking (Allee, 2003) between demand- and supply-side proactively using ICT. This a major shift in the tourism business ecosystem (Moore, 1993, 1996). All those transformations and shifts make the industry perceive the newly emerging role of tourist-consumers in responsible promotion, imaging, creation of a better place to visit and live in and championing the responsible practice against the overtourism phenomenon.

Conclusion

Beginning from the days of the emergence of modern tourism till the days of 2020-pandemic, the interaction between tourism and contemporary technology innovations remains crucial and critical. The outcome of interactions at different period contributed to the qualitative and quantitative progress of tourism. The pre-1950 interaction was better quality logistics requirement driven thus became crucial in the qualitative changes of tourism or to the qualitative transformation dependent tourism growth. The post-1950 interaction was largely quantitative growth and overgrowth oriented as an increase in tourist mobility is considered to be the tourism success indicator. Thus interaction, then, is largely shaped by overtourism and conversely, it also contributes to the fast growth of the overtourism phenomenon. Indeed some outcomes like the introduction of LCC in the aviation sector, automated reservation systems, emailing system, World Wide Web technology, near-instant communication, instant messaging, Voice over Internet Protocol (VoIP) telephone calls, two-way interactive video calls, discussion forums, blogs, YouTube, and other social networking systems contributed to the growth of overtourism phenomenon rapidly and substantially while the introduction of technology abetted luxury cruises, tourist charter flights, rail pass, luxury hotels stagger for a while before contributing to the growth or overgrowth of tourism. Since the understanding of overtourism grows out of contradictions between the limitless growth and its consequential-experience; the premise of growth principle and counter contentions; the affected and responsible for the effect; uncontrolled and controlled growth, ICT's role in making the contradictions visible and felt in the greater society and in revealing the consequential truth of overtourism phenomenon remains vital. Contradictory contributions of technology-tourism interaction outcome make overtourism and its counter phenomenon to be understood, prepare the industry to reject all the post-1950 growth approaches and its counters and promotes responsible practice as the countercultural reactions to the overtourism and its niche counter agendas.

References

Albattat, A. (2021). The impact of online Marketing in Travel Agency. In A. Hassan & S. Sharma (Eds.), *The Emerald handbook of ICT in tourism and hospitality* (pp. 85–98). Emerald Publishing Limited.

Allee, V. (2003). *The future of knowledge: Increasing prosperity through value networks*. Butterworth: Heimann.

Ansel, D., & Dyer, C. (1999). A framework for restaurant information technology. *Cornell Hotel and Restaurant Administration Quarterly, 40*(3), 74–84.

Arana, J. A., & Leon, C. J. (2013). Correcting for scale perception bias in tourist satisfaction surveys. *Journal of Travel Research, 52*(6), 772–788.

Breukel, A., & Go, F. M. (2009). Knowledge-based network participation in destination and event marketing: A hospitality scenario analysis perspective. *Tourism Management, 30*(2), 184–193.

Chathoth, P. K. (2007). The impact of information technology on hotel operations, service management and transaction costs: A conceptual framework for full-service hotel firms. *International Journal of Hospitality Management, 26*(2), 395–408.

Cheng, S., & Cho, V. (2010). An integrated model of employees' behavioral intention toward innovative information and communication technologies in travel agencies. *Journal of Hospitality & Tourism Research, 35*(4), 488–510.

Cho, W., Olsen, M. D., & Sumichrast, R. T. (1996). Expert system technology for hotels: Concierge application. *Cornell Hotel and Restaurant Administration Quarterly, 37*(1), 54–60.

Choi, Y., Hickerson, B., & Kerstetter, D. (2017). Understanding the source of online travel information. *Journal of Travel Research, 57*(1), 116–128.

Clarke, A. (2000). *The cultural tourism dynamics*. Business Education Publishing.

Condé Nast Traveller. (2019). *How technology can help us to tackle overtourism*. Retrieved from: https://www.cntraveler.com/story/how-technology-can-help-us-tackle-overtourism. Accessed: 28 Feb 2021.

D'Ambra, J., & Mistilis, N. (2009). Assessing the e-capability of visitor information centers. *Journal of Travel Research, 49*(2), 206–215.

Dewar, J. A. (1998). *The information age and the printing press: Looking backward to see ahead*. Retrieved from: http://www.rand.org/content/dam/rand/pubs/papers/2005/P8014.pdf. Accessed: 28 Feb 2021.

Eisenstein, E. (1979). *The printing press as an agent of change*. Cambridge University Press.

Goodwin, H. (2017). *The challenge of Overtourism*. Responsible Tourism Partnership.

Griffin, R. K. (1998). Data warehousing: The latest strategic weapon for the lodging industry? *Cornell Hotel and Restaurant Administration Quarterly, 39*(4), 28–35.

Hassan, A., & Sharma, S. (Eds.). (2021). *The Emerald handbook of ICT in tourism and hospitality*. Emerald Publishing Limited.

He, Z., Wu, L., & Li, X. (2018). When art meets tech: The role of augmented reality in enhancing museum experiences and purchase intentions. *Tourism Management, 68*, 127–139.

İştin, A. E. (2021). Social media transforming tourist behavior. In A. Hassan & S. Sharma (Eds.), *The Emerald handbook of ICT in tourism and hospitality* (pp. 113–130). Emerald Publishing Limited.

Jadhav, V. S., & Mundhe, S. D. (2011). Information technology in tourism. *International Journal of Computer Science and Information Technologies, 2*(6), 2822–2825.

Jung, T., Chung, N., & Leue, M. C. (2015). The determinants of recommendation to use augmented reality technologies: The case of Korean theme park. *Tourism Management, 49*, 75–86.

Kim, H.-B., Kim, T. (Terry), & Shin, S. W. (2009). Modeling role of subjective norms and eTrust in customers' acceptance of airline B2C e-commerce websites. *Tourism Management, 30*(2), 266–277.

Kim, J., Kim, P. B., Kim, J.-E., & Magnini, V. P. (2014). Application of construal-level theory to promotional strategies in the hotel industry. *Journal of Travel Research, 55*(3), 340–352.

Part II
Social Media Images

Chapter 3
Facing Overtourism: Social Media to Advise Tourists to Visit Unexplored Destinations

Anukrati Sharma and Venkata Rao Yedla

Abstract The purpose of this chapter is to find the status of different tourism destinations in terms of their carrying capacity in receiving tourists. A sort of imbalance leads to congestion, and it affects the destination and its local community adversely. While the famous destinations across the world are overcrowded, neither the Government nor the private players and media are looking for unexplored tourist destinations and allied activities. The research study was conducted in September 2019 during the visit to Sri Lanka by the two authors of this chapter, who belongs to India. The case study method of research was chosen to develop this chapter. For the same, the researchers surveyed through the questionnaire with residents and tourists of both the countries India and Sri Lanka. Thus, the generated primary data was analyzed and presented in this chapter. One of the primary reasons for using the term overtourism in a negative sense is that neither the host community nor the guest visited is happy with the overtourism on the destination. Tourism is a service industry, an industry by the people and the people for the prosperity of a place with a balanced ecosytem. But it is unfortunate that because of tourism promotion beyond its boundaries, both peace, prosperity, and public welfare are missing today. The consumption and mishandling of natural resources by the tourists are so high that there is nothing left for the residents. The situation is gradually heading towards a movement against the further advancement of tourism. The chapter attempts to find the answer to the question 'how to come out from overtourism by exploring the unexplored destinations in India and Sri Lanka, what is the awareness level of tourists about the tourist destination, and what role social media is playing in increasing overtourism? The chapter also gives details about the unexplored destinations and the overcrowded tourist destinations of both countries. More respondents from India were unaware of the tourist destinations of Sri Lanka.

A. Sharma (✉)
Department of Commerce and Management, University of Kota, Kota, Rajasthan, India

V. R. Yedla
Department of Tourism and Travel Management, Pondicherry Central University, Puducherry, India

A. Hassan, A. Sharma (eds.), *Overtourism, Technology Solutions and Decimated Destinations*, https://doi.org/10.1007/978-981-16-2474-2_3

Keywords Overtourism · Overcrowding · Social media · Tourist destination · Destination

Introduction

Overtourism is ever-growing and posing a threat to many renowned tourist destinations around the world. It is a new term in academia, research, and in general, for the public and practitioners. The term overtourism was first used on Twitter way back in August 2012. In 2017, at World Travel Market 2017, the Ministers' Summit of UNWTO's was addressed the causes and consequences of overtourism. This phenomenon prevalent is not limited to any specific area or continent of the world. Overtourism exists where all stakeholders like tourists, service providers, and the local community feel that there are too many activities over and above the limitations of the destination. As a result, the quality of life local community and tourists' quality of experience have deteriorated unacceptably. Often the phenomenon of overtourism has been taken as a negative and considered as harmful for tourist destinations. The negative impacts of overtourism have been studied for years. Issues such as environmental damage, cultural deterioration, security laps, impact on community welfare, and excess usages of natural resources have been researched many times. However, the consequences of tourism are rapidly visible now on tourist destinations. In view of these impacts and to divert the tourist flow towards the unexplored destinations with similar or more different types of attractions are very much required.

According to Sri Lanka Tourism Development Authority (2019), Asia and the Pacific developed into the primary source of tourist travel to Sri Lanka, with 58 percent of the overall travel received percent in September 2019. Europe accounted for 34 percent of the total travel, America 5 percent, Middle East 2 percent, and Africa 1 percent. India, China, United Kingdom, Australia, and Germany were Sri Lanka's top five international tourists creating markets in the month of September this year. India was the biggest tourist source to Sri Lanka, with 27 percent of the total traffic received in September 2019. China accounted for 8 percent of the total traffic, while United Kingdom, Australia, and Germany accounted for 8 percent, 6 percent (both), respectively. Sri Lanka includes eight world heritage sites, namely 1) Ancient City of Polonnaruwa; 2) Ancient City of Sigiriya; 3) Golden Temple of Dambulla; 4) Old Town of Galle and its Fortifications; 5) Sacred City of Anuradhapura; 6) Sacred City of Kandy; 7) Central Highlands of Sri Lanka and 8) Sinharaja Forest Reserve (ICRA Lanka and IMACS, 2011). The reason to take the Central Province of Sri Lanka in the study as it includes four world heritage sites out of these eight world heritage sites.

The Ministry of Tourism, Government of India (2019) rportd that foreign tourist arrivals in May 2019 were 610,590 compared to 606,522 in May 2018, registering a growth of 0.7 percent. The percentage share of Foreign Tourist Arrivals in India during May 2019 among the top 15 source countries was highest from Bangladesh (28.37 percent) followed by the USA (16.24 percent), UK (6.90 percent), Malaysia (3.75 percent), China (3.59 percent), Australia (2.81 percent), Canada (2.60 percent), Japan (2.47 percent), Sri Lanka (2.34 percent), Germany (2.30 percent), Singapore (2.01 percent), France (1.84 percent), Nepal (1.61 percent), Republic of Korea (1.61 percent) and Thailand (1.19 percent).

In the present chapter, the case study method has been used to present the real issues faced by India and Sri Lanka's selected tourist destinations. The reason to choose a case study method for the research is to provide extensive knowledge and information about the tourist destinations facing challenges because of overtourism. The case study is a method that is related to past studies and shows a new direction. Thus, it comprises past, present, and future, which can be imagined. The chapter's theme is overtourism, which needs in-depth study and related to community-based problems, economics, education, psychology, and sociology. The case study method is acknowledged as an instrument in several studies more specifically related to social sciences, education (Gulsecen & Kubat, 2006), community-related issues, likewise joblessness, the dearth of finance, lack of education(Johnson, 2006), and sociology (Grassel & Schirmer, 2006). This method provides a free hand to the researchers to conclude their observations, ideas, suggestions as solutions for a given problem. The more reasons to adopt the case study method are:

- To find out the issue of overtourism is naturally occurring, or it is somehow manipulated through some sources, likewise social media and online marketing.
- For drafting a theoretical framework that will be useful for further research in the area of overtourism more specifically related to India and Sri Lanka
- To bring the outcomes both positively and negatively of the selected tourist destinations of the aforesaid countries.

Case Study

This chapter is developed out of the case study method. This method was adopted to take out the findings from the selected two countries, namely India and Sri Lanka. The researchers are from India. The first researcher belongs to the University of Kota, which is under the State of Rajasthan. The State is very well known for tourism on a global scale. The second researcher belongs to the Central University of Pondicherry, which lies under the Union Territory of Pondicherry and is enclosed with Tamilnadu, Andhra Pradesh, Karnataka, and Kerala, which are again a well-known tourist destination in India with a record of receiving a large number domestic as well as foreign tourists. The reason to select Sri Lanka in this study is mainly as most of the tourists, and potential tourists are acquainted with the knowledge about tourism in Colombo and Kandy but not about Nuwara Eliya and other places of Sri Lanka. The increasing growth in tourist numbers created challenges in the form of "overtourism"in India's destinations, namely Shimla, Manali, Agra,

Jaisalmer, Darjeeling, Srinagar, Varanasi, Ooty, Pondicherry, Mamallapuram, Madurai, Visakhapatnam, Hyderabad and, Wayanad. In Sri Lanka, overtourism raised more during some particular seasons at Kandy and Colombo. In view of these facts, the chapter attempts to contribute to the literature through the in-depth research and case study of the chosen tourist destinations of both countries.

As stated in the report of UNWTO (2018), tourism is an industry that is for both tourists and communities both, and it needs to be build up stronger for mutual benefits. The researchers of this study who belongs from India observed during their three visits at Sri Lanka in the year 2017, 2018 and 2019 that the local community is not getting that much of benefits out of the tourism industry which they deserve. An imbalance found in the tourism development between Colombo and the Central Province which includes destinations, namely Kandy, Matale, Dambulla, Nuwara Eliya, Sigiriya, and Hatton. Although Kandy is getting well recognized by the tourists, still the other tourist sites under the Central Provinces are almost uncovered. On the other hand, the researcher belongs to India as Domestic tourists they experienced the issues related to traffic jams at the tourists' destinations, observed garbage on streets, increased pollution, the wastage of natural resources and deforestation, higher side of accommodation cost and lowest question of services at few tourist destinations. Marriott (2012) defended the stakeholders who play a vital role in supporting and shaping the sustainability strategy. It seems missing in the case of the studied tourist destinations because of the business gains and profit-making policies.

Methodology

The study is based on both primary and secondary data. For Sri Lanka, a thorough and detailed literature search was carried out. Thus, data for the Sri Lanka context was mainly generated from secondary resources, including articles, journals, both published/unpublished resources, newspapers, and the sources on the Internet. However, the required primary data also were collected from Sri Lanka. The analysis and findings were generated from the literature review, and it has rendered a summary with a pertinent thematic bibliography with valuable annotations. In the Indian context, results were mainly drawn based on a questionnaire survey and literature review. To know the awareness level of local people and tourists, the Likert scale was administered among the selected responds. The idea of measurement is principal to almost all experimental research studies (i.e., social science, data systems, marketing, and psychology), so the researchers used a Likert scale in this study. The main question of this survey was why the tourist inflows beyond its capacity at a few destinations and causing congestion? The questionnaire was designed to get the first-hand information about the awareness level of the tourists who have visited India, in particular Pondicherry and Colombo, Sri Lanka. Brief discussions outline the importance of this study that ends with recommendations.

The questionnaire includes questions that are vital to know the root cause of overtourism and imbalance of destination visitation by the tourist. The questionnaire covered questions; 1) Demographic factors of respondents of both the countries – Gender, Education Level, Marital Status, Employment status and Income, 2) The awareness level of the tourist destination of Sri Lanka at Kota City and Pondicherry, India. The researchers designed this question specifically for Indian local people and tourists who visited India during the time of this research study. The reason to ask this question is mainly that, as previously also mentioned, Sri Lanka is most visited by Indians. 3) Tourist has also been asked about the influential information sources to select a tourist destination. (Tables 3.1 and 3.2).

To generate the primary data, researchers asked randomly selected local people as a sample of the study (total n = 100), 50 from Kota City, Rajasthan, and 50 tourists from different countries who visited Pondicherry, India. From Sri Lanka also (total n = 100), 50 local people and 50 tourists were selected. Because of the time limitation, fewer respondents were chosen for the study. In India, when asked to the local people about the tourist spots of Sri Lanka, out of 50local respondents, 46 respondents were aware of only Colombo city as a tourist destination of Sri Lanka, and in total, 43 tourists at Pondicherry were aware of Colombo and Kandy. When asked at Sri Lanka about India 48 local respondents' destinations, Visa Versa confirmed that they are aware of Kerala, Bihar, Mumbai, and Himachal Pradesh hill station and 33 tourists were about Himachal Pradesh and Kerala. The rest of the

Table 3.1 Respondents' classification

Local people	Tourists
Survey and open-ended discussions were conducted at tourist places, railway stations, restaurants, malls, and shops in both countries. In India, for the local people, the survey was conducted at Kota City. In Sri Lanka, it was conducted at Colombo city.	Survey and open-ended discussions were conducted at tourist places and hotels of both countries. In India, the tourists were surveyed at Pondicherry. In Sri Lanka, tourists were surveyed at Colombo and Kandy.

Source: the authors (2020)

Table 3.2 Awareness level of the tourist destination of Sri Lanka at Kota City and Pondicherry, India

Tourist destination	Local people at Kota city	Tourist at Pondicherry
Colombo	46	32
Kandy	2	11
Nuwara Eliya	1	4
Sigiriya	0	0
Dambulla	0	0
Galle	0	2
Bentota	1	1

Source: the authors (2020)

tourists knew about India's other tourist states, likewise Rajasthan, Gujarat, Mumbai. Thus, in the Central Province of Sri Lanka, the tourist spots with 46 local people and 43 tourists' responses were found as almost no awareness and less important in compression of Colombo city. In addition to these responses from Sri Lanka residents and tourists, 48 local and 33 tourists only knew about Kerala and Himachal Pradesh as Tourist destinations. One of the reasons for overtourism at Colombo and Himachal Pradesh is lack of awareness about these countries' other tourist destinations. At this juncture, the researchers have looked into social media's role in gradual congestion at tourist destinations and how the same media can become a source to cover unexplored tourists' destinations in both countries.

Overtourism and Media: Myths and Realities

The incredible impacts of media communication can bring far-reaching developments in the style of working and code of conduct among the important stakeholders, national and worldwide, the travel industry for harmony, security, and sustainable improvement. According to Xiang and Gretzel (2010), social media's importance in the digital tourism field is increasing. The usages of social media are visible nowadays from the starting of the trip to the end of the journey (Zeng & Gerritsen, 2014). The visitation and promotion of tourist destinations have been leaning on online blogs'rising development (Banyai & Glover, 2011).

Further, Kim et al. (2013) said social media outline tourist's behavior, insight, and emotions during their trip. Different social media, likewise Twitter, Instagram, Facebook, and so on playing a vital role in attracting tourists towards a destination. The research was done by the aforesaid authors (Kim et al., 2013) also stressed the potential of Facebook specifically to influence tourists' emotions, which certainly helps to develop tourism experience. In the case of India's tourist destinations such as Shimla and Manali of the Himachal Pradesh State, it has been noticed that social media marketing previously worked to attract tourists towards these destinations. Still, now the local people and media are using social media to stop the tourists from visiting their places. Further, social media is used to divert tourists to different locations, including newly promoted attractions, hotels, and restaurants. For instance, the unexplored attractions like Art and Crafts Centre and Ousteri Lake are promoted through social media to divert tourists and avoid overcrowding in Pondicherry City Beach and White Town.

Sahin and Sengün (2015) rightly stated the significance of using social media as a tool for tourism marketing. This tool is impacting negatively on the over visited places. As most of the tourists, whether domestic or international, are visiting tourist destinations after taking all the information available online. Photographs/Images, blogs, and reviews are the more influence to attract potential tourists. The study done by Dohmen (2013) concluded that people prefer online sources rather than offline sources. Most of the studies and researches explained the importance of social media, online marketing, travel blogs in tourism promotion for affecting the tourist's emotions, and discussing the tourists' experience. The present study is done with the purpose of finding out two questions related to overtourism and social media.

The first question –excess usages of media are creating an issue of overtourism, a tourist destination, is it a myth or reality? And the second question is, how worst is the after-effects of overtourism propagated by social media? For the answer to the first question, the researchers thoroughly reviewed relevant literature. As defined by Koens et al. (2018) overtourism is occurred because of media communication. This is well said by Fotis et al. (2011) that the tourist usually formulates their choice on behalf of the feedback of the previous tourists' experience because in such case they can avoid hazard and feel more satisfied. Overtourism is a result of these sought-for feedback in photographs, blogs, reviews, Facebook, Twitter, and Instagram posts. In the twenty-first century, the Internet is accessed by most tourists for getting the information and most of them rely on the uploaded information, which attracts the tourists towards the destination. The creation of destination images these days depends on social marketing. The pattern and choices of tourist destinations are more or less similar because of social media promotion. Karataş et al. (2013), the technological advances are transforming the patterns, selection methods of tourism and the approach of customers look for the tourist products and services. For sure, information received from outside sources, including destination promotion, creates tempted images (Gunn, 1972), which work wonders to catch the attention of potential tourists. In this support, the research undertaken by Buted (2014) looks fit that the often practice of social media in the tourism and hospitality sector is because of easy access and complete information. Although he stated that the improper presentation might mislead the customer's, this is more harmful. Travel blogs cannot be considered as individual acts as the readers automatically become part of these blogs. Blogging is an area which is still needed to be uncovered (Baumer et al., 2008). The potential tourist seeks the negative assessment specifically (Mangold & Faulds, 2009). They get this negative information quickly via online feedbacks from blogs, reviews, and ratings. Thus, social media is somewhat working against the destination image. Unless the stakeholders do not take the responsibility and connect emotionally with the destination's image, this propaganda cannot be stopped. For the same purpose, the researchers collected primary data for knowing the influential information sources.

Likert Five Point Scales was used for Measuring it. Table 3.3 shows the different information sources and their influence on selecting a tourist destination. Likert five point scales were used to measure the attributes as 'extremely influential', 'moderately influential', 'somewhat influential," slightly influential,' and 'not at all influential.' This Table reveals the noteworthy finding that social media is considered an extremely influential information source to influence tourists' choice to select the tourist destination. Out of 50 respondents from India 40 (80 percent) respondents from India, 40 respondents were agreed that their choice of tourist destination was influenced by Social media. Followed by it, Blogs moderately influenced the selection of a destination, and videos were not at all influential sources in choosing the tourist destination. Strangely the researchers got the more or less same sought of

Table 3.3 Influential Information Sources to select a Tourist Destination

Information source	Influence on tourists	Tourist at Pondicherry	Tourist at Colombo
Videos	Not at all influential	1	2
Word of Mouth	Slightly influential	2	4
News and travel show on TV	Somewhat influential	3	5
Blogs and Reviews	Moderately influential	4	7
Social Media	Extremely influential	40	32

Source: the authors (2020)

responses from the tourist who visited Sri Lanka. Thirty-two respondents were in favor of social media, 07 respondents were in favor of blogs and reviews, 05 respondents said that news and travel shows on TV channels somewhat influence their selection, tourists slightly influenced by word of mouth and videos, not at all influenced their choice. Thus, on the basis of the generated primary data, it has been concluded that Social media is the most influential source of tourists to choose a tourist destination. Perhaps this is the reason why the Government needs to think and make some mechanism to use the same social media to promote explored destinations as a strategy to divert tourists and avoid the overtourism on specific tourists' destination. In this direction, governments need to formulate policies to reorient social media to promote all tourist destinations, keeping in mind social responsibility and ecological sensitivity. Otherwise, these days' social media are becoming, directly and indirectly, apart and partial in destroying the tourist destination image from the local community's perspective. As a whole, it leads to the unsustainability of the tourists' transaction on the destination.

Destination Image and Tourists Visitation

One of the biggest challenges in the tourism sector is to create unique destination image (Alqurneh et al., 2010). The unique brand image of the tourist destination attracts tourist and meet the expectations and left with a meaningful mark. Thus, studying destination marketing, destination branding, and destination image are the crucial issues. According to Stepchenkova and Mills (2010), for more than four decades, destination image has been a vital area of tourism research. Overtourism is also a part of any destinations'image. In the way, social media and media, in general, are creating and spreading the information and awareness about a particular destination makes the image of the tourist place in the mind of the tourist and potential tourists. Boulding (1956) image is all about the shared opinions of the people about anything that they are aware of. The problem of overtourism is gradually growing as the repeat and rapid promotion is happening of the famous and well-known tourist destinations. It is creating the insight of a group of individuals, and that makes the destination image (Zeng & Gerritsen, 2014). Tapachai et al. (2000) said the same which has been promoted is taken by the tourists respectfully as an insight or idea about a destination.

Tourist visits a destination, keeping in mind the overall image of the tourist destination. Sometimes tourists are visiting the destinations, which are on top of the world as a part of social status and image-making (Gartner, 1986). Architectural pattern, local events, archaeological riches, natural scenic beauty, cities, museums and art galleries, exploration, cultural heritage, ample places to stay far from mass, cuisine, convenient hotels, rituals, and the infrastructure are the factors to measure the image of a destination (Sonmez & Sriakaya, 2002). The tourist wants a full package, which is value for money and satisfying his desires. In this rat race, many of the tourist destinations are losing the peace of the local people and damaging the place. In the case of destinations like Manali, Shimla, Ooty, Goa, Pondicherry, and Mamallapuram of India, where tourists use to visit for soothing environment and peace is gradually destroyed.

Deforestation is another critical issue raised in the hill stations like Manali and Shimla to make hotels, restaurants, and cafeterias, whereas cultural and local sentiments are being affected due to overtourism in Ooty, Goa, Pondicherry, and Mamallapurm. In the winter season, the well-known State of Rajasthan for the heritage and culture does not even look safe and secure for the local people to visit the tourist spots. Because of the high number of tourist visits, the local community is so much engaged in fulfilling their services towards tourists in the states like Kerala, Jammu & Kashmir, Pondicherry that they are forced to ignore their physical and health-related issues. Creativity could be a part of destination marketing, but there is no sense to market and promote a destination if this image started the exploitation of the local community in different means. The factors to promote the destinations should be fair enough for community development. Nine factors that make destination image were stated by Chi and Qu (2008): 1) The environment in which a tourist travels to a tourism destination, 2) The natural beauty of a destination, 3) Festivals, Fairs, and events of a destination, 4) Culture and Heritage of the destination, 5) Infrastructure and necessary facilities, 6) Accessibility, 7) Recreation, 8) Adventure and 9) Value for money.

Through the case study of Sri Lanka Central Province, the researchers also reached on the findings that for making a balance and diversifying the tourist from Colombo, it is high time to develop the destination image of the Central Province. For this purpose, other than promoting the natural scenic beauty, it is also required to focus on fostering spiritual, religious, heritage, tea tourism and need to organize different events and focus on outdoor activities. The image-making of a destination can be a possible solution for outreach from the challenges of overtourism. Earlier, the main factors which tourists want to see in a destination are related to travel, accommodation and food. These days because of the growing awareness and changing lifestyle, tourists need events, easy accessibility, shopping, and environment (Chi et al., 2008). Thus, continuously the authorities of tourism need to add new attractions across the destination and recreate the image for different destinations to divert tourists to less-visited destinations. However, unexpected incidents like

April 2019 bomb attacks in Sri Lanka also one of the incidents to destroy the image of the country as a tourist destination. To overcome these types of crises, the Government needs to keep a contingency strategy for the quick revival of its image.

In Sri Lanka, the Government machinery has geared up to recover and restore normal conditions within a year with its vibrant machinery. Yes, indeed, the travel environment and accessibility are the most important attributes of a destination image (Alqurneh et al., 2010). India is still lying behind in providing the necessary facilities at some tourist destinations, still getting the number of tourists at these destinations. Although it is not in favor of the country as it creates a poor image of the country and brings dissatisfaction in tourists. Coban (2012) observed the factors which influence tourist satisfaction are the tourist attractions of the destination, cultural attractions, accessibility, diversity, and monetary benefits. The world is facing the challenge of overtourism and removing it. The tourist organizations need to work together so they can put forward an attractive image of the destination (Ejarque & Orengo, 2009). One of the serious issues on the part of the tourist for taking over tourism as a negative term is the irresponsible behavior of the tourist (Gallarza & Saura, 2006). Tasci et al. (2007) stated that destination image is indented with respect to a destiny. The challenge of overtourism can be controlled by personal involvement and affection from the destination (Prayag & Ryan, 2012). Tourist satisfaction is affected by the destination image, motivation, and entrainment facilities (Lee, 2009).

Conclusion

The primary aim of the tourism sector is all about the overall wellbeing of the world community, which is derived out of memorable experience and satisfaction of the tourist from attractions of the destination. The current situation in many of the famous tourist destinations is just the opposite of it due to the gradual move towards mass tourism. The overcrowded tourist destination is creating any value addition to tourists on their arrival neither and stay nor satisfying with the quality of services. On the other, local community participation and gaining advantages out of tourism activities are not happening. Indeed, to make a balanced and fit between the demands and supply, the Government and private players need to search the hidden treasures of their respective countries to promote the alternative destination to divert tourists towards new destinations. Especially, Government intervention is a must in using social media positively for the destinations which are unexplored in the countries. Some rewards to the travel agents, tour operators, event's organizer, and hoteliers can be offered to promote the unexplored destinations of the countries. To address the challenges of overtourism, the Government also needs to formulate, implement and evaluate the strategies of marketing, finance, and human resources. Social media may be used to promote unexplored destinations to avoid congestion at an aged, overcrowded, and popular destination. Besides Government, the local community and other service providers need to join hands in the formulation and proper implementation of the strategies to face the negative impacts of overtourism.

This research offers significant recommendations for policymakers and planners; to focus on the promotion of less-known tourist destinations, to link religious tourism with other types of tourism for giving multiple rewarding experiences to the tourists. It is taken for granted that tourism in Himachal Pradesh faces challenges in terms of significantly increasing the number of tourists in the duration of holidays (i.e., May and June, in summers, and Rajasthan from October to February the favorable climatic conditions). Most of the Europeans have been visited Pondicherry during the winter session. Government and private players of both countries need to develop new tourists' products in order to provide greater access to the tourist in non-seasonal duration. It is indispensable to provide information through different media, including virtual tours, but over usage of media or misuse of media must be stopped. Developing new tourist circuits of the unexplored tourist destinations can be a possible solution to face the issues of overtourism. This will reduce the disappointments and frustrations in tourists and enhance the meaningful experience for their visit.

Moreover, the urge for tourism is believed in making a peaceful planet, people proposer and a healthy environment. Tourist destinations could come out from the challenges of overtourism if the stakeholders change their outlook with the socially responsible and ethical mind in promoting destinations without any disparity and redistribute tourists among all the destinations. Stakeholders can play a vital role in controlling the issues related to overtourism by educating people, motivating entrepreneurs, and protecting the environment. Thus, on the whole, the study reflects that the tourist destinations in every country need to receive tourists within their capacity to advance the tourism sector for its sustainable growth.

References

Alqurneh, M., Isa, F., MD., & Othman, A. R. (2010). *Tourism destination image, satisfaction and loyalty: A study of the Dead Sea in Jordanian curative tourism*. Retrieved from: http://repo.uum.edu.my/2505/1/Majed_Al-Qurneh_-_Tourism_Destination_Image.pdf. Accessed 19 Sept 2020.

Banyai, M. & Glover, D. (2011). *Evaluating Research methods on travel blogs*. Retrieved from: https://www.academia.edu/5010243/Evaluating_research_methods_on_travel_blogs. Accessed 20 Sept 2020.

Baumer, E., Sueyoshi, M., & Tomlinson, B. (2008). *Exploring the role of the reader in the activity of blogging*. Retrieved from: http://citeseerx.ist.psu.edu/viewdoc/download?doi=10.1.1.151.6092&rep=rep1&type=pdf. Accessed 20 Sept 2020.

Boulding, K. (1956). *The image – Knowledge in life and society*. University of Michigan Press.

Buted, D. (2014). *Promoting the tourism industry of Calatagan Batangas, Philippines*. Retrieved from: http://www.questjournals.org/jrbm/papers/vol2-issue5/A250104.pdf. Accessed 20 Sept 2020.

Chi, C.G-Q., & Qu, H. (2008). Examining the structural relationships of destination image, tourist satisfaction and destination loyalty: An integrated approach. *Tourism Management, 29*, 624–636.

Coban, S. (2012). The effects of the image of destination on tourist satisfaction and loyalty: The case of Cappadocia. *European Journal of Social Science, 29*(2), 222–232.

Dohmen, A. (2013). *Effectiveness of Online Travel Recommendations; Determining the importance of and Preference for Online Travel Recommendations*. Retrieved from: https://thesis.eur.nl/pub/15625/Dohmen-A.S.M.-314325ad. Accessed 20 Sept 2020.

Ejarque, J., & Orengo, H. A. (2009). Legacies of change: The shaping of cultural landscapes in a marginal Mediterranean mountain range, the Garraf Massif, North-Eastern Spain. *Oxford Journal of Archaeology, 28*(4), 425–440.

Fotis, F., Buhalis, D. & Rossides, N. (2011). Social media impact on holiday travel planning: The case of the Russian and the FSU markets. *International Journal of Online Marketing, 1*(4), 1–19.

Gallarza, M. G., & Saura, I. G. (2006). Value dimensions, perceived value, satisfaction and loyalty: an investigation of university students' travel behaviour. *Tourism Management, 27*(3), 437–452.

Gartner, C. (1986). Temporal influences on image change. *Annals of Tourism Research, 13*, 635–644.

Grassel, E., & Schirmer, B. (2006). The use of volunteers to support family careers of dementia patients: results of a prospective longitudinal study investigating expectations towards and experience with training and professional support. *Zeitschrift Fur Gerontologie Und Geriatrie, 39*(3), 217–226.

Gulsecen, S., & Kubat, A. (2006). Teaching ICT to teacher candidates using PBL: A qualitative and quantitative evaluation. *Educational Technology & Society, 9*(2), 96–106.

Gunn, C. A. (1972). *Vacationscape: Destination tourist regions*. The University of Texas.

Johnson, M. P. (2006). Decision models for the location of community corrections centers. *Environment and Planning B-Planning and Design, 33*(3), 393–412.

Karataş, M., & Babür, S. (2013). Gelişen Dünya'da Turizm SektörününYeri. *KMU SosyalveEkonomik AraştırmalarDergisi, 15*(25), 15–24.

Kim, J. J., Fesenmaier, D. R., & Johnson, S. L. (2013). The effect of feedback within social media in tourism experiences. In *International conference of design, user experience, and usability* (pp. 212–220). Springer.

Koens, K., Postma, A., & Papp, B. (2018). Is overtourism overused? Understanding the impact of tourism in a city context. *Sustainability, 10*(12), 4384.

Lee, T. H. (2009). A structural model for examining how destination image and interpretation services affect future visitation behavior: A case study of Taiwan's Taomi eco-village. *Journal of Sustainable Tourism, 6*(17), 727–745.

Mangold, W. G., & Faulds, D. J. (2009). Social Media: The New Hybrid Element of the Promotion Mix. *Business Horizons, 52*(4), 357–365.

Marriott. (2012). 2011–2012 *Sustainability Report*. Retrieved from: http://www.marriott.com/Multimedia/PDF/CorporateResponsibility/MarriottSustainabilityReport_2011and2012c. Accessed 20 Sept 2020.

Ministry of Tourism, Government of India. (2019). *A compiles monthly Foreign Tourist Arrivals (FTAs) & FTAs on e-Tourist Visa on the basis of information received from Bureau of Immigration (BOI)*. Retrieved from: http://tourism.gov.in/sites/default/files/FTA/Briefpercent20Notepercent20FTAspercent20Maypercent2019.pdf. Accessed 20 Sept 2020.

Prayag, G., & Ryan, C. (2012). Antecedents of tourists' loyalty to Mauritius: The role and influence of destination image, place attachment, personal involvement, and satisfaction. *Journal of Travel Research, 3*(51), 342–356.

Sahin, A. P. D. G. G., & Sengün, G. (2015). The effects of social media on tourism marketing: A study among university students. *Management and Administrative Sciences Review, 4*(5), 772–786.

Sonmez, S., & Sriakaya, E. (2002). A distorted destination image? The case of Turkey. *Journal of Travel Research, 41*(Nov), 185–196.

Sri Lanka Tourism Development Authority. (SLTDA). (2019). *Monthly tourist arrivals report*. Retrieved from: http://www.sltda.lk/monthly-tourist-arrivals-reports. Accessed 20 Sept 2020.

Stepchenkova, S., & Mills, J. E. (2010). Destination image: A Meta-Analysis of 2000–2007 research. *Journal of Hospitality Marketing & Management, 19*(6), 575–609.

Tapachai, N., & Waryszak, R. (2000). An examination of the role of beneficial image in tourist destination selection. *Journal of Travel Research, 39*(1), 37–44.

Tasci, A., Gartner, W., & Cavusgil, S. (2007). Measurement of destination brand bias using a quasi-experimental design. *Tourism Management, 28*, 1529–1540.

UNWTO. (2018). *Overtourism? Understanding and managing urban tourism growth beyond perceptions*. UNWTO.

Xiang, Z., & Gretzel, U. (2010). Role of social media in online travel information search. *Tourism Management, 31*(2), 179–188.

Zeng, B., & Gerritsen, R. (2014). What do we know about social media in tourism? A review. *Tourism Management Perspectives, 10*, 27–36.

Chapter 4
Social Media Transforming Tourist Behavior: Leading Towards Overtourism

Manisha Solanky and Sachin Gupta

Abstract Most admired and exotic tourist destinations around the globe are distress from challenges posed by overtourism, which in turn affect the well-being of nature and humans. The causes of overtourism are many but social media can be considered as the foremost reason for overtourism in the era of web 2.0. This chapter aims to address the role of social media in transforming the behavior of tourist by descriptive research approach using secondary sources of information collected from e-resources, journals, articles, and books. It has been concluded from the literature analysis that social media plays a tremendous part in transforming the behavior of tourist due to the craze of user-generated content like reviews, selfies, photographs, visuals, and lots more. A positive attempt has been made to suggest ways to combat overtourism. Tourist behavior has been changed due to the novel coronavirus. Impact of COVID-19 on overtourism has also been touched to provide insight.

Keywords Overtourism · Social media · Tourist behavior · Instagram · Blogs · User-generated content · COVID-19 · Pandemic

Introduction

Overtourism is the most growing complex phenomenon that needs to be addressed immediately. Individuals become tourist when they willingly leave their natural habitual and surrounding to visit another environment for recreational activities, health, sport, holiday, and religious purposes. This is known as leisure among tourists, as per the investigation of Camilleri (2018). Tourism is known as a highly

M. Solanky (✉) · S. Gupta
Department of Business Administration, Mohan Lal Sukhadia University, Udaipur, Rajasthan, India

A. Hassan, A. Sharma (eds.), *Overtourism, Technology Solutions and Decimated Destinations*, https://doi.org/10.1007/978-981-16-2474-2_4

knowledge-based industry, which allows extensive use of the transferring and sharing, reusing and storing, and generation of knowledge through the usage of Information and Communication Technology. Tourism is considered to be an information-intense industry, according to the views of Nezakati et al. (2015). Tourism provides wide opportunities like job creation, foreign exchanges, business development as well directly indirectly support the overall economic development of the countries. While tourism provides many broad spectra of opportunities, it also invites a mass population in the country that probably overburden's the tourist destination as well as create a nuisance for localities. Tourism is seen as one of the sustainable economic growth approach for cities at the same time create multifarious disturbances in the cities. Tourism has experienced rapid growth and diversification worldwide, as suggested by Perkumienė and Pranskūnienė (2019). Milano et al. (2018) emphasized the growing phenomena of overtourism and defined it "as the excessive growth of visitors leading to overcrowding in areas where residents suffer the consequences of impermanent and seasonal tourism peaks, which have imposed everlasting changes to their lifestyles, access to amenities and general well-being". The researchers have alleged that overtourism is destroying the natural landscape settings, polluting beaches, positioning infrastructure under massive strain, and charging residents out of the property market. Overtourism is considered to be an immensely complex issue that is often oversimplified. Overtourism has multiple impacts on the local population as well as on the traveler themselves as they diminish the ambiances of the destination, lead to overcrowding and waste management pressures. Goodwin (2017) defined overtourism as those destinations where local residents and local's visitors feel that the area is too overcrowded and that the quality of life and the quality of the experience has depreciated insufficiently. In general, overtourism can be understood as too many tourists visiting a particular area and the part of technology play in facilitating the blight. Koens et al. (2018) the study focused on the overtourism and emphasized bringing clarity to the term overtourism. Overtourism is a negative impact on host communities and on the natural environment, which gradually leads to harm to the confined environment and depressing and negative attitudes among local residents in urban and rural areas. Rosenow and Pulsipher (1979) recognized three different causes of overtourism. First, there is too much unpleasant tourist impact like noise pollution, misbehaving other annoying behavior, second too many visitors possibly aggravated by seasonality and last physical impact of the tourist on the economy like obliteration to natural resources. Over the last years, the tourism industry and its representatives are trying hard and their best to raise public awareness of overtourism, its impact, economic significance, and developmental potentials.

Literature Review

Wen et al. (2020) emphasize on the potential effect of COVID-19 on Chinese citizens' lifestyle and travel and claimed in the findings that COVID–19 have changed the overall tourist behavior, choices, attitudes, preferences in the short and long term. Due to the fear of the novel coronavirus, tourist shall avoid traveling to overcrowded destinations and shall remain aloof of public transport, etc. Thus the study concludes that COVID-19 has left a long-lasting impact on overall tourist behavior and consumption patterns.

Gössling et al. (2020) study focuses on pandemics, tourism, and global change, a rapid assessment of COVID-19 and emphasizes on consequences of COVID–19 on the global tourism industry due to lockdowns, restriction on movement, travel restrictions, air transport restrictions, restrictions on sports, events, festivals, and meeting, etc. This global crisis can be an opportunity for the global tourism industry as policymakers. Tourism authorities can rethink and reframe the tourism policies and learn striking lessons from the global crisis.

Alonso-Almeida et al. (2019) emphasize the effect of social media in pushing towards overtourism and analyze the case of Barcelona. This study evaluated the persuasion of social networks on overtourism. The main impact of overtourism pointed out where the negative behavior of travelers, overuse of natural resources, and the irruption of low-cost transportation. Social media plays an important role in creating opportunities through the internet and thousands of people. Social media allows user-generated content that is posted by users like blogs, vlogs, social media posts on various platforms, consumer opinion platform, digital content full of emotions, personal opinions, eWOM. Thus it can be concluded that social media impacts or encourages travelers and tourist, hence contributing to congestions and creating problems for local communities and surroundings.

Capocchi et al. (2019) view that overtourism has been an emerging observable fact that needs to be fixed effectively. This research study aims to assess the implication of tourism and future perspectives of tourism. The problems of overtourism have been caused due to the development of new technology, low–cost airlines, uncontrolled demand, etc., affecting the destination and local communities negatively. There is a need to educate the tourist and visitors towards overtourism.

Reynold (2019), in the blog, spotted out that social media can be blamed for the global overtourism crisis. Apps like Instagram, Facebook are driving people crazy to travel across the globe but at the same time increase the overtourism effects on the destinations. Overtourism is affected by social media platforms causing many damages to society and the environment as well as creating hurdles for local communities.

Milano et al. (2019) explained how overtourism is evolving as a complex phenomenon and attempted to describe as undue growth of visitors in peak season leads to overcrowding and causes permanent changes and disturbance in the lifestyles of the local communities and the general well-being of people and environment. Overtourism continues to be dissected, interrogated, and critiqued. Multifarious issues have emerged with no consensus on how best to respond, plan and manage to undergo design, implementation, and testing. There is a need to understand the drivers of overtourism and, if continued, can possibly become a more pressing issue globally.

Oklevik et al. (2019) analyze in the case study of Norway the emerging concern of overtourism. This study emphasizes destination management-based optimization as a tool to combat the growing issue of overtourism. Every year, Norway faces the problem of overtourism and overcrowding effect due to a greater number of tourist arrivals causing many problems to localities and the environment. There is a need to compress overtourism through effective ways like restriction on numbers of tourist visits, hike in the process of services, amenities and to keep a keen eye on the activities of tourist through information and communication technology.

Milano et al. (2018) evaluate overtourism as a global phenomenon that needs to be fixed. Overtourism is the unnecessary growth of tourist, which gradually leads to over-crowding in particular destinations, causing permanent changes in local residents' lifestyle. The overtourism causes social and environmental damages, reducing housing affordability to the local communities, diminishing the destination's inimitable ambiance, and leading to over-crowd and waste-management pressures. The ways to combat overtourism are to follow demarketing strategies, high taxation policies, etc.

Hall (2014) explained social marketing as the utilization of marketing principles and methods to change behavior to benefit individuals and communities. The concept of social marketing is growing continuously and has been recognized as extremely important for tourism b creating a better balance between tourism and host community, conserving the cultural heritage and natural attraction, improving visitor experiences as well as in assisting in developing new tourism opportunity. Social media marketing has the potential to influence personal, community, and business behavior throughout the tourism system.

Živković et al. (2014) in the study analyzed the impact of social media on tourism. The objective of the article was to evaluate the effectiveness of social-media on tourism. Information and communication technology are now a day's relied upon to promote sales, develop management relationships with customers and promote tourism destinations directly. Web 2.0 has an immense impact on tourist behavior. Digital marketing communication such as website approach, social media, mobile marketing, internal specific ads, blogs, and online communities plays an important role in influencing tourists worldwide. Social media is considered one of the most expressive and influencing platforms that encourage experienced travelers to post and communicate their travel experience and comments and suggests new destinations that can be explored. Electronic word of mouth (e- WOM) is also very effective in marketing tourism destinations, thereby enhancing visitors' satisfaction; solve problems and doubts about travel. Therefore, social media impact and influence tourist behavior around the globe.

Cox et al. (2009) investigate the user-generated content and its trustworthiness and credibility and evaluate its impact on tourist behavior through an online survey. The results claim that 51% of the respondents trust the user-generated content to decide the final destination. The result also suggests that all user-generated content cannot be trusted and are not reliable except for the content provided by state tourism sites and airlines sites, and other official sites. Therefore, the information provided by user-generated content influence the tourist behavior, and thereby their decision on planning the final decision for the travel also widely depends on the user-generated content.

In the study, Volo (2010) emphasizes the blogs success factor and evaluates it as an opportunity for destination management. Key features that have been evaluated are usability of blogs and sociability of blogs that is one-to-one communication. The blog is considered the most effective user-generated content that enhances users' ability to create tailored information and easily share content. Blogs provide the simplicity of browsing, reliability, trustworthiness, creditability, sense of community, social interaction, and customer feedback. Hence it can be said that traveler blog influences the tourist and their behavior.

Akehurst (2009) elaborates on the user-generated content with special reference to web blogs for the promotion of tourism. According to the author, user-generated content is mostly trusted by consumers and is considered trustworthy and credible. User-generated content like reviews of professionals and customers and blogs written by travelers provides information, educates consumers, and facilitates tourism. Blog content provides actual information with little detailed information about expectations, satisfaction with products and services, and expectations. Blogs are utilized for consumer profiling, acquirement, engagement, commitment, creating brand awareness, and strengthening reputation management and customer services. Therefore, user-generated content reflects more genuine opinions and attitudes, which influence tourist behavior and travelers.

Dodds and Butler (1990) analyze the enabler of overtourism. The enabler of tourism is the greater number of tourist, traveling becoming cheaper and affordable, young groups of tourist, growth-focused mindset, the greater struggle for spaces, facilities, and services, wider access to social media and information & communication technology, lack of management over the restriction on the number of tourist. All these enablers affect the local residents & communities, over competition, high housing cost, and social and environmental issues. Hence there is a need to potentially take actions and measures to counter the problems of overtourism.

From the literature review, it can be concluded that overtourism is an emerging phenomenon that peaked due to social media's adverse effect. These platforms are used globally and influence many youth and adults to plan their travel widely depending upon the information provided by the user-generated content.

Research Methodology

Objective

The aim of the chapter is to address the role of social media in transforming the behavior of tourists or travelers and to initialize the ways to combat the issue of overtourism.

Research Design

This study adopts a descriptive research design. Descriptive research attempts to spotlight the current issues and problems by collecting data either through primary or secondary data that describes the complete situation or phenomenon. The purpose of descriptive research is to Describe, Elucidate and Validate Research Findings.

Type of Data

Secondary data has been opted to complete the study. Secondary data refers to data that has already been collected and analyzed by someone else.

Sources of Data

Secondary data has been collected from the various journals, articles, websites, E – libraries. Resources accessed are IEEE Xplore, Procedia, JSToR, Springer link, Google scholar, etc. have been accessed to gain quality journals, articles. E- libraries like Wiley online Library, ACM Digital Library, NDL.

Findings

Social Media an Overview

Social media is the most powerful tool in providing wider access to travel information. Unlike other forms of communication media, social media provides users with a variety of information and allows them to identify an information provider's source, as suggested by Chung and Han (2017). Kuofie et al. (2015) enlightened on overtourism and social media and quoted that in the era of competition, the environment is social media driven. It is the most essential to understand when to utilize social media. Social media is a family of online platforms and mobile applications that take advantage of web 2.0 technologies. These platforms provide smooth progress in creating and sharing content like videos, audios, images, and live streaming by the users with user-friendly interface. Social media includes blogs like WordPress, Blogger, etc., content sharing like YouTube; Flick; Photobucket; professional networks like LinkedIn, Xing, and micro- blogging like Twitter, Plazes, etc. Social media allows one-to-one, one-to-many, one-to- particular others, few-to-few and many-to-many, as well as synchronous and asynchronous conversations. Social media make it easy to create and post content and share content posted by others as per Gretzel (2019) views. By and large, social media refer to web platforms that allow different users to post and share different types of content. More ideologically, the concept of social media enables users to access the internet for sharing, updating content like videos, audios, images etc. without the limitation of time or natural environment, according to Lange-Faria and Elliot (2012).

Kietzmann et al. (2012) suggest that maximum consumers utilize these platforms such as sharing sites, blogging, creating social networks, accessing content worldwide through social media. Identity, conversation, sharing, relationships, presence, reputation, and groups are the seven fundamental building blocks of social media. While social media were largely text-based initially, they quickly evolved to include audio, visual, and animated content. Social media like Facebook, Instagram, and Pinterest allow various unique characteristics that allow interaction like hashtags, tagging, ratings, reviews, "likes", which constitute various people interacting. Social media have a remarkable impression on how people act on online platforms, search and seek information, play, communicate, form and join communities, construct and uphold relationships, and create, tag, modify, and share content across different sites devices. The concepts of selfies gaze tourism, Fear of Missing Out (FOMO), and You Only Live Once (YOLO) social comparisons are analyzed for providing an enhanced understanding of how the social media persuade why, where and how tourists execute and experience tourism and the impact of the latter on their well-being as per the suggestion of Sigala (2019). The role of social media has brought drastic transformation with special reference to tourist behavior.

Excessive literature is evidence that social media has influenced the maximum number of users worldwide. According to a recent study, 30–40% of travelers below the age of 30 plan their destination and pack their bags based on Instagram hashtagged famous destinations. Almost 97% of millennials share their experiences by posting images and visuals of the trip or tour online. Social media have altered the way tourists search for information, make arrangements for their travels, encounter their vacations, connect with different tourists, travelers and localities, and easy ways to communicate with tourism service providers and share word-of-mouth accordingly to the suggestions of Gretzel (2019). Social-media has transformed the tourism sector and has affected all facets of tourism and hospitality services varying from research and development, to booking travel and reviews and suggestions on these platforms. Changes in information and communication technology and the surfacing of new media, the advent of social media platforms also contribute to overtourism as they are considered the important means of promoting famous destinations worldwide. Facebook is one of the most accessed social media platforms that had 2.27 billion monthly active users as of the third quarter of 2018. Twitter and Instagram had 300 million and one billion, respectively, as Dodds and Butler (2019) mentioned. Therefore, from the above study, it has been proved that social media plays an active role in influencing tourist and social media behavior can also be utilized as a platform to combat overtourism in affected destinations.

Social Media and Tourist Behavior Leading Towards Overtourism

"Man is by nature a social animal; an individual who is unsocial naturally and not accidentally is either beneath our notice or more than human" quoted by Aristotle.

Nowadays, modern tech–oriented travelers plan their trips, vacations, and holidays by the recommendations of friends, relatives, and third parties. The World Tourism Organization (UNWTO) annual report claimed that tourists and travelers worldwide increased to 1.3 billion in 2017 from 25 million in 1950. This tremendous growth had advantages of the massive increase in tourists, which gradually contributes to social and economic development and promotes cultural development of the destinations. The disadvantage of tourism focuses on the local communities'negative attitude towards tourists due to over-crowding, noise pollution, environmental changes, and nuisance, thereby impacting their natural lifestyles, as investigated by Acampa et al. (2020).

Social media has made it easy to acquire such knowledge and information without any pain and just through a random click. Traveling to different destinations is the most passionate things loved by almost every one and when the most powerful tool is by the side, people can conquer the whole world. Social media is considered the most important source of information that tends to influence the tourism industry's decision-making behavior. Acquisition of travel-related information is the foremost step in planning the whole process of travel. Sharing experiences, photographs with different dimensions, capturing nature, selfies at various locations provide insights or knowledge to other people through social networking. One of the exceedingly knowledge-based industries is the tourism industry Nezakati et al. (2015).

User-generated Content (UGC) is gaining attention along with the wide usage of social media. User-generated content on social media platforms is the real evidence of tourists, emotions, beliefs, and Preferences, and taste about tourists' travel-related experiences. It is also considered to be one of the most reliable and relevant sources of information generated by the tourist directly, according to the insight provided by Conti and Lexhagen (2020). User-generated content has an influential bearing on the attitude, opinions and choice of tourists who have experienced it. User-generated content has more credibility than information shared by the destinations as tourists post real experiences.

The word of mouth information through social media from friends, relatives, and other network maximizes the effect decision and opinions of people who aspire to travel soon, according to the suggestions of Huertas (2018). The information acquired through user-generated positively influences tourists' behavior and paves the way to organize their future trips accordingly. User-generated content can be of any form of information like blogs, videos, recommendations, photographs, audios, or any media structure shaped by users. This content created by the user has the ability to change the perception of visitors towards any destination.

Social Platforms That Influence Tourist Behavior

Blogs and Vlogs

Travel blogs and vlogs have a tremendous impact on tourist behavior. Researchers and professionals strongly agree that travel and tourism blogs and vlogs have a strong influence on tourists' minds. According to the social media and tourism statistics, 70% of travelers believe in online recommendations. 50% of blog readers read travel-related blog, and 57% of aspiring tourist read reviews on travel websites. The role of social media has provided a different level of experience of planning trips and booking travel instantly. Travel blogs provide written content generated by users to encourage others to consider those destinations while planning their next vacations. As per the findings of Mirasol & Cunanan (2017), travellers and blog readers have positive attitudes towards travel blogs as they meet the readers expectations, provide relevant and appropriate information about their dream destinations.

Blogs are popularly known as the platform for sharing real-time experiences and building a strong, appealing network to interact with other people sharing the same interest. The interaction can be done through posting comments, receiving feedback, and engaging the audience in the communication. User-generated content is considered to be the most reliable and trustworthy source of information by researchers and professionals. Information provided by the bloggers can alter the customer's final decision making on the destinations. Sharda and Ponnada (2008) defined blogs as virtual diaries stored on the web and made available to anyone who shares a common interest. Blogs are a combination of textual information, visuals, and audios.

Volo (2010) investigated that blog allows accessing to a wide range of information quickly and instantly, blogs are the source of information provided by the bloggers who disseminate information quickly and instantly to those who are seeking factual, real and true information. This platform enables the consumer to consumer interaction effectively. Tourists and travelers worldwide have command over online resources to search for information regarding destinations, transportation, accommodations, and leisure activities. Content uploaded directly by users on websites through blogs is the travelers' personal views and thoughts who have experienced the moments. Hence, the personal travel stories, details of trips of specific destinations, videos, and audios have an immense impact on tourist and hotspot destinations, which gradually result in overcrowding as per Akehurst's (2009) views. Online communities like tripadvisor.com give users the opportunity to interact with other users through written comments and reviewing recommendations and feedback. Word-of-mouth and electronic word-of-mouth are regarded as one of the best and reliable sources of information that strongly impact the decision of tourists.

Micro-Blogging

As a recently emerged social media, micro-blog provides information and helps tourists make travel decisions and share experiences and provide public services and tourism destinations. According to Huang and Xie (2013), micro-blogging has adopted the concept of blogging in compact mode. Micro-blogging is usually paired with mobile phones. Twitter was launched in 2006. Twitter is the most popular micro-blogging platform and free social networking services that allow users to share short posts limited to 140 characters called tweets. Tweets can include hyperlinks and hashtags. According to the statistics, Twitter has 313 M monthly active users. 82% are active mobile users. Twitter plays a positive role in sharing information and experiences through mobile phones. It also plays a prominent role in promoting destinations through interconnections with other websites. Sotiriadis et al. (2015) in study mentioned that micro-blogging is a tool for incorporated exchange of ideas with tourism consumers. Twitter is regarded as a power source or medium for interactive communication with other consumers. It potentially contributes and helps in acquiring customer feedback and recommendations.

Instagram

The most trending social media app that has over 800 million users across the world is Instagram. It is a platform that allows users to create content by posting photos of visited places, historical locations, and also adventurous locations. People make photos or videos to create memories and moments that last forever. Thus, social media intervention is constantly growing as per the investigation of Mukhina, et al. (2017). Photography, videos, live videos, and audio-based content convey and appeal the value of real experiences and thereby motivate the other tourist to add that destination to their wish list. Users assign unique and common hashtags, captions, and geo-tagging to their pictures and videos to describe contents, emotions, and beliefs. They may be self-evident in the photos, according to Conti's views Lexhagen (2020).

Instagram allows users to access unique geo-tagging features that allow users to search for the location and view the recent and related feeds shared. According to the survey, 48% of Instagram users use the social network to pick the next vacation destination, and 35% use Instagram to discover new places. If we talk about Instagram statistics, 102MM+ numbers of images are hashtagged #travel, and 38MM+ number of images is hashtagged #vacation. 4 out of 5 millennials typically view travel articles and photos online while 2 out of 3 millennial view travel videos online. 39% of millennial travelers inspired to plan travel activities based on the information available on the social media sites, according to the report of Apollo Agency (2020). Many people, especially millennials, widely depend on social media platforms like Instagram and Facebook to explore new and happening travel destinations. According to the survey, 84% trust online reviews to plan their next vacations.

Instagram and the travel feeds also allow users to instantly book their plan on the spot by "book now" option. This appealing feature allows users to book their travel instantly and finalize them adopted from Mendes (2018). Over one billion people travel each year internationally, and Instagram is determining where and how to travel. For many, travels have become a contest. It's all about the best picture, exclusive location, adventure, number of likes, views, and followers. Bali, Indonesia, is one of the destinations that face overtourism as massive tourists come every year to capture magical water reflection, temples, and other popular sites. The culture of best photos is in trends and invites massive tourist from across the world. This culture promotes and motivates flocks from different countries to visit popular destinations and, in return, harm the destination's ambiance, natural environment, and disruption to local communities and environment.

In their investigation, Berhanu and Raj (2020) concluded that social media users trust the travel information provided and shared on social media platforms. The young generation is more inclined towards social media. This implies that maximum tourist believes on the user-generated content and make sure to follow their footprint. The priority heavily causes Overtourism to take photographs, live videos, audios, and more important, selfies. Social media has made clicking pictures more important than visiting and enjoying the places, moments, and the byproduct of congestion and overcrowding. Reynold (2019), in the blog, claimed that social media could be blamed for the global overtourism crisis.

Apps like Instagram and Facebook are driving a massive number of people to travel across the world but also increase the overtourism effects on the destinations. To understand the concept of social media, Kaplan and Haenlein (2010) initiated a categorization design depending on the theories of media affluence, social-presence, and social-processes (self-presentation, self-disclosure), which are considered to be the base of social media platforms. The concept of social presence indicates the superior the degree of social-presence, the higher social stimulation communicators have on each other's behavior. Media richness theory presumes that any communication goal is the "resolution of ambiguity and the reduction of uncertainty (Kaplan & Haenlein, 2010). The theory of self-presentation and self-disclosure presumes the virtual worlds allowing for high social- presence and self-presentation levels. In the era of web 2.0 technologies, information can be easily disclosed via images and visuals shared by travelers, according to the study of Kaplan and Haenlein, (2010).

Social media emerged as the powerful tool to persuade and influence and encourage users to share posts of their real time experiences, visuals, audios using any popular platforms. The most important element of life is fun, excitement and the ideology of "enjoying the life to the fullest". This ideology inspires maximum to people to travel more, explore new destinations to impress their social network. Social media induced FOMO "Fear of Missing Out" and YOLO "You Only Live Once" Thus, while "social media use is not the only, and likely not the most important, causes of overtourism, it certainly promotes behaviors of tourist that can pave way to overcrowding and it perpetuates photographs, images, visuals that influence others to travel to certain places and, once there, behave in certain ways" as concluded by Gretzel (2019).

Ways to Combat Overtourism

From the above literature, it has been proved that social media's role is that of a destroyer and provokes overtourism at many destinations. But initiatives can be taken to reduce the impact of overtourism through different innovative ways.

Hashtag Culture

Through the unique features of social media like the power of hashtags can be used to promote the adoption of eco-friendly hacks like eco – friendly accommodations, using sustainable products like refillable water bottles, encouraging under tourism destinations, traveling in the offseason, and lots more. #eco-friendly can be a trending hashtag among the tourists. Similarly, hashtag of least visited destinations should be promoted to be diverted from the hotspot destinations.

Anti-hashtag

Vienna Tourist Board initiated this anti hashtag campaign to address tourist behavior. In this campaign, Vienna tourist board urged every visitor to enjoy Vienna's moments instead of documenting posts for social media. The slogan carrying "See Vienna, Not #Vienna" and were placed in and around airports, according to the blog of Buckley (2018). This initiative can be adopted by every destination facing the growing complexity of overtourism.

Visitors Guiding System

Visitors guiding system in the form of apps or through social media can be implemented to guide visitors about the peak season, off-peak season, the best time to visit the destination so that space, time, and quantity can be maintained without hurting the environment and the local inhabitants, as suggested by Erschbamer et al. (2018).

Government Interventions

The government's role in reducing overtourism can be effective like redesigning the tourism policy, limiting the number of visitors per day, and imposing heavy tourist taxes. Governments can control development through regulation and licensing by developing tourism strategies. To be effective, they need to determine what constitutes an optimum capacity and identify when the limits to change that are acceptable to residents and visitors are approaching. As suggested by Joppe (2019) Peru, which is facing the problem of overtourism has limited the number of visitors per day to the ancient site of Machu Picchu, while the Maldives charges an environmental "Green Tax" per person per day, according to the blog of Birtles (2018). Setting up an upper limit and restriction on access can be implemented to combat overtourism effectively. For example, Dubrovnik historical points of attraction, which attract 1.5 million visitors every year while the population of the city is just 40,000, the government along with other boards, planned to restrict access in peak season and set the upper limit to avoid harm to the environment and local ambiances according to the suggestion of Erschbamer et al. (2018).

Demarketing

Demarketing is a proactive strategy that can be utilized to reduce the demand and influences of overtourism. The most considerable demarketing strategies that can be implemented are no marketing of those destinations which faces overtourism, redirection/marketing strategies of alternative places where the crowd can be shifted, informational place demarketing, restricting access that is allowing the limited number of tourist in a particular area or fixing the number of entries beforehand, and pricing mechanisms (Olokesusi et al., 2019). The study results indicated that demarketing had been adopted in the tourism industry to address problems of overtourism in several attractions across many countries around the world. The most important underlying principle of demarketing is to Preserve ecological integrity by limiting excessive demand, b) setting an upper limit for visitors visiting ecologically sensitive tourism assets, curtail socially unacceptable consumption in the form of rowdiness and misconduct at destinations. The study's overarching proposition is that demarketing constitutes a vigorous approach for realizing and upholding sustainable tourism development. Various strategies of demarketing can be adopted to reduce the impact of overtourism at hotspot destinations.

Instagram has a unique feature that allows the users to "book travel", while scrolling down their feeds. When the user clicks on the book travel option, Instagram can provide insights to the user about the number of bookings of that particular destination, whether it's the peak season or off season, whether overcrowded or congested lots more. This type of insight can persuade the tourist to drop the plans to visit such places and can divert them to those spots which the tourists may like.

Through social media, knowledge management processes can be used as a tool to share experiences and provide insights to travelers or tourists about the most crowded destinations, peak season, off-season, and other destination-related knowledge so that massive booking can be massively avoided initially. The potential of knowledge management in social media can be utilized to cope with the global crisis of overtourism.

Impact of COVID-19 on Overtourism

The outbreaks of pandemic COVID-19 have immensely affected humankind in many ways, and no doubt its impact will stay forever on humans' minds. The outbreak of COVID-19 worldwide has posed many critical issues in front of humankind, therefore imposing threat of financial crisis, health crisis, and overall posing various threats to society. Every nation's economic conditions have been declined and heading towards the great depression, and many sectors and businesses have been crushed due to the pandemic. The hospitality and tourism industry has come to a sudden halt due to the outbreak of COVID-19 as many nations have imposed emergency lockdown within and outside therefore sealing international boundaries, restriction on mobility, closure of international flights, restaurants, lodges, hotels, malls, etc. These restrictions have crushed travelers' wishes and overall crushed the tourism industry and shall leave a long-term impact on this industry. The fear of the virus has also changed travelers' attitude and their love and affection towards traveling and exploring new destinations. The traveling trends have been drastically changed, and it's difficult to predict how long these trends will remain. COVID-19 spreads via human to human transmission, therefore, proving to be the most contagious virus. Social distancing is the only solution to avoid and prevent the infection. Therefore, this has changed the tourist behavior towards well-known tourist destinations. Tourists will avoid traveling through public transport and try to prefer destinations that are not famous and will highly avoid overcrowded destinations to stay safe. The novel coronavirus has immensely affected every individual's lifestyle and the taste and preferences of tourist-related behavior, as Wen et al. (2020) suggested. Thus, the change in the tourist behavior towards overcrowded destinations will halt the issue of overtourism.

Pandemics'fear is likely to exist on travelers' minds and might rethink twice before making travel plans to major attracted cities. Pandemic has certainly shuttered the tourism industry but has also beaten the critical issue of overtourism. Due to the fear of pandemic mobility restrictions, overcrowding in over- tourist destinations has been gradually decreased to a great extent. It's time for over- tourist cities like Dubrovnik, Venice, and Barcelona, to rethink and reset the overtourism policies and strategies and reclaim the over – tourist cities. This pandemic impact on the tourism sector is just a shift and for the time being. Overall, the fears of virus have stopped overcrowding issues, the decline in tourist volume, shift in travelling trends, changing attitude of travelers, and lots more. Gössling et al. (2020) emphasized mobility restriction globally, which has gradually affected the whole tourism system like international and domestic travels, air transports, hotels, lodges, restaurants and café, events, festivals, and lots more. Mobility restrictions in overcrowding destinations will help in reviving of natural ambiance of the cities. Ecological imbalance now desires for environmental healing, and therefore, this global crisis has paved the way for environmental healing through lockdowns and movement restrictions globally. Overtourism caused environmental distress, and restriction on movement and lockdown has helped this distress to heal. Therefore, it can be concluded that COVID-19 has paved the way for tourism authorities to rethink and revive the existing overtourism policies and strategies to the earliest as this movement restriction is not going to last long, hopefully.

Conclusion

The creative reviews point out that the "buzz" of social media access is positively influencing how travelers plan their travel and make a final decision, as mentioned by Lange-Faria and Elliot (2012). But with the enlargement of web 2.0 technologies, it has become necessary for tourism industries to establish those technologies that have the potential to serve their long-term strategic goals. More and more travel experiences are shared via social media. Social media allows experienced users to share various information about destination, travel plans with outsized audiences, and enable them to share their behavior at destinations that are often copied or tried by following visitors. Thus, social media hype of a location shows the way to massive numbers of travelers and invites inappropriate behaviors of visitors, which causes damage to the environment and the inhabitants of the destinations and affects the destination's natural ambiance. The increasing issue of overtourism and the role of social media in fueling the issues but also help fix the global issue of overtourism. Social media accelerates overtourism, but social media can also be a mediator in reducing the complex phenomena of overtourism.

Social media acts as a mediator for information search and helps people hail from different times, gender, cultures, and other demographics and socio-graphic characteristics. Social media persuades tourist choice regarding the destination. Users have the option to actively contribute in communities to find information, socially interact, and make ways wider. To prevent overtourism at hotspot destinations all the related stakeholders of the tourism industry have to come forward and take responsibility and initiative towards responsible tourism. Social media can be actively used to create an online forum where all stakeholders can come together and exchange views, ideas, feedback, recommendation, and discussions to solve the emerging phenomena. The pandemic COVID−19 has positively solved overtourism globally due to lockdowns, seal of international borders, and closure of air transports. This is an opportunity for over tourist authorities to review and maintain the balance of footfalls through effective overtourism policies and strategies. Mobility restrictions in overcrowding destinations will help in reviving of natural ambiance of the cities. Ecological imbalance now desires for environmental healing, and therefore this global crisis has paved the way for environmental healing through lockdowns and movement restrictions globally. Overall, there is a shift in tourist-related behavior towards overcrowded destinations, therefore naturally resolving overtourism. Consequently, it can be concluded that COVID-19 has paved the way for tourism authorities to rethink and revive the existing overtourism policies and strategies to the earliest as this movement restriction is not going to last long, hopefully.

This study is based on a descriptive approach and content analysis. To make this study more effective, the researcher needs to collect empirical data from social media users and prove that social media transforms tourist behavior. This study has focused on only a few known social media platforms, and many other platforms can be explored to understand the overall effect of social media in transforming tourist behavior. The researcher has touched only a few user-generated contents that tourists widely use to share their emotions and feelings on social media. Many other such user-generated content is left untouched and explored to learn about social media tourist behavior.

References

Acampa, G., Grasso, M., Marino, G., & Parisi, C. M. (2020). Tourist flow management: Social impact evaluation through social network analysis. *Sustainability, 12*(2), 731.

Akehurst, G. (2009). User generated content: The use of blogs for tourism organisations and tourism consumers. *Service Business, 3*(1), 51.

Alonso-Almeida, M. D. M., Borrajo-Millán, F., & Yi, L. (2019). Are social media data pushing overtourism? The case of Barcelona and Chinese tourists. *Sustainability, 11*(12), 3356.

Apollo Agency. (2020). *How social media has transformed tourism?*. Retrieved from: www.apollo.agency/blog/. Accessed 12 May 2020.

Berhanu, K., & Raj, S. (2020). The trustworthiness of travel and tourism information sources of social media: Perspectives of international tourists visiting Ethiopia. *Heliyon, 6*(3), e03439.

Birtles, K. (2018, April 3). Instagram and overtourism. *Wayfairtravel*. Retrieved from: www.wayfairtravel.com. Accessed 12 May 2020.

Buckley, J. (2018). Vienna Instagram holiday tourism campaign hashtag dress unhashtag. *I news*. Retrieved from: www.inews.co.uk. Accessed 12 May 2020.

Camilleri, M. A. (2018). *Travel marketing, tourism economics and the airline product: An introduction to theory and practice*. Springer.

Capocchi, A., Vallone, C., Pierotti, M., & Amaduzzi, A. (2019). Overtourism: A literature review to assess implications and future perspectives. *Sustainability, 11*(12), 3303.

Chung, N., & Han, H. (2017). The relationship among tourists' persuasion, attachment and behavioral changes in social media. *Technological Forecasting and Social Change, 123*, 370–380.

Conti, E., & Lexhagen, M. (2020). Instagramming nature-based tourism experiences: A netnographic study of online photography and value creation. *Tourism Management Perspectives, 34*, 100650.

Cox, C., Burgess, S., Sellitto, C., & Buultjens, J. (2009). The role of user-generated content in tourists' travel planning behavior. *Journal of Hospitality Marketing & Management, 18*(8), 743–764.

Dodds, R., & Butler, R. W. (1990). The enablers of overtourism. *WORLD, 2010*, 2030.

Dodds, R., & Butler, R. (Eds.). (2019). *Overtourism: Issues, realities and solutions* (Vol. 1). De Gruyter Oldenbourg.
Erschbamer, G., Innerhofer, E., & Pechlaner, H. (2018). *Overtourism: How much tourism is too much*. Eurac Research.
Goodwin, H. (2017). *The challenge of overtourism*. Responsible Tourism Partnership.
Gössling, S., Scott, D., & Hall, C. M. (2020). Pandemics, tourism and global change: A rapid assessment of COVID-19. *Journal of Sustainable Tourism, 29*, 1–20.
Gretzel, U. (2019). The role of social media in creating and addressing overtourism. In R. Dodds & R. Butler (eds.), *Overtourism: Issues, realities and solutions* (pp. 62–75). de Gruyter.
Hall, C. M. (2014). *Tourism and social marketing*. Routledge.
Huang, L., & Xie, L. (2013, April). A study of tourism destination official microblog marketing based on content analysis. In 2013 *International Conference on Service Sciences* (ICSS) (pp. 170–175). IEEE.
Huertas, A. (2018). How live videos and stories in social media influence tourist opinions and behaviour. *Information Technology & Tourism, 19*(1–4), 1–28.
Joppe, M. (2019). *The roles of policy, planning and governance in preventing and managing overtourism*. De Gruyter.
Kaplan, A. M., & Haenlein, M. (2010). Users of the world, unite! The challenges and opportunities of social media. *Business Horizons, 53*(1), 59–68.
Kietzmann, J. H., Silvestre, B. S., McCarthy, I. P., & Pitt, L. F. (2012). Unpacking the social media phenomenon: Towards a research agenda. *Journal of Public Affairs, 12*(2), 109–119.
Koens, K., Postma, A., & Papp, B. (2018). Is overtourism overused? Understanding the impact of tourism in a city context. Sustainability, 10(12), 4384.
Kuofie, M., Gholston, K., & Hakim, A. C. (2015). An overview of social media for marketing. *International Journal of Global Business, 8*(2), 65–82.
Lange-Faria, W., & Elliot, S. (2012). Understanding the role of social media in destination marketing. *Tourismos, 7*(1), 193–211.
Mendes, M. (2018). How Instagram has taken overtourism marketing. *Pixlee*. Retrieved from: https://www.pixlee.com. Accessed 10 May 2020.
Milano, C., Cheer, J. M., & Novelli, M. (2018). Overtourism: a growing global problem. *The* Conversation. Retrieved from: https://theconversation.com/overtourism-a-growing-global-problem-100029. Accessed 10 May 2020.
Milano, C., Cheer, J. M., & Novelli, M. (Eds.). (2019). Overtourism: Excesses, discontents and measures in travel and tourism. CABI.
Mirasol, R. G., & Cunanan, M. E. R. R. (2017). Attitudes towards travel blogs and its effect over tour related decisions. *International Journal of Education and Research, 5*(5), 203–210.
Mukhina, K. D., Rakitin, S. V., & Visheratin, A. A. (2017). Detection of tourists attraction points using Instagram profiles. *Procedia Computer Science, 108*, 2378–2382.
Nezakati, H., Amidi, A., Jusoh, Y. Y., Moghadas, S., Aziz, Y. A., & Sohrabinezhadtalemi, R. (2015). Review of social media potential on knowledge sharing and collaboration in tourism industry. *Procedia-social and behavioral sciences, 172*, 120–125.
Oklevik, O., Gössling, S., Hall, C. M., Steen Jacobsen, J. K., Grøtte, I. P., & McCabe, S. (2019). Overtourism, optimisation, and destination performance indicators: A case study of activities in Fjord Norway. *Journal of Sustainable Tourism, 27*(12), 1804–1824.
Olokesusi, F., Agboola, G., Aina, C., & Ajayi, O. (2019). Evolution of Demarketing in the Tourism Industry and Implications for Sustainability. *International Journal of Innovative Science and Research Technology, 4*(10), 625–630.
Perkumienė, D., & Pranskūnienė, R. (2019). Overtourism: Between the right to travel and residents' rights. *Sustainability, 11*(7), 2138.
Reynolds, R. (2019). Social Media global Overvtourism crisis. *Aljazeera*. Retrieved from: www.aljazeera.com. Accessed 19 May 2020.
Rosenow, J. E., & Pulsipher, G. L. (1979). *Tourism the good, the bad, and the ugly*. Century Three Press.

Sharda, N., & Ponnada, M. (2008). Tourism blog visualizer for better tour planning. *Journal of Vacation Marketing, 14*(2), 157–167.

Sigala, M. (2019). The bright and the dark sides of social media in tourism experiences, tourists' behavior, and well-being. In D. J. Timothy (Ed.), *Handbook of globalisation and tourism* (pp. 247–259). Edward Elgar Publishing.

Sotiriadis, M. D., & Van Zyl, C. (2015). Tourism services, micro-blogging, and customer feedback: A tourism provider perspective. In J. N. Burkhalter & N. T. Wood (Eds.), *Maximizing commerce and marketing strategies through micro-blogging* (pp. 154–173). IGI Global.

Volo, S. (2010). Blog success factors: Opportunities for destination marketing. In *Conference on information and communication technologies in tourism.*

Wen, J., Kozak, M., Yang, S., & Liu, F. (2020). COVID-19: Potential effects on Chinese citizens' lifestyle and travel. *Tourism Review, 76*(1), 74-87.

Živković, R., Gajić, J., & Brdar, I. (2014). The impact of social media on tourism. In *Prva međunarodna konferencija Sinteza Conference.*

Chapter 5
Social Media and Transformation of Consumer Behavior in Tourism

Lakhvinder Singh

Abstract Social media enjoy an exceptional accomplishment in terms of acceptance and its uses for different purposes. Although sufficient efforts have been made to discover social media's application in consumer behavior for purchasing travel services from a different perspective, the usage of social media during different phases of travel starting from pre-trip to post-trip stage remains imprecise. Hence, the present chapter intends to fill this gap by identifying social media uses on consumer behavior in different phases of travel. The chapter is based on qualitative techniques to study and discover the different aspects of consumers' social media uses during the travel process and its effect on consumer behavior. The discussion with experts and reviews of websites, social media, and digital platforms related to the tourism and hospitality sector were accessed for the accomplishment of the aims of the chapter. The chapter noticed that consumers had extensively used social media during all the traveling process stages to make a purchase decision start from the response, and these end in the information sharing. From the analysis, it is detected that social media uses in travel decision making have a significant impact on consumer behavior and used extensively in information exchange, flexibility to adopt this dynamic technology, searching and discovering information related to travel. Moreover, several practical implications can also be drawn from the study from the perspective of professionals in the field of tourism.

Keywords Social media · Tourist behavior · Adoption · Search · Experience · Share

---

L. Singh (✉)
Government College, Kaithal affiliated to Kurukshetra University, Kurukshetra, Haryana, India

A. Hassan, A. Sharma (eds.), *Overtourism, Technology Solutions and Decimated Destinations*, https://doi.org/10.1007/978-981-16-2474-2_5

Introduction

The extended development and technological innovation forced people to take on the online channels to devise their routine life and events by synchronizing with internet source. For sure, since the last decade of nineteenth century, the appearance of internet revolution and especially social-media change individuals' way of looking, shopping, abundance of informal communication over social-media digital network, and even travel arrangement exercises (e-Marketer, 2014). The consistent development in the clientele of social media reflecting its popularity and suitability in the contemporary time period. As displayed in Fig. 5.1, the size of social media traffic across globe touch the high number of 3 billion as it has changed the system of communication, learning, entertainment and selection of different products and services such as travel (Statist.com, 2020).

In this situation, it becomes imperative to understand the emerging concept of social media and the implications associated with this. There are as yet numerous continuous discussions and conversations in regards to description of social-media, as this digital channel has been changing and converging into the advancing advancement of new media (Solis, 2010). Despite this discussion, sufficient literature exists that focus on description and tenacity of social-media. Safko and Brake (2009) have upheld the idea proposed by Kaplan and Haenlein (2010), that defined this digital platform as medium to express behavior, activities, communicating to masses by sharing views and opinions overall digital network. In every situation, social-media assists the progression of communication by authorizing assurances and involvement of everyone interested. This is a double-sided conversation when

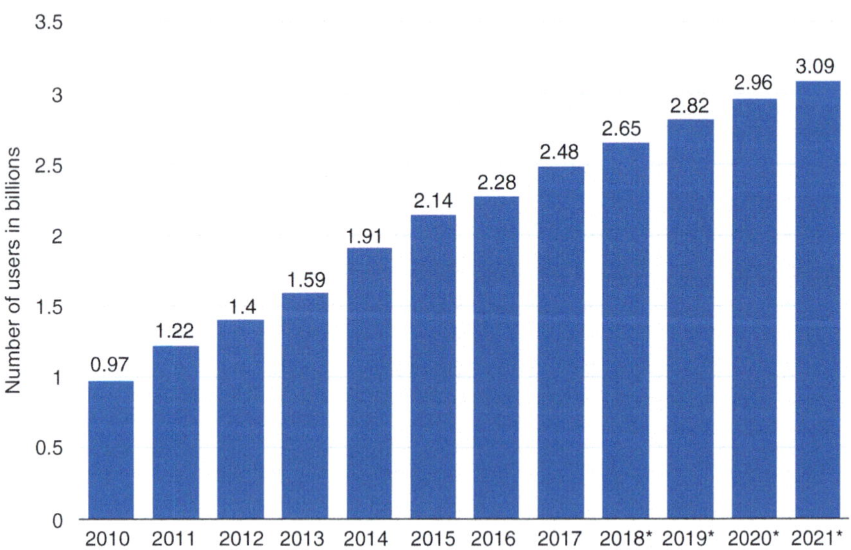

Fig. 5.1 Worldwide social media network users. (Source: Statista.com, 2020)

contrasted with the ordinary media since digital communication channel (social-media) are open up to condemnation and cooperation (Mayfield, 2008). At its quintessence, social-media portrays the amazing novel system people are depicting in with content on the internet, as well as the other way around; that is, as many have noticed, this media is customer ambitious in which it emphasizes and works with the most common way to make, remark, alter, share and overall it is the, "democratization of the substance".

As indicated by Fig. 5.2, it can be easily recognized that Twitter is by a long shot the most famous social-media juncture, trailed by Facebook, Vimeo, and Instagram. Evidently, different social-media points are utilized for various purposes by digital communities. For tourists, Twitter is especially well known and broadly used to discuss about general topics, whereas, Facebook is more normal when discussing innovation and digital implements (e-Marketer, 2014).

These days, social networking channels are indivisible from the existence of the expansive greater part of tourists. Continually, novel social-media are executed, for which by far most of the travelers will attempt to see if they contribute in a positive manner to make a difference according to life status. The tourism industry is facing certain deviations happened because of the reception of various innovation changes over the most recent couple of many years. It is in this way expected that social-media affect travellers' choice to book a travel service. This has been considered by researchers in the literature in travel and accommodation field (Gretzel & Fesenmaier, 2009) examine how various technologies are utilized all through the three phases of the travel process and decision making. Notwithstanding, significant exertion has been made to look at what social-media and its use in purchasing of travel services in entire travel process. An engaged way to deal with social-media can assist with building a solid brand, permit a more noteworthy piece around the world, support reservations, make and keep up with better associations with clients and, to wrap things up, give the ideal apparatus to communicate straightforwardly

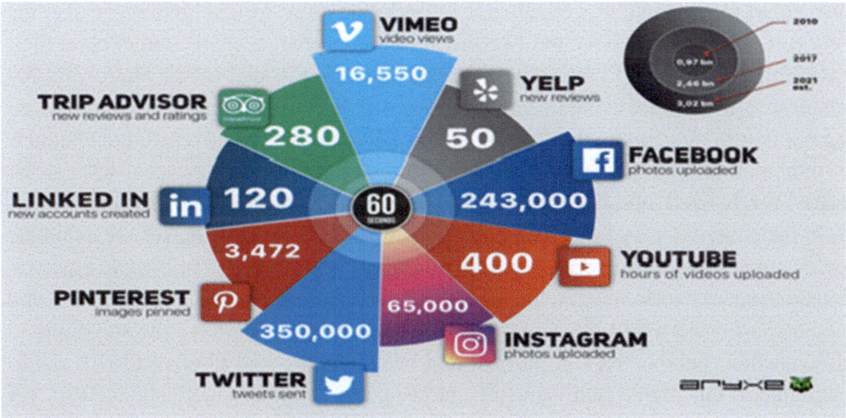

Fig. 5.2 Leading social media networks. (Source: Aryxe.com, 2017)

with them. Social-media channels could be seen as a channel to look and bookings for their facilities, visits, and exercises. Now, it is of paramount importance to overview the usage of social media during different phases of travel.

Taking into account that a response in terms of assessment of "how do social-media impact travel decision making process" is not recognized at this point, the current study focus on covering this limitation by examining the extent of effect and also understand in what direction social-media influence travelers' decision making cycle. Because of the way that very few examination article available in literature that investigate the phenomena of travel decision making is associated with social-media, the current research explanatively and subjectively examines this issue.

The research approach is concerned with the associated theories on 'consumer behavior' in pursuit of social-media. More explicit discussion is worked from the conceptual perspective, and it is the association of related notions. Hence, the conjectural improvement can be summarized in the theme of present study. Current study used to discuss the concept with experts from the field of industry and academician. The online tourism and hospitality booking portals, including the top tourism and hotel enterprises' websites, car rentals, escorting and guiding companies, were accessed and reviewed by authors to get in-depth information regarding the latest trends in the field. The researcher also consulted with tourists from peer groups and different traveler community, including different stakeholders.

Social Media in Tourism Sector

Now a days, it is inordinately challenging to imagine a world without social-media. They impact individual's day to day routines and effect different enterprises, particularly the tourism & hotel sector, in aspects like expansion, handling trade, and investigation mechanism because of their continuous universal growth (Zeng & Gerritsen, 2014). Because of this enormous utilization of social-media channels, like sharing and remarking on content, consumers' force put on more weight amazingly over the long run. Travellers do have the likelihood to look, arrange, share, and explain their travelling experiences (Leung et al., 2013) using various social-media platforms, that numerous travelers are using in entire journey process. Subsequently, an ever increasing number of consumers think about word of mouth publicity over digital channels as indicator of authentic informational source, explicitly for services that can be looked for, booked, and bought on the digital platform (Sparks & Browning, 2011).

In the contemporary time period, huge range of social-media daises are available for its users. The existing incorporate sites/digital broadcasts, informal organizations/online networks, social bookmarks surveys, social news, microblogs, and media-sharing-networks. During these phases, consumers do have the substitute to blowout their perspectives, inquiries, and needs. Xiang and Gretzel (2010) delineated that as this digital platform since blowout information effectively and they get taken on by the travelers, one can say that it is inescapable for the travel and accommodation industry to focus on any new mechanization over digital platform.

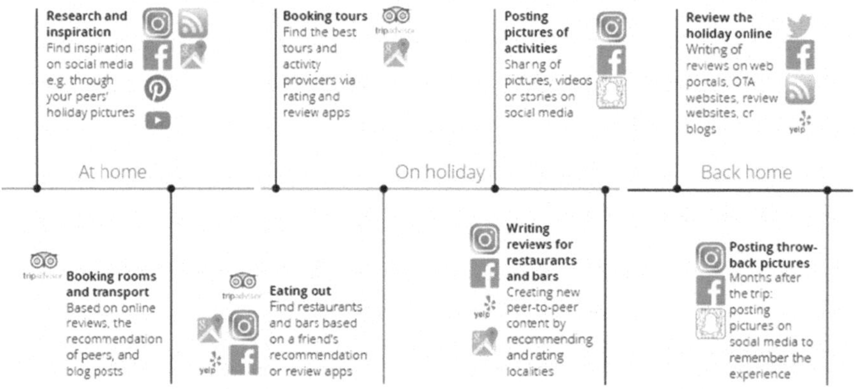

Fig. 5.3 Use of social media and holiday consumer behaviour. (Source: e-Travel)

Nonetheless, Leung et al. (2013) found accommodation firms actually did not utilize social-media to its full degree. Cox et al. (2009) see the prospect for the tourism business industry and allied sectors to work on digital "B2C" systems by executing customer fashioned product/information in the firms' web portal so that travelers are keen on contribute feedback/experience related information. Likewise, added analyses on the convenience can be acquired by entreating the email address from travelers. Hence, lodgings and travel services can send them a solicitation for creative input.

The travel and accommodation industry experts required the consideration about the way tourists assess the facts, information and purchasing highpoints over different advanced stages like social-media (Murphy et al., 2016). It is contended that travel industry firms utilize social-media to make interest in their assets, present expansions, draw the media consideration, and make mindfulness, acknowledgment, and advancement about associated products (Eroz & Dogdubay, 2012). With sharing of information and facts via social-media tools by the travel industry organizations are the significant mechanism of publicizing and advancement, a somewhat regrettable word or tweet, post and remark partook in social-media may handily sully the standing and appearance about organization (Kim et al., 2016; Curkan, 2013). Consequently, it is recommending that it is precarious for tourism superiors to give focused contemplation to their clients' necessities for information in social-media so they can react effectively and indulgenced the impression of useful advantages (Fig. 5.3).

Social Media and Tourism Behavior

Consumers go through the various phases of buying decision procedure when settling on a buy choice, including need acknowledgment, information search, assessment, buy choice, and post-buy choice (Kotler et al., 2003). As a data source, the significance of social-media for the travel industry has been reported by different creators and is an arising field that has gotten expanded consideration from the travel industry researchers of different disciplines to comprehend the outcomes on travelers' behavior, decision making, and corporate correspondence (Minazzi, 2015).

Various investigations can be found in the writing about social-media and its impacts on the consumer purchasing selection. Kaplan and Haenlein (2010) demonstrated that social-media could be utilized as a chance in understanding the purchaser choices making measure. Atadil (2011) analyzed the view of interpersonal organizations of the travel industry clients, and his review indicates about accuracy of the information of accommodation units through social-media. It was likewise tracked down that this arrangement of information assembled over social-media assumes a basic part in their determination of accommodation premises. Eroz and Dogdubay (2012) researched the elements of social-media corresponding to showcasing standards. The reviews analysis clarified that social-media worked as an important method for the determination of the travel industry supervisions.

As advanced investigations, Munar et al. (2013) examined the groundbreaking force of the travel industry social-media in their digital book. Zivkovic et al. (2014) clarified that the tourists are intensely affected by social-media and they change information acquired from social-media into the process of making travel decision.

Likewise, the majority of the examinations highlighting social-media is generally encompassed by tourists to look, sort out, sharing, and proclaiming travelling experience using digital channels such as social media (Hays et al., 2013). The investigations unmistakably report this digital platform (social-media) is a powerful instrument for advertising, particularly for the travel industry, and it tends to be stated that the customers' buying procedure over digital channel is straightforwardly connected to social-media showcasing.

The process of purchase decision of consumers is the extent of study on the consumer behavior. It tends to be seen according to an exploration viewpoint on the consumer behavior in service sector. It expresses that specialists consider consumers to be the frontrunners. According to the point of view, the buying brings about which a consumer has some issue, and afterward through the judicious interaction, he/she tackles the issue. The viewpoint depicts that a shopper certainly does a progression of steps when he/she completes buying exercises. The means incorporate the recognizable proof of the issue, search, elective assessment, choice, and assessment after the buying. The point of view's root is intellectual and mental encounters and other monetary elements (Umar, 2003). In particular, the utilization of facts available on social-media, as expressed by McCann (2008) happens in a sort of association, coordinated effort, and information sharing. The substance gave in the section depicts social-media utilizes in the dynamic of the travel industry. As clarified already, client's dynamic on the travel industry items is partitioned into three stages: previously, when, and after the travel journey.

The significant role of information in the process of making a decision not just shows up in the information assortment stage. Information exists and has a significant influence likewise in each phase of travel process. The information alluded to here is that made and transferred by different consumers, that is likewise known as,

"User Created Content (UCC)". Social-media as a juncture and its different tools that permits clients to transferring and sharing UCC is among major social media that is available to help and "accompany" the traveling decision making process embraced by the tourists. Present study discovering about use of social media by travellers by; (1) at pre-trip stage traveler using UCC via social-media over some type of phases; reacting, gathering, assessing, and affirming, (2) at during-the trip phase travellers evolving the data, and (3) at post-trip phase, tourist tend to share among different groups and individuals.

Social Media and Consumer Behavior in Pre-trip Stage

In this absolute first phase of the travel decision making process, the discoveries show that the arrangement of data search can change across various individuals strikingly.

In the beginning phases of pre-trip, explicitly in perceiving the movement needs/ wants, the witness reacting to "UCC" in social-media as a boost that helps them in perceiving the necessity to accomplish traveling exercises. The witness admittance information via social-media as an everyday action. While getting to social-media with content on the travel industry, the source reacts to the substance by seeing pictures, watching recordings, or understanding text. Such exercises assist the sources with distinguishing or perceive the need and longing for movement and travel to specific objections. In the following stage, tourists gather the information to have the option to address the issues/wants to travel. Information assortment is an interaction by which shoppers noticed and overviewed their current circumstance to get exact information as a reason for settling on the best choices. At the phase of collecting information, travelers inquiring about experience and feedback shared by previous travellers and facts about different the travel industry items that will be utilized/burned-through on travelling activity through different social-media stages and different sites. At the phase of collecting information, the principle prerequisites saw by witnesses are ampleness and culmination of information. The source is prepared to lead the assessment and choice of information as a reason for choices with respect to the traveling activities.

Subsequent to gathering information, the following stage is the elective assessment stage. Here the observer assessing the data that has been acquired from different sources. Audit, remarks, and appraisals transferred by different clients can help the assessment interaction before the traveler making a choice. Subsequent to assessing information, the vacationer moves to the phase of finalizing a decision. Here, they are affirming convictions on the information that has been assessed beforehand. The information is affirmed by settling on choices about explicit objections and the travel industry items for activities undertaken during journey. Making a decision is finished by booking/buying on the web through the Internet. The most widely recognized buy done through the Internet is intended for transportation and boarding & lodging services. By building up this movement choice, the

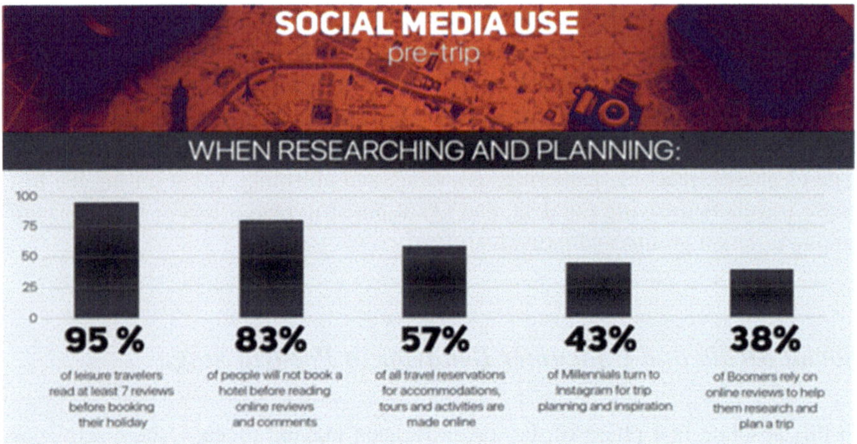

Fig. 5.4 Social media and consumer behavior during pre-trip. (Source: Aryxe.com, 2017)

holidaymaker becomes prepared to travel or enter the "during trip" stage in his/her travel decision making process.

After the place to be visited has been chosen, the inquiry keeps on assessing different boarding and loading arraignments. It is a crucial stage for different tourism organizations as it deals with the tourism business's positive survival by increasing the searching and selection of destination, accommodation, and other services and facilities. According to a survey than 80 percent observed as compulsory use of digital review platform for selection of accommodation. Moreover, 85 percent clients related to Trip Advisor revealed the image of an accommodation firm is strongly affected from its review available on digital sources (Aryxe.com, 2017) (Fig. 5.4).

Consequently, one cannot deny the inescapable need of online reviewing channel for vacationers during pre-trip phase of the travel journey process. By far most uses them normally when arranging an excursion and puts an extreme spotlight on valid audits. In this way, analysts, particularly on TripAdvisor, make a momentous commitment to the accommodation's last reserving choice. For substantial information about transportation, Checkfelix seemed, by all accounts, to be the most well-known site for travelers to illuminate themselves about the best value deals and exceptional flight bargains the whole way across the world. Besides, YouTube recordings for sightseers given by objections will in general turn out to be progressively significant and utilized as a future travel data basis regarding famous locale on tourist place. During this stage, it very well may be learned that digital channels are mostly utilized including social-media by the travelers. As the defendants generally have a thought of where they need to go, they begin searching for hospitality and mobility immediately, for which digital evaluates are a critical information channel. In this review, no respondent expressed to examine the lodging's own site with respect to costs, area, and others. This demonstrates that it is fundamental for an accommodation to either be pleasing on these social-media stages or have extraordinary informal publicity. Else, it very well may be a disconcerting to remain competitive in accommodation sector.

Social Media and Consumer Behavior During the Trip Phase

Tourists are going through the travel experience, and this is the point at which the utilization of the travel industry services like transportation, food, boarding & lodging, and site seeing happens. Travelers at this stage additionally look to explore about different attractions and mobility options in a specific region (Dwityas & Briandana, 2017). At this point, tourists need to participate in different activities and to look for more attractions around that destination. Tourists depend on the feedback provide by their peer groups on digital media about destination and other services that they wish to visit. Subsequently, digital communication channels like Facebook are used to interact with peer groups (Shyle, 2015).

During this stage, the tourists are going through the experience from consumption of travel services like food, stay arrangements, mobility, locale and entertainment. It likewise incorporates looking for extra information needed around the tourism products' extension when they are in one place.

This includes tourists' exercise to seek ample information and travel experience from their visit. While doing this, travellers are looking to find out every required information using social-media that supposed to be consumed during the journey. The requirement for colossal and fast data in "during-trip stage" drives tourists towards an interactive digital channel that reliable to provide quick and accurate information, for instance, by utilizing explicit catchphrases in the searching information. Likewise, in this phase of, "during trip", the tourists additionally interchange UCC to other people. This movement is remembered for the step tour experiences concentrating by travelers.

"Exchange of information" is carried by dialogue with acquaintances or the individuals who did not take part in the travel experience. The topic of discussion in such situation is related to sharing information about "eye witness report" of travel services and activities to be planned to carry out at the destination. By and large, tourist is sharing posts, travel experience, photographs or status besides getting feedback over social-media. Enrichment and sharing of information at the stage "during trip", includes different activities that are undertaken altogether. On arrival to the destination, tourists seek to find out more information and share their opinions to other individuals in companion group over social-media channels. This action should likewise be possible all the while. For instance, content sharing exercises completed by tourists can likewise part & parcel of process to improve the information about tourism industry.

The tourists are experiencing travel products and services, and it involves all the activities during journey process, searching about destination and travel services remain persistent among travellers. The analysis of COMSCORE (2013) uncovers that during this phase of travel nearly sixty percent of travelers are looking for travel activities, followed by search for food venues and cafeterias contributing nearly forty percent, and search for accommodation and hospitality by more than thirty-five percent travelers. Besides, tourists also sharing information and holiday experience over social-media channels. Facebook is one of the most favored digital tool for sharing life occasions and travel-related substance, and from a quantitative

Fig. 5.5 Social media and consumer behavior during-trip phase. (Source: Aryxe.com, 2017)

perspective it came to notice that more than fifty percent of travelers were inspired from their peer group's photo shared over social-media. In continuation to this more than seventy percent tourists reported willingness to share their photos and experience over social-media channels. With regards to Instagram, with in excess of 200 million posts labeled food and 23 million with beverages, food, and drink photographs are effectively probably the most famous sorts of content there (Fig. 5.5).

This result permits the suspicion that the innovation with which the travellers grew up continually impacts significant choices during their journeys. They seem to consider online feedback and reviewing system as a solid base of social-media and along these lines use it when essential.

Social Media and Consumer Behavior in Post-trip Stage

After the trip stage is the point at which every activity has been finished, it is when tourists end their exercise of travel process and are returned to homeland on completion of holiday tour. This is when travellers assess their experience to check if they were fulfilled and conclude whether to drop a few travel elements, or the information sources utilized during the travel process (Dwityas & Briandana, 2017).

The social-media platforms, for example, TripAdvisor and Facebook would now be able to be seen as new electronic publicity channels (e-WOM), which can be characterized collectively by individuals who participate in discussions utilizing electronic channels basically through social-media, and making a frame a social relationship. This paradigm shifts convert social-media as suitable channels for a creating and seeking publicity over digital platform (Rodriguez, 2009). Since individuals trust the word of mouth publicity and marketing, social-media channels have turned into important information searching platform for travelers.

The post-trip stage is the stage when a progression of activities during journey have been refined. So, it very well may be expressed that it is the stage when the tourists have been back home again from their voyaging. In the stage, their service encounters are assessed, and the tourists can finish up the contentment level in response to consume of different tourism services or about the destination and source of information used during entire travel process. This assessment of tourists is important source of information of future tourist. It also supports travelers to share their experience and keep the photos/video and status over social-media as a memory of their holiday.

Concerning period of travel service providers, rhyming to the use of social-media during travel purchasing cycle, Evans (2008) has introduced the exemplary buying channel model as a pointer of how social-media has affected on customer decision making stages.

The last stage of the travelers' travel experience is viewed as when they are once again at home and ponder their holiday journey. With this, planned to recognize is social-media is as yet used to share information and experience about the excursion they embraced, discover to which degree they are utilized, in contrast with the pre-trip stage and during the journey.

At the post-trip stage, it is opinioned that on completion of holiday travel process and returning to residence, the travellers tend to share their experience through photos/videos over social-media sites. This phenomenon also includes sharing experience from different service providers such as travel intermediaries, transporters, accommodation units, food & restaurants and other professional such as guides, escorts, photographers and so on. The tourists after completion of their holiday also seems interested to give feedback about various products/services of firms over websites, booking portals and social-media pages of concerned agency. This feedback and review system can be important source of information for future travellers to seek about destination and make a final decision to purchase different tourism services. Thus, travellers in this way went about as a co-maker who make content and offer it to help other people, particularly the individuals who have comparable interests (for example travel planning). The consequences of the assessment are imparted to others in front of an audience "post-trip", not simply the assessment of the experience and level of fulfillment with the travel services, yet additionally the information source that have been utilized all through the travel decision making process. Likewise, the impacts of this investigation additionally discovered about travel decision-making process by the utilization of social-media can be seen more as a crescentic practice.

Post-trip stage activities have been accomplished in this phase of travel. So, it very well may be expressed that it is the stage when the travelers have returned to journey starting point from destination. In the stage, their experienced from tour are assessed, and the they can finish up the satisfactory feedback in terms of different products and services consumed in entire holiday phase. The feedback given by tourists can be important driver for future tourist behavior. Unexpectedly, when getting back, the tourists begin assessing the nature of their experience. An online review led by Statista shows that excursion, social-media clients are roused by seeing potential objections via social-media, or they effectively research suggestions on the web. While holiday, many individuals post pictures and offer their experience with their friends through social-media applications or destination experience, along these lines making credible distributed substance, which is utilized by others for additional motivation. The impact of content introduced by loved ones online on the genuine travel choices is particularly solid with twenty to thirty year olds. Nearly ninety percent of them say that they plan their holiday dependent on recently seen pictures posted by their friends, and almost similar number of individuals offer photographs or recordings while travelling.

In the aforementioned discussion, it is tracked down that the employments of information through social-media in travel process are a direct cycle wherein the interaction comprises of organized stages. The stages start from the responding, and these end in the sharing information and experience (travel decision making process).

Conclusion

The current study exhibits the travel decision making process as a roundabout interaction and the role of society in changing consume behavior. The study noticed the usage of social-media by the consumer in the searching and selection of destination is base for the travel decision making process. Social-media goes about as a transformer of the consumer behavior in all phases of travel process including arranging, booking, and evaluation of the travel products and services. After a careful investigation of social-media's effect on consumer behavior during different phases of travel process, it becomes apparent that the pre-trip stage is most significant in regards to the use of social-media. Because of the frequent enquiry about destination and other travel services, thusly invest a great deal of energy via social-media applications all together.

Nonetheless, it also involves the assessment about products and services consumed during tour. At this stage, along with the assessment, the tourists also evaluate the believability and precision of the information sources that have been utilized already. On the off chance that the evaluation outcome exhibits the positive course on consumer behavior as travellers unmistakably expressed the possibility of re-utilization of social media channels as information source to assist in forecasting and making an upcoming travel decision. It might occur on the basis of usefulness and association of existing information with past travel experiences, or then again if the data sources effectively give admittance to the right substance extent either for the information from an marketer or travel organizations and furthermore the substance put together by different customers.

During the post-trip stage, social-media is utilized by tourists as a stage to share personal travelling experiences through photos and recordings. Concerning the use of review over digital platform in the last phase of travel process (i.e., "post-trip"), it came to knowledge that it is based on contentment rating of traveler reported over social-media channels. In case tourists were reported high satisfaction or report disappointed from travelling experiences, tend to compose feedback/review over social-media sites that will be useful to potential tourists. Tourists also persuaded to post experiences over social-media. Moreover, it was depicted that blogging portals, microblogging channels, social- newscast and bookmarks, and digital sharing applications do not need critical catch the tourists interest therefore.

The review came to realized that verbal exchange is more powerful than any promoting effort; social-media is utilized usually for sharing pictures, recordings by voyagers. The voyager likewise composes surveys of their convenience, exercises, or cafés to impart their encounters to other people, hence making new motivation and content for different explorers. Social-media locales are turning into a virtual list of must-dos of spots that can be visited and that give the scene to incredible pictures. Being available via social-media can assist with drawing in vacationers to specific areas, and the travel industry normally helps neighborhood economies, which is the reason numerous districts are anxious to draw in via social-media. Notwithstanding, some regrettable sides additionally saw during utilization of social-media, particularly for this situation it is "over the travel industry". This happens when certain areas or simply little spots become popular via social-media and draw in a greater number of guests than can be dealt with by the neighborhood specialists or the climate.

The investigation of the concentrate likewise gives reasonable ramifications; one might determine a few viewpoints that actually should need to improve by tourism professionals and service providers for the sake of boosting the benefit acquired from use of this digital communication tool (i.e., "social-media") on the grounds that the vast majority of the vacationers are currently a days are emphatically depend via social-media stages for determination their location. Subsequently, the travel industry specialist organizations or social-media chiefs might have to screen positive and explicitly regrettable remarks intently and manage them speedily. The creator likewise gives grounds to future studies to understand tourist behavior in the era of digital technology that will be helpful in formulation of appropriate social-media based marketing techniques.

References

Aryxe.Com. (2017). *Digital transformation role of social media in hospitality industry*. Retrieved from:https://aryxe.com/en/blog/digital-transformation-role-social-media-hospitality-industry. Accessed 31 Jan 2020.

Atadil, H. A. (2011). Otel Isletmelerinde Sosyal Medya Pazarlamasi: Turizm tuketicilerinin sosyal paylasim sitelerine iliskin algilari uzerine bir alan calismasi. *Yuksek Lisans Tezi*. Dokuz Eylul Universitesi, Sosyal Bilimler Enstitusu, Turizm Isletmeciligi Anabilim Dali.

COMSCORE. (2013). *Base: Leisure travelers who searched for information through social media during the trip*. Retrieved from: https://aryxe.com/en/blog/digital-transformation-role-social-media-hospitality-industry. Accessed 5 Feb 2020.

Cox, C., Burgess, S., Sellitto, C., & Buultjens, J. (2009). The role of user-generated content in tourists' travel planning behaviour. *Journal of Hospitality Marketing & Management, 18*(8), 743–764.

Curkan, S. C. (2013). Sosyal Medya ve Turizm: Turkiye'deki Turistik Destinasyonlarin Degerlendirilmesi. *Yuksek Lisans Tezi*. Balikesir Universitesi Sosyal Bilimler Enstitusu Turizm Isletmeciligi ve Otelcilik Anabilim Dali.

Dwityas, N. A., & Briandana, R. (2017). Social media in travel decision making process. *International Journal of Humanities & Social Sciences, 7*(7), 193–201.

e-Marketer. (2014). *Tourists use different social networks to post, buy differentproduct types*. Retrieved from: http://www.emarketer.com/Article/Tourists-Use-Different-Social-Networks-Post-Buy-Different-Product-Types/1010609/2. Accessed 1 Dec 2019.

Eroz, S., & Dogdubay, M. (2012). Turistik Urun Tercihinde Sosyal Medyanin Rolu ve Etik ilişkisi. *Dokuz Eylul Universitesi, Iktisadi ve Idari Bilimler Fakultesi Dergisi*. Cilt.27, 133–157.

Evans, D. (2008). *Social media marketing: An hour a day*. Wiley.

Gretzel, U., & Fesenmaier, D. R. (2009). Information technology: Shaping the past, present and future of tourism. In T. Jamal & M. Robinson (Eds.), *The SAGE handbook of tourism studies*. Sage.

Hays, S., Page, S. J., & Buhalis, D. (2013). Social media as a destination marketing tool: its use by national tourism organizations. *Current Issues in Tourism, 16*(3), 211–239.

Kaplan, A. M., & Haenlein, M. (2010). Users of the world, unite! The challenges and opportunities of social media. *Business Horizons, 53*, 52–61.

Kim, S., Lee, K. Y., Shin, S., & Yang, S. B. (2016). Effects of tourism information quality in social media on destination image formation: The case of Sina Weibo. *Information and Management, 5*(1), 2–9.

Kotler, P., Bowen, J., & Makens, J. (2003). *Marketing for hospitality and tourism*. Prentice Hall.

Leung, D., Law, R., van Hoof, H., & Buhalis, D. (2013). Social media in tourism and hospitality: A literature review. *Journal of Travel & Tourism Marketing, 30*(1-2), 3–22.

Mayfield, A. (2008). *What is social media*. i-Crossing.

McCann, U. (2008). *Power to the people: Social media tracker wave 3*. Retrieved from: http://www.slideshare.net/mickstravellin/universalmccann- international-social-media-research-wave-3. Accessed 3 Nov 2019.

Minazzi, R. (2015). *Social media marketing in tourism and hospitality*. Springer.

Munar, A. M., Cai, L., & Gyimóthy, S. (2013). *Tourism social media: Transformations in identity book on community and culture*. Emerald Group Publishing Limited.

Murphy, H. C., Chen, M. M., & Cossutta, M. (2016). An investigation of multiple devices and information sources used in the hotel booking process. *Tourism Management, 52*, 44–51.

Rodriguez, I. L. (2009). *Social media in tourism behavior*. Bournemouth University. Université de Savoie.

Safko, L., & Brake, D. K. (2009). *The social media bible. Tactics, tools and strategies for business success*. Wiley.

Shyle, I. (2015). Social media and its impact on decision making for trip. *European Journal of Interdisciplinary Studies, 1*(1), 8–15.

Solis, B. (2010). *Introducing the conversation prism version 3.0*. Retrieved from: http://www.briansolis.com/2010/10/introducing-the-conversationprism-version-3-0/. Accessed 3 Jan 2020.

Sparks, B. A., & Browning, V. (2011). The impact of online reviews on hotel booking intentions and perception of trust. *Tourism Management, 32*(6), 1310–1323.

Statista.com. (2020). *Number of global social network users 2010–2021*. Retrieved from: http://www.statista.com/statistics/272014/global-social-networks-ranked-by- number-of-users/. Accessed 22 Jan 2020.

Umar, H. (2003). *Metode Riset Perilaku Konsumen Jasa*. Ghalia Indonesia.

Xiang, Z., & Gretzel, U. (2010). Role of social media in online travel information search. *Tourism Management, 31*(6), 179–188.

Zeng, B., & Gerritsen, B. (2014). What do we know about social media in tourism? A review. *Tourism Management Perspective, 10*(2), 27–36.

Zivkovic, R., Gajic, J., & Brdar, I. (2014). The Impact of Social Media on Tourism, Singidunum. *Journal of Applied Sciences Supplement, 39*(1), 758–761.

Part III
Marketing and Promotion

Chapter 6
Information and Communications Technology for Overtourism Management: An Exploration

Priyakrushna Mohanty, V. S. S. Shreya, and Rashmi Ranjan Choudhury

Abstract With increasing attention being paid to aspects like carrying capacity and sustainability of destinations, overtourism has emerged as one of the biggest and most fiercely debated issues across the world. In the last decade, overtourism has become a buzzword and a social movement in tourism, gaining a lot of media attention. Effective, efficient, and innovative solutions to address the problem of overtourism at various destinations have been called for. One such solution is the use of advanced ICT tools for destination management that can contribute to reducing stress on overbooked destinations, planning for alternative destinations, targeting an alternate group of tourists, and most importantly to have control over the tourist inflow through systematic and planned destination management strategy. While literature concerning the usage of ICT for tourism promotion is in plenty, the aspect of destination management through the use of advanced technology has been rarely dealt with at a holistic level. This paper attempts to fill that gap by developing a framework for destination marketing organization (DMOs) in the quest to mitigate the problems of overtourism through the usage of new advanced technology. Based on the review of secondary literature, this paper tries to highlight the challenges associated with overtourism and then deals with the ways by which the problems can be mitigated by the usage of advanced and updated technological tools.

Keywords Overtourism · Technology · ICT · Tourism management · Overtourism management

P. Mohanty (✉)
Department of Tourism & Travel Management, Jyoti Nivas Autonomous College, Bengaluru, India

V. S. S. Shreya
Department of Tourism Studies, Pondicherry University, Pondicherry, India

R. R. Choudhury
North-Eastern Hill University, Shillong, India

A. Hassan, A. Sharma (eds.), *Overtourism, Technology Solutions and Decimated Destinations*, https://doi.org/10.1007/978-981-16-2474-2_6

Introduction

With increasing attention being paid to aspects like carrying capacity and sustainability of destinations, overtourism has emerged as one of the biggest and most fiercely debated issues across the world (Higgins-Desbiolles et al., 2019; Smith et al., 2019). In the last decade, overtourism has become not just a buzzword but also a social movement in tourism, gaining a lot of media attention (Milano et al., 2019). Defined as "the impact of tourism on a destination, or parts thereof, that excessively influences the perceived quality of life of citizens and/or quality of visitors experiences in a negative way" (Koens & Postma, 2020), overtourism can hinder sustainable social development by deteriorating residents' quality of life (Song et al., 2020) and give rise to cultural clashes (Goodwin, 2017).

Overtourism occurs when tourism expansion fails to acknowledge that there are limits for growth (Milano et al., 2018). Overtourism is multifaceted in a way that its effects are multidimensional (i.e., on residents, on tourist experience, on the economy, and the destination as a whole). Subsequently, there are multiple reasons for the rise of overtourism. Still, the most mentioned include the technological advancements, increase in the affordable middle-class group, low-cost airlines, wider access to the internet and media (Gretzel, 2019), resulting in the growth of tourism in emerging countries as well (Capocchi et al., 2019; R. Dodds & R. Butler, 2019a; Erschbamer et al., 2018; Lawton, 2017; Peeters et al., 2018).

It is argued that if a place wants to avoid the problems of overtourism, it is highly important to start proper tourism development even before a place becomes a destination (Song et al., 2020). Further, some experts believe that countering multidimensional overtourism can be done through policies, organizations, institutions, and behavior (Benner, 2019). All in all, effective, efficient, and innovative solutions to address the problem of overtourism at various destinations have been called for. One such suitable solution is the use of ICT tools.

Investment in technological innovations is most recommended for long term sustainability (Mrsic et al., 2020). Modern technologies can not only help the development of new tools for the tourism industry but also aid in managerial decision making through the use of tools like big data analysis and the Internet of Things (IoT). While literature concerning the usage of ICT for tourism promotion is in plenty, the aspect of destination management through the use of advanced technology has been rarely dealt with at a holistic level (Eckert et al., 2019). This paper attempts to fill that gap by developing a framework for DMOs in the quest to mitigate the problems of overtourism through the usage of new advanced technology. Based on the review of suitable studies, a Framework for Technology-driven Solutions for Overtourism (FTSO) has been proposed in the study.

Methodology

The study can be classified as conceptual research, and it has adopted a systematic review of secondary data present in both online and offline format. The authors initiated the drafting of the paper by searching the mentioned six keywords in different valid combinations on three academic search engines, namely Web of Science, Scopus, and Google Scholar. Suitable articles and other scientific literature about the subject were collected and thoroughly read to develop fundamental themes present in the literature. Following the enlisting of the themes, non-scientific forms secondary data accessed from verified websites, reports, gazettes, newspapers, magazine pieces, and documents of different private, government, and NGO organizations were accessed to trace out current implications of technology in managing overtourism. The blueprint of the paper was then prepared after a brainstorming session among the three authors. Each author was assigned to specific parts of the work. Lastly, the corresponding author assimilated all the parts of the paper to prepare the first draft, which was then reviewed twice to remove any errors left.

Overtourism: An Overview

Overtourism has come a long way from being just a word in tweets to being shortlisted as the "word of the year" in 2018, taking its place in Oxford English Dictionary. Right from its inception, the term gained rapid momentum be it because of its self-explanatory nature or relatability triggered by the media coverage. Overtourism is defined as a visible but unacceptable negative deterioration of residents' quality of life and/or visitors' quality of experience (Duignan, 2019; Goodwin, 2017). It indicates the exceeding of physical, social, economic, or psychological carrying capacity thresholds (Peeters et al., 2018). The development of tourism comes with the cost of overcrowding. Many experts view overtourism as a consequence of unplanned tourism development (R. Dodds & R. Butler, 2019a) and failure in estimating the impacts of a large number of tourists (Ivars-Baidal et al., 2019). Some view it as just a new word for the long-existing problem of mass tourism (R. Dodds & R. Butler, 2019a; Namberger et al., 2019; Perkumienė & Pranskūnienė, 2019). The impact related studies date back to tourism capacity study by O'Reilly (1986, p. 254) followed by sustainable tourism, which later emerged into new concepts of soft tourism, responsible tourism, and off late overtourism. Doxey's Irridex Model (1975) stands as the first study to explain the resident's reactions to the impacts.

The impacts of overtourism are largely visible off late as the tourism wave hit some major tourist destinations like Barcelona, Prague, Lisbon, Berlin (Koens et al., 2018; Milano et al., 2018), forcing other destinations to consider it as a serious threat to the tourism industry (Eckert et al., 2019). Due to this, increasing studies on the causes and strategies to overcome the impacts of overtourism have come up in academia. The resident resistance leading to protests has paved the way for a debate on overtourism. The consideration for tourists per resident ratio is on the rise, especially in cities where the problems of overtourism seem to be higher (R. Dodds & R. Butler, 2019a; Ivars-Baidal et al., 2019; Koens et al., 2018). The phenomenon of overtourism is not just numbered anymore. It is much complex than the term (Namberger et al., 2019).

Also, contrary to the popular belief overtourism is no more a European phenomenon and is widely spreading into Asian and developing countries (Goodwin, 2019). While some of the worst-hit destinations started taking the initial measures like entry restrictions and pricing strategies, other destinations are still in a wait and watch approach (Eckert et al., 2019) while considering overtourism as "other city's problem" overlooking the negative consequences for the sake of economic benefits (Ivars-Baidal et al., 2019). The multidimensional effect of overtourism is questioning the traditional ways of management (Eckert et al., 2019). These issues cannot be solved overnight. Every destination is unique, and so are the problems in them. A holistic approach is required to manage overtourism. Monitoring will help in understanding the actual problem and to take concrete measures (Eckert et al., 2019). There is a need for a system to examine, measure, and predict tourism to control it. Technology is often offered as a solution to tackle the issues of overtourism (Mahizhnan, 1999).

Background of Technology and Overtourism Linkage

Capocchi et al. (2019) describes the phenomenon of overtourism within three broader and connected parameters called growth, concentration, and governance. He argues that in the last few decades, there has been higher dependence on tourism for economic development, which has resulted in a *growth* of the sector; this growth has then brought in the massive *concentration* of tourists in particular destinations giving rise to undesired impacts like a breach of carrying capacity, overcrowding, and concerns of environmental sustainability. Lastly, Capocchi et al. (2019) connect overtourism with the concept of *governance* of community resources and infrastructure.

Consequently, there is an extraordinary dependence on technology to be which is characterized by being futuristic, holistic, and equitable (Mohanty, 2020). Technology is a great equating agent as it does not discriminate between the rich and the poor and is available to everyone with less or almost no resources required (Olson, 1996). It has therefore been argued that technology's role in mitigating the issues in overtourism ventures is perennial.

As can be seen in Fig. 6.1, there exists an intersection between the issues of overtourism and technology-driven solutions. This paper focuses to explore this area and develop a framework that can serve a tool for various DMOs and policymakers while designing structural plans to eradicate the impacts of overtourism.

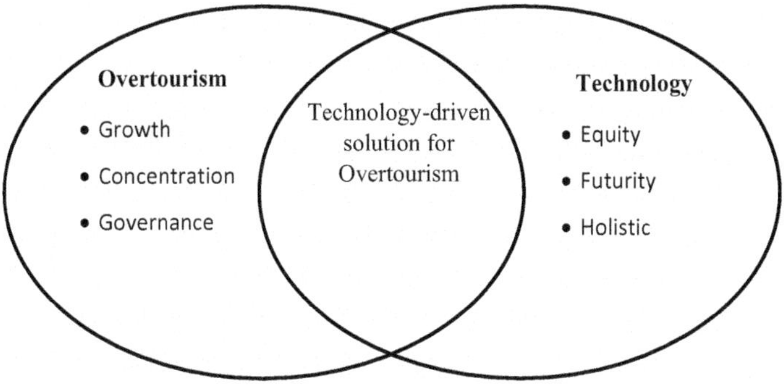

Fig. 6.1 Overtourism-technology synergy. (Source: Authors' own work, 2020)

Framework for Technology-Driven Solutions for Overtourism (FTSO)

Advanced ICT tools for destination management can contribute to reducing stress on overbooked destinations, planning for alternative destinations, targeting an alternate group of tourists, and most importantly to have control over the tourist inflow through systematic and planned destination management strategy. While literature concerning the usage of ICT for tourism promotion is in plenty, the aspect of destination management through the use of advanced technology has been rarely dealt with at a holistic level. This paper attempts to fill that gap by developing a framework for DMOs to mitigate the problems of overtourism through the usage of new advanced technology.

Based on the extensive review of the pertinent literature, a Framework for Technology-driven Solutions for Overtourism (FTSO). At first, the framework proposes to enlist generalized solutions and strategies to mitigate the problem of overtourism. It then links those strategies to various ICT tools and highlights the fact how the latest technology can provide holistic solutions to the majority of the issues involved in overtourism. To provide validation, the framework enlists several suitable pieces of literature (not all). The elements of the framework have been discussed below (Table 6.1):

Managing Visitors

Overtourism is a complex issue, and it varies from destination to destination. Not a simple solution can overcome the concern of overtourism, which needs long term planning at the destination level. Though uprooting the issue of overtourism is still quite challenging, regular monitoring and precautionary measures can slow down the issue. Overtourism can be managed through collaborative actions taken by stakeholders and tourists.

Table 6.1 Framework for Technology-Driven Solutions for Overtourism (FTSO)

Possible solutions	Strategies involved	Prospective technological interventions	References
Managing visitors	Monitoring	Artificial intelligence-based management	Zsarnoczky (2017)
		Location-based service	Yu and Chang (2009)
		Crowd dispersal systems	WTTC (2018)
	Work on tracking	GPS based tracking	Shoval and Ahas (2016)
		Geotagging	Edwards and Griffin (2013)
	Movement data	Tourist segmentation by e-data	Katsoni et al. (2013)
		Visitor movement data	Mckercher and Lau (2008)
	Analyzing visitor data	Big data analysis	Cillo et al. (2019)
Managing attractions and periphery	Alternative routes	Use of blogs and m-apps for alternate route promotion	Hospers (2019) and Vargas-Sánchez (2019)
	Places outside the center	Visitor management systems, GIS, 3D geospatial visualization	Hospers, (2019) and Zubiaga et al. (2019)
	De-seasonalization	Social media campaigns, market intelligence software	Ivars-Baidal et al. (2019)
	Safety and security	Express and secure service with biometrics	Leong and Li (2011), Rao (2018), and Nam et al. (2019)
Managing community resources	Identifying community resources	Geo-localized information and flow management system,	Ivars-Baidal et al. (2019) and Panayiotopoulos and Pisano (2019)
	Managing traffic	Intelligent transportation system	Zimmerman et al. (2004)
	Common development plans	IoT, big data analysis, and GIS	Erschbamer et al. (2018)
	Equalization	M apps for visitor allocation	Pasquinelli and Trunfio (2020)
Managing experiences	Creating immersive experiences	Augmented reality	Dewailly (1999) and Yung and Khoo-Lattimore (2017)
	Providing alternate experiences	Virtual reality	Hospers (2019) and Zubiaga et al. (2019)
	Providing remote experiences	VR and extended reality	Neuhofer et al. (2013)

(continued)

Table 6.1 (continued)

Possible solutions	Strategies involved	Prospective technological interventions	References
Managing promotions and creating awareness	Sensitizing stakeholders	Through smartphones and mobile applications	Dickinson et al. (2014) and Lu et al. (2015)
		E-commerce to m-commerce	Buhalis and O'Connor (2005)
	Developing actions	Social media campaigns	Sigala (2020)
		Marketing campaigns for overtourism awareness	Gowreesunkar and Thanh (2020)
Managing environmental impacts	Accessing the impacts on the environment	ICT enabled EIA, eGain forecasting	Du Cros and McKercher (2020)
	Preservation and protection of the environment	Green technology, green building, green energy	Gavrilović and Maksimović (2018)
	Enhancement of the environment	Smart electrical grids, energy-saving gains	van Zanten and van Tulder (2020)

Source: Authors' own work (2020)

Monitoring

The ICT development has altered the preference, search procedure, and decision-making process of the tourism consumers, mostly accomplished over the digital platform. The innovation of artificial intelligence (AI) helps in providing suggestions, and recommendations over the past search preferences are trending in the tourism industry as well (Zsarnoczky, 2017). When the industry demands efficiency, sustainability, and productivity AI plays a significant role in providing personalized solutions for visitors about tourism services, forecast tourism demands, and monitoring ways out for the stakeholders. AI-based services like automatic entry caps, no go zones, an alternate route to avoid congestion, and similar production recommendations are few monitoring measures for overtourism (Kuflinski, 2020).

The mobile revolution has developed in providing real-time services or location-based tourism systems in the industry base on GPS services, which can be used as a monitoring tool for overtourism management (Yu & Chang, 2009). Crowd dispersal is another significant way for overtourism management while monitoring the overcrowded destinations. AI audio guides, alternate route suggestions, and tourism gamification are such illustrations of crowd dispersal while speeding and managing the flow of tourists (WTTC, 2018).

Tracking

In monitoring and resource management, tracking technology gives numerous possibilities in tourism management and smart destination development (Shoval & Ahas, 2016). Technologies like smartphone-based GPS, social media geotagging, and hashtag analytics provides information about visitor's movement, behavioral intentions of tourists (Hardy et al., 2017). The visual tracking data provides extensive insights for the destination management organizations about the spatial use of the destination. The tracking technology can further be used as a diagnostic tool for overtourism to develop destinations, training organization staff, and providing the consumer with a better travel experience (Edwards & Griffin, 2013).

Visitor's Movement Data and Analysis

The visitor movement pattern is crucial for tourism stakeholders to plan for destination development, provide quality service to customers, and sustainable management of resources (Mckercher & Lau, 2008). GPS based visitors' movement is quite helpful in overtourism management by regulating the carrying capacity of destination and reduce conflicts between tourist and the locals. The tourist movement data can further be classified into different segments like tourist demographics, information search, trip purpose, and traveler's behavior to succeed long run tourism (Katsoni et al., 2013). The Digital revolution has made the segmentation more efficient, and further effective use of the data is found in marketing, policy management, and destination development. Tourism is a multifaceted industry associated with several sectors; hence the data management in tourism is a little complex in nature. The recent trend of big data analytics can be a solution for overtourism by collecting data on several factors associated with the domain. Big data analytics provides the opportunity for DMOs to develop strategies for any tourism situation (Cillo et al., 2019) similarly for overtourism as well.

Managing Attractions and Periphery

Overcrowding at particular attractions along with uneven distribution of visitors at various attractions stands as one of the biggest causes of overtourism. In this regard, different mechanisms to manage the attractions have been discussed.

Alternative Routes

Popular sites often suffer from long queues and traffic problems in the connecting roads. Most people prefer to visit the site from the popular routes. This forces local people to avoid these routes. Promoting secondary routes (Hospers, 2019) can be a solution to overcome such problems. Most visited tourist sites can be connected to multiple routes to disperse the tourists among different directions and connect them to navigation systems to access the less crowded routes to reach the destination. Making use of bloggers to promote these new accessible routes will make the job much easier. All available resources have to be used to divert tourists to less-visited places and hidden gems. The technology adopted by Valencia to promote alternative routes is unique and effective at the same time. They developed the "Alter Eco App" that shows new routes and local traditions to the tourists, thereby ensuring reduced crowding and better behavior of visitors (Vargas-Sánchez, 2019).

Places Outside Center

Every destination is famous for some specific honeypot sites. These are the places where most consequences of overtourism are faced. To achieve spatial distribution, promoting places outside the city centers is necessary. Though the primary solution is to distribute tourism to different places (Weber, 2017), diverting tourists from honeypot sites is not feasible (R. Dodds & R. Butler, 2019a) as tourists feel incomplete without visiting the popular sites. Distributing tourists to other places via promotions, new attractions will reduce seasonality (Peeters et al., 2018) Offering outside centers to tourists (Eckert et al., 2019), marketing unknown places (Hospers, 2019), diversified attractions like special events and incentives for target markets (Lee et al., 2008), de-centralization of tourists (Ivars-Baidal et al., 2019) are some of the solutions of spatial and seasonal distribution of tourists. Technology can be a major tool to distribute tourists among different locations while managing popular sites. Data-driven visitor management systems (Hospers, 2019) can help manage the visitors at sites. GIS is another tool that provides data on pedestrians' spatial distribution and 3D geospatial visualization, making interpreting geographic information easy for planning and managing the visitors (Zubiaga et al., 2019). Trento has come up with the technological innovation of placing Some Smart Points of Interest away from crowded hotspots to encourage visitors to enjoy the popular honeypots using physical web technology. London has also introduced Free gaming app that shows different attractions around the city and gives vouchers and discounts when those places are visited (Vargas-Sánchez, 2019).

De-seasonalization

While seasonality is an inseparable part of tourism, the issues with peak season are too many to be ignored. While complete closure of sites during peak season is not a viable option, de-seasonalization can be a strategy to mitigate the problems of seasonal peaks. De-seasonalization is nothing but promoting tourism in the offseason while limiting the numbers in peak season. But it is argued that tackling seasonality by promoting in non-peak season does not reduce the demand in the peak season (Dodds & Butler, 2019a, b). Strategies to promote off-season tourism include offers in shoulder seasons (Erschbamer et al., 2018), organization of events in off-peak season (Hospers, 2019), promoting low season, attracting new targeted groups (Eckert et al., 2019), Offering off-season holiday packages, developing new attractions activities and events, increase year-round visitation (Lee et al., 2008) and off-season incentives. Also, Campaigns can be organized to attract different markets in different seasons (Lee et al., 2008). De-seasonalization and development of products and events outside high season help in the better seasonal distribution of tourists (Ivars-Baidal et al., 2019). Social media-based marketing can be utilized to market off-seasons providing attractive offers. Market intelligence software can be used for the de-seasonalization of destinations and de-centralization of tourists (Ivars-Baidal et al., 2019).

Safety and Security

Exponential growth in the number of travelers can be observed globally every year, demanding fast-moving and secure services in the tourism industry. The tourism industry has been transformed into a digital medium; hence the issue of trust and security is a major concern at present. Several services like biometric recognition technology, digital identity services have overcome the traditional identification and authorization service, taking the tourism domain in a new direction (Rao, 2018). Radio Frequency Identification (RFID) technology is used in China to manage a large number of tourists providing quality of experience, security, and sustainable usage of resources during the peak tourism season (Leong & Li, 2011). Keeping given digital security in tourism a recent trend blockchain technology is getting attention and adopted by various organizations. The decentralized, anonymous, and peer to peer network-based blockchain technology provides secured, trusted, and cost-effective services (Nam et al., 2019). Though the technology is in the stage of infancy, the trend could influence smart tourism and smart city development in the recent future (Willie, 2019).

Managing Community Resources

Unplanned tourism development often comes with a cost of overtourism. The repercussions of overtourism are felt the most by the residents as it results in a deteriorated quality of life among residents. Hence residents' quality of life and well-being became an important factor due to increasing social movements against tourism. The actual phenomena of overtourism gained momentum with the protests of residents against tourism and tourists. Hence, managing the community resources plays a critical role in mitigating the negative impacts of overtourism (Erschbamer et al., 2018). Some of those aspects have been discussed below;

Identifying and Allocating Community Resources

This is an important measure to combat overtourism and promote tourism acceptance (Eckert et al., 2019). Though there is no specific explanation on how and on whom there should be no limitation, it is evident from the literature that limitation on resident usage of infrastructure should be avoided to control the negative perception of residents. Limiting resident's role in tourism develops a sense of avoidance among them. Plans like 'urban armature' can divide the traffic and promote the use of shared spaces by locals and tourists (Panayiotopoulos & Pisano, 2019). Geolocalized information can be used for accommodation and flow management at destinations so that it does not hinder the usage of spaces by residents (Ivars-Baidal et al., 2019).

Common Development Plans

Developing a destination solely to cater to tourists will never be successful until it is supported by the residents. Residents are an inseparable part of destination development. Overtourism occurs when tourists' priorities override the interests of local communities (Goodwin, 2017) and hence local urban planning (Goodwin, 2017) and keeping residents' welfare must above the needs of the global tourism supply chain (Milano et al., 2018) ensures social sustainability of the destination. A transparent and balanced approach with sustainability initiatives will make a positive reputation for tourism in public. (Alexis, 2017). A balanced destination should be attractive to tourists while being habitable to locals (Milano et al., 2019). The residents' concerns and ideologies of future tourism development must be considered to ensure common development plans for both the tourists and residents (Erschbamer et al., 2018). While cooperation among stakeholders is prominent in designing common plans (Eckert et al., 2019; Goodwin, 2019), destinations often suffer from an imbalance of power among stakeholders (R. Dodds & R. W. Butler, 2019b). Residents complain about having to compete with tourists for spaces, services, and amenities (R. Dodds & R. W. Butler, 2019b). Henceforth, development plans should ensure a better quality of life for residents. A smart destination strategy can be used for sustainable and inclusive development to improve residents' quality of life as well as visitor experience (Ivars-Baidal et al., 2019). Bottom-up technical and political approaches are needed to balance the economic benefits and social costs (Milano et al., 2018). This equalizing of supply and demand can be done by using timed tickets and dynamic pricing (Goodwin, 2019). If the destination management strategy has to be holistic, more technology inclusion is needed (Ivars-Baidal et al., 2019). IoT, Big Data, and Geographic Information Systems together provide all information for managing destinations (Zubiaga et al., 2019).

Equalization

The problems of overtourism are caused mainly due to the concentration of a large number of tourists at a particular place. This affects both the tourists' experience as well as the residents' perception of crowding. Overcrowding at destinations can be mitigated by the equalization of tourist flows among different sites of a destination (Namberger et al., 2019), thereby distributing tourist traffic (Erschbamer et al., 2018). Dynamic pricing strategies to attract target markets (Leeet al., 2008) and product diversification to distribute tourists (Peeters et al., 2018) can equalize the tourists to some extent. Smart solutions can manage the tourist flows, establish limits, and de-growth but should not be used for demand growth (Ivars-Baidal et al., 2019). Distribution of visitors to other regions through clear target marketing not only eases crowding and improves tourism satisfaction and quality but also achieves qualitative sustainable growth and balanced seasonal development (Song et al., 2020). Current and future technological advancements like apps to guide the visitors can be used to spread manage tourists (Ivanov & Webster, 2017) without compromising growth (Erschbamer et al., 2018).

Managing Experiences

As the Internet-based tourism or cyber tourism goes forward, the industry demands an innovative form of presentation. At present the artificial demonstration like virtual reality (VR) and augmented reality (AR) are becoming a more important tool in tourism marketing and destination management, entertainment, accessibility, and heritage preservation as well (Dewailly, 1999; Guttentag, 2010). Though AR and VR are related to each other, the major difference is the level of immersion. The environment in VR is completely virtual, whereas in AR the user still experiences the real world (Yung & Khoo-Lattimore, 2017). Virtual experience generates more positive emotion among customers, which influences their behavioral intention to visit a destination (Jung et al., 2017). Various tourism organizations like British Columbia, Tourism Australia, and the city of Valladolid in Spain have already been adopted AR and VR technology in their websites to provide virtual experience to attract customers (Yung & Khoo-Lattimore, 2017).

AR, VR, and ER can provide solutions to cater to the changing realities post-COVID-19. To begin with, one may discuss the power of AR and VR to provide a virtual experience of a place without physically going there. AR and VR can promote a sustainable configuration of tourism after the pandemic that will require tourists to abandon all unnecessary excessive movements given their negative impacts and experience them virtually (Loureiro et al., 2020). Lastly, these ICT tools can reduce the burden on overbooked attractions or the ones needing protection.

Managing Promotions and Creating Awareness

Technology has influenced the tourism industry in several manners, such as the practitioner's representation of destination and services, information search, decision making, and customer experience sharing. Every segment of tourism is advancing with technological innovations. Of those innovations, web-based promotions and dissemination of information have emerged as one of the crucial areas for managing overtourism. Some of the aspects have been discussed below;

Sensitizing Stakeholders

In the technological revolution, information and communication technology (ICT) is the most prolific innovation of the twenty-first century. ICT gets a rapid uptake through compatible smartphone technology associated with mobile apps. Along with voice service, messaging, internet, portable media player, and the compact camera makes the smartphone an all-in-one travel gadget (Dickinson et al., 2014). Travel becomes more unpretentious through real-time information search, while mid-travel experience sharing makes the travel memorable through smartphones and mobile apps. Ease of use, convenience, and compatibility influence travelers to adopt smart mobile apps in travel (Lu et al., 2015). Information, as well as other services, can be availed through smartphones and dedicated apps smoothly and expressly than the traditional way (Shoval & Ahas, 2016). The smart and secure services turned the facet of the travel industry towards e-commerce decades back when information, bookings, and few other services were availed to customers over the internet (Buhalis & O'Connor, 2005). As a result of the rapid digital revolution, traditional e-commerce has been transformed into mobile e-commerce (m-commerce) through wireless data, smartphones, and mobile applications from any corner of the world. Travelers use mobile phones for information seeking and entertainment purposes throughout their travel; hence, mobile-ready websites and user-friendly applications act as driver and marketing tools in the m-commerce era (Lei & Law, 2019).

Developing Actions

Social media plays a significant role in tourism by providing a medium for promotional activities, a communication platform between service providers and customers, and many more. Social media presentation has grown up to the next level with internet 2.0 through end to end communication and promotes tourism in a faster way. In the technological era, the video-sharing channels and blog contents promote tourism destinations to the glare of publicity in less time.

The nature of the tourist experience was influenced by the revolution of the ICT in the tourism industry. Previously technology plays a supplementary role in tourism experience building, whereas in the present scenario, technology became an integral part of tourism experience and, in some cases becoming the experience itself (Neuhofer et al., 2013). The technology has been developed from information sharing web content to interactive websites, social media, and mobile applications that are influencing the customer experience. The experience has been transformed from a passive experience to a technology-empowered experience. Technology has changed the scenario of information creation, processing, and exchange, creating tourism experience and shaping traditional e-tourism in the direction of smart tourism.

Managing Environmental Impacts

With a sudden surge in the number of tourists in overtourism, issues surrounding environmental sustainability begins to emerge (Zhong et al., 2008). ICT tools like E- Environmental Impact Analysis (EIA) and ICT Footprint measurement tool [see Bull and Kozak (2014)] can enhance the understanding of the destination carrying capacity. Further, advanced ICT-based tools can facilitate demarketing of the destinations (Clarke, 2005) and eco-labeling, Environment Management Information System (EMIS), and carbon calculators can help restore the environment (Ali & Frew, 2014). Last but not least, with the new technologies like Smart electrical grids, energy-saving gains, a lot of energy can be preserved, which will act towards reducing/balancing the negative impacts of higher tourist inflow with overtourism.

Conclusion

Overtourism is emerging out to be one of the biggest problems faced by the tourism industry, with negative impacts spread on to the majority of the stakeholders and especially the residents of the place. The problem of overtourism is multifaceted, and it calls for holistic solutions to reduce its negative impacts and to achieve the bigger goal of destination sustainability. Subsequently, ICT tools are also showing promising signs in driving the sustainability movement forward. These tools are integrated, holistic, futuristic, and ensures equity (Mohanty, 2020). Therefore, the possibilities of ICT tools reducing the negative impacts of overtourism are endless. While the aspect of ICT for tourism promotion has been thoroughly discussed in the existing literature, the aspect of technology-driven destination management has been rarely deliberated at a holistic level. This paper attempts to fill that gap by developing a framework for DMOs to mitigate the problems of overtourism through the usage of new advanced technology. Based on the review of secondary literature, this paper tries to highlight the challenges associated with overtourism and then deals with the ways by which the problems can be mitigated by the usage of advanced and updated technological tools. It is argued that this framework can act as a guiding material for various policymakers and DMOs in need of overcoming the issues of overtourism. (Du Cros & McKercher, 2020).

References

Alexis, P. (2017). Over-tourism and anti-tourist sentiment: An exploratory analysis and discussion. *Ovidius University Annals, Economic Sciences Series, 17*(2), 288–293.

Ali, A., & Frew, A. J. (2014). Technology innovation and applications in sustainable destination development. *Information Technology & Tourism, 14*(4), 265–290.

Benner, M. (2019). From overtourism to sustainability: A research agenda for qualitative tourism development in the Adriatic. *Zeitschrift für Wirtschaftsgeographie = German Journal of Economic Geography, 2*(6). Retrieved from https://ideas.repec.org/p/pra/mprapa/92213.html. Accessed: 15 Nov 2020.

Buhalis, D., & O'Connor, P. (2005). Information communication technology revolutionizing tourism. *Tourism Recreation Research, 30*(3), 7–16.

Bull, J. G., & Kozak, R. A. (2014). Comparative life cycle assessments: The case of paper and digital media. *Environmental Impact Assessment Review, 45*, 10–18.

Capocchi, A., Vallone, C., Pierotti, M., & Amaduzzi, A. (2019). Overtourism: A literature review to assess implications and future perspectives. *Sustainability, 11*, 3303.

Cillo, V., Rialti, R., Giudice, M. D., & Usai, A. (2019). Niche tourism destinations' online reputation management and competitiveness in big data era: evidence from three Italian cases. *Current Issues in Tourism, 24*(2), 177–191.

Clarke, J. (2005). Effective marketing for rural tourism. In D. R. Hall, I. Kirkpatrick, & M. Mitchell (Eds.), *Rural tourism and sustainable business* (Vol. 26, pp. 87–102). Channel View Publications.

Dewailly, J. M. (1999). Sustainable tourist space: From reality to virtual reality? *Tourism Geographies: An International Journal of Tourism, 1*(1), 41–55.

Dickinson, J. E., Ghali, K., Cherrett, T., Speed, C., Davies, N., & Norgate, S. (2014). Tourism and the smartphone app: Capabilities, emerging practice and scope in the travel domain. *Current Issues in Tourism, 17*(1), 84–101.

Dodds, R., & Butler, R. (2019a). The phenomena of overtourism: A review. *International Journal of Tourism Cities, 5*(4), 519–528.

Dodds, R., & Butler, R. W. (2019b). *The enablers of overtourism*. Retrieved from https://www.degruyter.com/document/doi/10.1515/9783110607369-002/pdf. Accessed 12 Dec 2020.

Doxey, G. V. (1975). *A causation theory of visitor-resident irritants: Methodology and research inferences*. Paper presented at the Travel and tourism research associations sixth annual conference proceedings.

Du Cros, H., & McKercher, B. (2020). *Cultural tourism*. Routledge.

Duignan, M. (2019). 'Overtourism'? Understanding and managing urban tourism growth beyond perceptions: Cambridge case study: Strategies and tactics to tackle overtourism. In *'Overtourism'? Understanding and managing urban tourism growth beyond perceptions: Case studies* (pp. 34–39). UNWTO.

Eckert, C., Zacher, D., Pechlaner, H., Namberger, P., & Schmude, J. (2019). Strategies and measures directed towards overtourism: A perspective of European DMOs. *International Journal of Tourism Cities, 5*(4), 639–655.

Edwards, D., & Griffin, T. (2013). Understanding tourists' spatial behaviour: GPS tracking as an aid to sustainable destination management. *Journal of Sustainable Tourism, 21*(4), 580–595.

Erschbamer, G., Innerhofer, E., & Pechlaner, H. (2018). *Overtourism: How much tourism is too much*. Eurac Research.

Gavrilović, Z., & Maksimović, M. (2018). Green innovations in the tourism sector. *Strategic Management, 23*(1), 36–42.

Goodwin, H. (2017). *The challenge of overtourism*. Retrieved from https://haroldgoodwin.info/wp-content/uploads/2020/08/rtpwp4overtourism012017.pdf. Accessed: 12 Dec 2020.

Goodwin, H. (2019). *Overtourism: Causes, symptoms and treatment*. Retrieved from https://responsibletourismpartnership.org/wp-content/uploads/2019/06/TWG16-Goodwin.pdf. Accessed: 12 Dec 2020.

Gowreesunkar, V. G., & Thanh, T. V. (2020). Between overtourism and under-tourism: Impacts, implications, and probable solutions. In H. Séraphin, T. Gladkikh & T. V. Thanh (Eds.), *Overtourism* (pp. 45–68). Springer.

Gretzel, U. (2019). The role of social media in creating and addressing overtourism. In R. Dodds & R. Butler (Eds.), *Overtourism: Issues, realities and solutions* (pp. 62–75). De Gruyter.

Guttentag, D. A. (2010). Virtual reality: Applications and implications for tourism. *Tourism Management, 31*, 637–651.

Hardy, A., Hyslop, S., Booth, K., Robards, B., Aryal, J., Gretzel, U., & Richard, E. (2017). Tracking tourists' travel with smartphone-based GPS technology: A methodological discussion. *Information Technology & Tourism, 17*, 255–274.

Higgins-Desbiolles, F., Carnicelli, S., Krolikowski, C., Wijesinghe, G., & Boluk, K. (2019). Degrowing tourism: rethinking tourism. *Journal of Sustainable Tourism, 27*(12), 1926–1944.

Hospers, G.-J. (2019). Overtourism in European cities: From challenges to coping strategies. *Paper presented at the CESifo Forum, ifo Institut – Leibniz-Institut für Wirtschaftsforschung an der Universität München, München, 20*(3), 20–24.

Ivanov, S. H., & Webster, C. (2017). Adoption of robots, artificial intelligence and service automation by travel, tourism and hospitality companies–a cost-benefit analysis. *Prepared for the International Scientific Conference "Contemporary Tourism – Traditions and Innovations"*. Sofia University: The 19th-21st October, 2017.

Ivars-Baidal, J. A., Hernández, M. G., & de Miguel, S. M. (2019). Integrating overtourism in the smart tourism cities Agenda. *e-Review of Tourism Research, 17*(2), 122–139.

Jung, T., Dieck, C. T., Moorhouse, N., & Dieck, D.t. (2017). *Tourists' experience of virtual reality applications*. Paper presented at the IEEE International Conference on Consumer Electronics (ICCE), Manchester.

Katsoni, V., Giaoutzi, M., & Nijkamp, P. (2013). Market segmentation in tourism: An operational assessment framework. In Á. Matias, P. Nijkamp and M. Sarmento (Eds.), *Quantitative methods in tourism economics* (pp. 329–352). Physica.

Koens, K., & Postma, A. (2020). *'Overtourism'? – Understanding and managing urban tourism growth beyond perceptions*. UNWTO. Retrieved from https://www.e-unwto.org/doi/book/10.18111/9789284420070. Accessed: 12 Dec 2020.

Koens, K., Postma, A., & Papp, B. (2018). Is overtourism overused? Understanding the impact of tourism in a city context. *Sustainability, 10*(12), 4384.

Kuflinski, Y. (2020). *10 AI travel trends for 2020 and beyond*. iflexion.com. Retrieved from https://www.iflexion.com/blog/ai-travel. Accessed: 12 Dec 2020.

Lawton, T. C. (2017). *Cleared for take-off: Structure and strategy in the low fare airline business*. Routledge.

Lee, C., Bergin-Seers, S., Galloway, G., O'Mahony, B., & McMurray, A. (2008). *Seasonality in the tourism industry: Impacts and strategies*. Australia: Sustainable Tourism CRC.

Lei, S., & Law, R. (2019). Functionality evaluation of mobile hotel websites in the m-commerce era. *Journal of Travel and Tourism Marketing, 36*(6), 665–678.

Leong, A. M. W., & Li, X. (2011). A study on tourist management in China based on radio frequency identification (RFID) technology. In A. M. W. Leong & X. Li (Eds.), *Global hospitality and tourism management technologies* (pp. 12–26). IGI Global.

Loureiro, S. M. C., Guerreiro, J., & Ali, F. (2020). 20 years of research on virtual reality and augmented reality in tourism context: A text-mining approach. *Tourism Management, 77*, 104028.

Lu, J., Mao, Z., Wang, M., & Hu, L. (2015). Goodbye maps, hello apps? Exploring the influential determinants of travel app adoption. *Current Issues in Tourism, 18*(11), 1059–1079.

Mahizhnan, A. (1999). Smart cities: the Singapore case. *Cities, 16*(1), 13–18.

Mckercher, B., & Lau, G. (2008). Movement patterns of tourists within a destination. *Tourism Geographies, 10*(3), 355–374.

Milano, C., Cheer, J. M., & Novelli, M. (2018). Overtourism: A growing global problem. *The conversation*. Retrieved from https://theconversation.com/overtourism-a-growing-global-problem-100029. Accessed: 12 Dec 2020.

Milano, C., Novelli, M., & Cheer, J. M. (2019). Overtourism and tourismphobia: A journey through four decades of tourism development, planning and local concerns. *Tourism Planning & Development, 16*(4), 353–357.

Mohanty, P. (2020). ICT and sustainable development: Implications for the tourism industry. . In A. Hassan & A. Sharma (Eds.), *The emerald handbook of ICT in tourism and hospitality* (pp. 259–272). Emerald Publishing.

Mrsic, L., Surla, G., & Balkovic, M. (2020). Technology-driven smart support system for tourist destination management organizations. In A. Khanna, D. Gupta, S. Bhattacharyya, V. Snasel, J. Platos & A. Hassanien (Eds.), *International Conference on Innovative Computing and Communications. Advances in Intelligent Systems and Computing* (pp. 65–76). Springer.

Nam, K., Dutt, C. S., Chathoth, P., & Khan, M. S. (2019). Blockchain technology for smart city and smart tourism: latest trends and challenges. *Asia Pacific Journal of Tourism Research*, 1–15.

Namberger, P., Jackisch, S., Schmude, J., & Karl, M. (2019). Overcrowding, overtourism and local level disturbance: how much can Munich handle? *Tourism Planning & Development, 16*(4), 452–472.

Neuhofer, B., Buhalis, D., & Ladkin, A. (2013). A typology of technology-enhanced tourism experiences. *International Journal of Tourism Research, 16*, 340–350.

O'Reilly, A. M. (1986). Tourism carrying capacity: Concept and issues. *Tourism Management, 7*(4), 254–258.

Olson, M. (1996). Distinguished lecture on economics in government: Big bills left on the sidewalk: Why some nations are rich, and others poor. *Journal of Economic Perspectives, 10*(2), 3–24.

Panayiotopoulos, A., & Pisano, C. (2019). Overtourism dystopias and socialist utopias: Towards an urban armature for Dubrovnik. *Tourism Planning & Development, 16*(4), 393–410.

Pasquinelli, C., & Trunfio, M. (2020). Reframing urban overtourism through the Smart-City Lens. *Cities, 102*. https://doi.org/10.1016/j.cities.2020.102729

Peeters, P., Gössling, S., Klijs, J., Milano, C., Novelli, M., Dijkmans, C., Eijgelaar, E., Hartman, S., Heslinga, J., Isaac, R., Mitas, O., Moretti, S., Nawijn, J., Papp, B., & Postma, A. (2018). *Research for TRAN committee – Overtourism: Impact and possible policy responses*. European Parliament, Policy Department for Structural and Cohesion Policies.

Perkumienė, D., & Pranskūnienė, R. (2019). Overtourism: Between the right to travel and residents' rights. *Sustainability, 11*(7), 2138.

Rao, K. (2018). The promise of biometric identification in travel industry. *Mindtree*. Retrieved from: https://www.mindtree.com/blog/promise-biometric-identification-travel-industry. Accessed: 19 Nov 2020.

Shoval, N., & Ahas, R. (2016). The use of tracking technologies in tourism research: the first decade. *Tourism Geographies, 18*(5), 587–606.

Sigala, M. (2020). Tourism and COVID-19: impacts and implications for advancing and resetting industry and research. *Journal of Business Research, 117*, 312–321.

Smith, M. K., Sziva, I. P., & Olt, G. (2019). Overtourism and resident resistance in Budapest. *Tourism Planning & Development, 16*(4), 376–392.

Song, H., Kim, M., & Park, C. (2020). Temporal distribution as a solution for over-tourism in night tourism: The case of Suwon Hwaseong in South Korea. *Sustainability, 12*(6), 2182.

van Zanten, J. A., & van Tulder, R. (2020). Towards nexus-based governance: defining interactions between economic activities and Sustainable Development Goals (SDGs). *International Journal of Sustainable Development & World Ecology, 28*(3), 210–226.

Vargas-Sánchez, A. (2019). Overtourism: A challenge for smart cities. Paper presented at the *Fomento de la Cultura Científica, Tecnológica y de Innovación en Ciudades Inteligentes.*

Weber, F. (2017). Overtourism. An analysis of contextual factors contributing to negative developments in overcrowded tourism destinations. *Paper presented at the BEST EN Think Tank XVII: Innovation and Progress in Sustainable Tourism.* University of Mauritius: the 14th–17th June 2017.

Willie, P. (2019). Can all sectors of the hospitality and tourism industry be influenced by the innovation of Blockchain technology? *Worldwide Hospitality and Tourism Themes, 11*(2), 112–120.

WTTC. (2018). *How can new technologies help deal with overcrowding?* Retrieved from: https://medium.com/@WTTC/how-can-new-technologies-help-deal-with-overcrowding-dda554fb164b. Accessed: 12 Dec 2020.

Yu, C.-C., & Chang, H.-P. (2009). *Personalized location-based recommendation services for tour planning in mobile tourism applications.* In T. Di Noia & F. Buccafurri (Eds.), E-Commerce and Web Technologies. EC-Web 2009 (pp. 38–49), Springer.

Yung, R., & Khoo-Lattimore, C. (2017). New realities: A systematic literature review on virtual reality and augmented reality in tourism research. *Current Issues in Tourism, 22*(17), 1–26.

Zhong, L., Deng, J., & Xiang, B. (2008). Tourism development and the tourism area life-cycle model: A case study of Zhangjiajie National Forest Park, China. *Tourism Management, 29*(5), 841–856.

Zimmerman, C. A., Daigle, J., & Pol, J. (2004). Tourism business and intelligent transportation systems: Acadia National Park, Maine. *Transportation research Record, 1895*(1), 182–187.

Zsarnoczky, M. (2017). How does artificial intelligence affect the tourism industry? *VADYBA, 31*(2), 85–90.

Zubiaga, M., Izkara, J. L., Gandini, A., Alonso, I., & Saralegui, U. (2019). Towards smarter management of overtourism in historic centres through visitor-flow monitoring. *Sustainability (Switzerland), 11*(24), 7254.

Part IV
Psychology, Community and Environmental Issues

Chapter 7
Residents' Perception of Overtourism, Tourism Impacts and Economic Dependence in Gökceada Island

Neşe Kafa and Halil Korkmaz

Abstract The aim of this study is to identify and investigate local inhabitants' perceptions of tourism's socio-cultural effects in Gökceada, Turkey. A total of 273 Gökceada residents were included in the study. From June through December 2019, survey questionnaires were used to collect data for the study in Gökceada. Eight questions about tourism's contribution to Gökceada inhabitants, as well as a 23-item scale measuring residents' perceptions of tourist impacts, were included in the survey's 24 questions. According to the findings of the study, inhabitants of Gökceada see particular positive and bad consequences of tourism in their village. The supply of jobs, increased personal incomes, stimulation of the local economy, and better security in the destination region are only a few of the beneficial effects of tourism. The majority of inhabitants of Gökceada want to see development tourism grow in the future. As a result, the paper will aid local government and tourism planners in bolstering local support for tourism.

Keywords Overtourism · Residents' perception · Impacts · Economic dependence · Gökceada Island

Introduction

Many visitors come to tourist areas in every country in the world. Thus, the increase of tourists causes more crowd in touristic places. This situation is seen as positive for tourist places. Nevertheless, once the destination's carrying capacity is

---

N. Kafa (✉)
Gökçeada School of Applied Sciences, Çanakkale Onsekiz Mart University, Çanakkale, Turkey
e-mail: nesekafa@comu.edu.tr

H. Korkmaz
Faulty of Tourism, Çanakkale Onsekiz Mart University, Çanakkale, Turkey

A. Hassan, A. Sharma (eds.), *Overtourism, Technology Solutions and Decimated Destinations*, https://doi.org/10.1007/978-981-16-2474-2_7

surpassed, too many tourists might cause significant difficulties. This issue, which is frequently referred to as "overtourism", appears to have increased in recent years (Weber et al., 2017). Overtourism was mentioned by UNWTO at World Travel Market "Ministerial Summit" in 2017 (Oklevik et al., 2019).

> *"Overtourism describes destinations where hosts or guests, locals or visitors, feel that there are too many visitors and that the quality of life in the area or the quality of the experience has deteriorated unacceptably"* (Goodwin, 2017: 1). On the other hand, over-tourism is defined *"as the excessive growth of visitors leading to overcrowding in areas where residents suffer the consequences of temporary and seasonal tourism peaks, which have enforced permanent changes to their lifestyles, access to amenities and general well-being"* (Milano et al., 2018: 2).

The causes of overtourism are the falling cost of travel, affordable accommodation enabling more people to travel, the public realm is free, distribution strategies, whether the spreading of tourism impacts is a result of local government action, seasonality bunches tourism concentrates numbers, tourism creates jobs, new originating markets, successfully marketed, established destinations, which attract more tourists, destination marketing organizations, transport is a larger scale than it was ten years ago (Goodwin, 2017). Aside from that, cities are growing economically in order to fulfil the demands of tourists. As a result, local communities' living costs rise (Milano et al., 2018). Residents find it impossible to do things like relax outside, go local shopping, use public transit, or just walk around the city due to the consequences of overtourism (Pearce, 2018). Locals have both good and negative opinions of tourism's effects as a result of these factors (Jacobsen et al., 2019). As a result, residents' support for tourism is contingent on how good and bad impacts are seen by locals (Park et al., 2012; Peric et al., 2016). Understanding locals' opinions and attitudes is therefore critical for the long-term growth of tourism (Gutiérrez-Taño et al., 2019).

The quality of life is linked to how residents perceive tourism's influence (Andereck & Nyaupane, 2011; Shani & Pizam, 2012). Reduced life quality as a result of congestion, bad working conditions, excessive costs, or improper visitor behaviour, environmental consequences (trash, noise, pollution, etc.), leaks or unilateral benefits, and poor quality of life and security issues as a result of crime and corruption. Aesthetic degradation, misuse of infrastructure, traffic issues, and insufficient execution of tourist initiatives are all typical issues (Weber et al., 2017). Abdool (2002)'s study determined that there are three perceptions of the resident. The first perception is community attachment. This is a measure of residents' sense of belonging to their community. The second one is personal and community benefits. It is an indication of the value residents attaches to the benefits of the tourism industry. The third one is socio-environmental impacts. This is an assessment of the consequences residents attribute to tourism development. Andereck and Vogt (2000) and Javari discussed these effects in three groups as social, economic, and environmental.

A wide variety of destinations are now described as experiencing overtourism (Express Newspaper, 2017). Palma de Mallorca, Paris, Kyoto, Berlin, Reykjavik, Venice, Majorca, Barcelona, Boracay Island, Amsterdam, Angkor Wat Temple,

Galapagos Islands, Macchu Picchu Ancient City, Iceland, Santorini, Dubrovnik, Bali, some national parks in the USA, Thailand's Islands and Mount Everest are negatively affected by overtourism (Cnt Editors, 2018; Milano et al., 2018). Turkey's largest island of Gökçeada is an example of the tourist season for tourism because of the extreme intensity. Gökçeada is a lively place in terms of both internal and external migrations, and a number of national and international decisions have determined the population density of the island. The island has housed a large number of seafarers, invaders, and immigrants since hundreds of years ago. The dynamism in this population change continues today as a demographic feature of the island (Kahraman, 2005). Additionally, the population of the island has increased to 9440 (Gökçeada Municipality, 2020). Gökçeada is a place frequented by local and foreign tourists, especially during the summer months. Ancient Greek villages, organizing festivals in Tepeköy every year, benefiting from the Salt Lake in Aydıncık for health, cleaning the beaches, sea sports activities such as sailing, surfing, diving are among the interesting touristic opportunities of Gökçeada (http://www.Gökçeada.com). Accordingly, the aim of this study is to determine the perceptions of the residents of Gökçeada towards the development of tourism.

Literature Review

Many studies emphasize the concept of overtourism and examine different issues. These include spreading tourists to regions, measuring segment tourists according to their income, purpose, etc., measuring the perception of local people towards crowding, focusing on overtourism and alternative accommodation methods, and offering solutions (Bresson & Logossah, 2011; Duyar & Bayram, 2019; Gutiérrez-Taño et al., 2019; Kuščer & Mihalič, 2019; Oklevik et al., 2019; Panayiotopoulos & Pisano, 2019; Pinke-Sziva et al., 2019; Şahin, 2018). At the same time, the reaction of local people to tourists is a previously discussed topic in the literature (Jacobsen, 2000; Pinkster & Boterman, 2017; Rouleau, 2017; Sommer & Helbrecht, 2017; Pinke-Sziva et al., 2019).

According to many studies, local people who are disturbed by the negative consequences of excessive tourism, such as rising house rents, environmental pollution, and difficult daily life, organize mass protests (Alexis, 2017; Coldwell, 2017; Peter, 2017; Milano et al., 2018; Kuščer & Mihalič, 2019). Jacobsen (2000) focused on the "Anti-Tourist" behavior developed by the local people against the tourists concentrated in destinations such as Greece and Majorca. Some studies found that participants perceived the social effects of tourism negatively (Tatoglu et al., 2002; Öztürk et al., 2015; Boğan & Sarıışık, 2016; Rouleau, 2017; Koens et al., 2018; Martín et al., 2018; Pinke-Sziva et al., 2019). These effects are urban sprawl and population growth, development- land-use conflicts, availability of retail outlets, crowding and congestion, peace and quiet, and drug and alcohol addiction. The factors negatively evaluated by Barcelona's people are the deterioration in the quality of personal life due to tourism development, peace and tranquillity, and the protection of the lifestyle (Martín et al., 2018:10). These are due to negative effects on the livelihood of citizens (Martín et al., 2018). Many studies found that participants perceived the economic effects of tourism negatively (Rouleau, 2017; Koens et al., 2018; Martín et al., 2018; Pinke-Sziva et al., 2019). These negative economic impacts increase in the price level of rental houses, taxis, shops, restaurants and cafes, public transportation, and entertainment facilities (UNWTO, 2018). Otherwise, in Boğan and Sarıışık (2016)'s study, it was determined that the participants have a positive view of the economic effects of tourism.

In many studies, the relationship between socio-demographic characteristics of local people and the effects of perceived tourism has been examined. Several studies have discovered a link between gender and perceptions of tourism's effects (Mason & Cheyne, 2000; Harvey et al., 1995; Williams & Lawson, 2001; Ritchie & Inkari, 2006). Many research have found that women have more unfavourable opinions regarding tourism's influence than men (Harvey et al., 1995; Mason & Cheyne, 2000; Williams & Lawson, 2001; Ritchie & Inkari, 2006). Tourism, according to women, has a detrimental influence on leisure options (Harvey et al., 1995), entertainment opportunities, and infrastructure issues (Mason & Cheyne, 2000). On the other hand, Korça (1998) and Akdu and Ödemiş (2018) could not find a significant difference in the participants' attitudes towards tourism according to their gender. Some researchers discovered age-related variations in impact perception (Smith & Krannich, 1998; Teye et al., 2002). King et al. (1993) found that older participants had a more positive opinion of tourism impacts than younger participants Haralambopoulos and Pizam (1996), Weaver and Lawton (2001), and Martín et al. (2018) found that young people perceive the effects of tourism more positively than the elderly. Otherwise, Korça (1998) and Akdu and Ödemiş (2018) could not find a substantial difference in the attitudes of the participants towards tourism according to their ages. Several studies determined that individuals with higher education levels have more positive attitudes towards tourism (Haralambopoulos & Pizam, 1996; Korça, 1998; Teye et al., 2002; Kuvan & Akan, 2005). Teye et al. (2002) discovered that when people' educational levels improved, so did their opinions regarding cultural influences. In contrast, Martín et al. (2018) did not found the difference between the effects of tourism and the education level of residents. Teye et al. (2002)'s study indicated a positive relationship between income and attitude scores. Martín et al. (2018) found that low and middle-income residents are more negatively view of the tourism impacts. On the other hand, Korça (1998) could not find a significant difference in the participants' attitudes towards tourism according to their income.

Korça (1998) could not find a significant difference in the participants' attitudes towards tourism according to their professions. Korça (1998) could not find a significant difference between resident's marital status and perceptions of tourism impacts.

Many studies accepted that perceived personal benefits from tourism are positively associated with an attitude that supports additional tourism development (Lankford & Howard, 1994; Haralambopoulos & Pizam, 1996; Vargas-Sánchez et al., 2011; Gonzalez et al., 2018). Haralambopoulos and Pizam (1996) found that the employed residents supported more arrivals and further development. On the other hand, Martín et al. (2018) found that the local people reject tourism due to the economic effects of tourism. Additionally, Ritchie and Inkari (2006)'s study determined that there were not any significant results in impact perception relate to the willingness to accept more tourism. Additionally, residents, who participate in tourism-related decisions, perceive the effects of tourism more positively (Choi & Murray, 2010).

Many studies determined that residents who benefit economically from tourism perceive the effects of tourism more positively. (Korça, 1998; Andereck et al., 2005; Kuvan & Akan, 2005; Inbakaran & Jackson, 2006; Sharma & Dyer, 2009; Ward & Berno, 2011; Garcia et al., 2015). Additionally, many studies have also confirmed that the employed residents in tourism had positive perceptions of tourism impact. (Williams & Lawson, 2001; Gursoy et al., 2002; Brent & Incari, 2006; Andereck & Nyaupane, 2011; Gonzalez et al., 2018). Martín et al. (2018) found that the differences between the social and economic impacts of tourism and residents who earn income from tourism.

Research Methodology

The main population of this research is the individuals living in Gökceada. Gökceada's population is approximately 9780 people, according to 2018 figures. In addition, it is known that the population on the island decreases in winter due to the difficulty of transportation from the mainland to the island and the difficult living standards in winter participants.

The survey method was determined as a data collection technique. The convenience sampling method was used in the study. A questionnaire was applied to 276 local people who wanted to participate in the survey between June 2019 and December 2019. Due to the fact that 3 questionnaires were left blank among the surveys, they were excluded from the sample. So that, the sample of the research consists of 273 people. Attitudes toward tourism development scale were used in the study. Twenty expressions of attitudes toward tourism development scale were adapted from the study of Wang and Pfister (2008). The two expressions about overall impact were adopted from Andriotis and Vaughan (2003), and one expression about grand satisfaction was gotten from Jani (2018). The statements to measure Attitudes toward tourism development statements were measured with five-point Likert scale (1: strongly disagree and 5: totally agree). Before starting the data analysis, the skewness and kurtosis values were checked to determine whether the data showed normal distribution. In social sciences, when exceeding 30 samples, it is assumed that the data approach to a normal distribution. Also, it has been accepted that the data used in the study show a normal distribution since the skewness level is between ±2 and the kurtosis level is between ±7 (West et al., 1995; Kline, 1998).

In the research, data were obtained with the help of the questionnaire, the SPSS 21 package program, which is used in the analysis of quantitative research in the field of social sciences. In this context, explanatory factor analysis, independent sample t-test, one-way analysis of variance (ANOVA), and linear regression analysis were performed.

Results

As seen in Table 7.1, the descriptive characteristics of the participants included in the study were evaluated using the statistics of frequency and percentage values. Approximately 57% of the 273 respondents are male, and approximately 43% are women. It is seen that approximately one-third of the participants are between the ages of 18–24. Approximately 29% of the people participating in the research are from the private sector. It was determined that the majority (55.6%) of the subjects in the study were married. More than one-third of the participants (35.2%) reported that they were high school graduates. The majority of the respondents (56.4%) stated that they earned less than 2500 TL. Again, it is seen that the majority of the participants (53.4%) have lived in Gökceada for more than ten years (Table 7.1).

Participants were asked about their views on the development of tourism in Gökceada, and the answers are given in Table 7.2. According to this Table, approximately two-thirds of the participants think that tourism has developed in Gökceada in the last 5 years. Again, the vast majority of the participants (73.6%) stated that tourism in Gökceada should develop further in the future.

An explanatory factor analysis (EFA) was conducted to determine the construct validity of the Residents' Attitudes toward Tourism Development scale used in the study. KMO value (0.801) Bartlett's Sphericity test results were found as significant ($x^2 = 931.431$; $p < 0.000$). When looking at the chi-square and KMO values, it can be said that the sample is suitable and sufficient for factor analysis.

During the analysis, three expressions with common variance values below 0.50 were removed from the scale. In addition, four expressions with factor load less than 0.5 were not included in the analysis result Tables. As a result of the explanatory factor analysis, it was seen that 16 expressions for Residents' Attitudes toward Tourism Development scale were grouped under four dimensions and explained 52.28% of the total variance (Table 7.3).

The first dimension, called "support for tourism development", has 2.67 eigenvalues and explains 16.7% of the variance. The second dimension, called "Contributions to community", has an eigenvalue of 2.07 and explains 12.9% of the variance. The third dimension, called "destination living standards" and the fourth dimension, called "Grand satisfaction" have each eigenvalues of 1.81 and explain 11.3% of the variance.

In addition, reliability analysis for the scale was made. Cronbach's Alpha value was found to be moderately reliable as 0.793. Considering the results of the reliability analysis regarding the dimensions, the Support for the tourism development dimension was found to be reliable at a medium level. Other dimensions were found

Table 7.1 Demographic profile of respondents

Variables		n	Sample %
Gender	Female	117	42.86
	Male	156	57.14
Age	18–24 years	85	31.14
	25–34 years	68	24.91
	35–44 years	54	19.78
	45–54 years	36	13.19
	55 or older	30	10.99
Occupation	Private sector	79	28.94
	Self-employed	49	17.95
	Public services	40	14.65
	Student	37	13.55
	Retired & housewife	35	12.82
	Unemployed	33	12.09
Marital status	Single	152	55.68
	Married	121	44.32
Education level	Elementary school	60	21.98
	High school	96	35.16
	Associate degree	56	20.51
	Undergraduate	48	17.58
	Graduate	13	4.76
Monthly income	Less than 2500 TL	154	56.41
	2500–5000 TL	84	30.77
	Over than 5000 TL	35	12.82
Duration of residence	Not a resident	30	10.99
	Less than 5 years	63	23.08
	5–10 years	34	12.45
	More than 10 years	146	53.48
Total		*273*	*100.00*

Table 7.2 Respondents' opinion about improvement of tourism in Gökceada

Variables		n	Sample %
Has tourism in Gökceada improved in the last five years?	Yes	185	67.77
	No	88	32.23
How would you like tourism in Gökceada to change in the future?	Remain the same	72	26.37
	Increase	201	73.63
Total		273	100.00

Table 7.3 Dimensions of residents' attitudes toward tourism development

Dimension/Items	x̄	Factor loadings	Eigen value	Variances explained (%)	Cronbach's alpha
Support for the growth of tourism (x̄ = 3.89)					
Tourism should be extensively promoted in Gökceada, in my opinion	3.80	0.774	2.67	16.69	0.756
Tourism should be aggressively promoted in my neighbourhood, in my opinion	4.01	0.703			
It is critical to create plans to control tourism expansion.	3.89	0.673			
I support tourism and want to see it grow into a significant component of my town.	4.12	0.658			
The local administration was proper in its support for tourism development.	3.75	0.585			
Contributions to community (x̄ = 3.14)					
As a result of increased tourism in my community, the quality of public services has improved.	3.09	0.805	2.07	12.94	0.644
Because of the inflow of tourists in my neighbourhood, local recreation activities have grown.	3.20	0.696			
Since the arrival of tourists, I have had additional recreational options.	3.00	0.606			
The tourist amenities in my community have enhanced the quality of life in my community.	3.26	0.579			
Destination living standards (x̄ = 3.55)					
Long-term planning by local leaders can mitigate the harmful environmental effects of tourism.	3.70	0.781	1.81	11.34	0.616
Because of the money spent by tourists, our household quality of living has improved.	3.42	0.563			
One of the most significant advantages of tourism is how it may raise local living standards.	3.63	0.545			
My neighborhood should become more of a tourism attraction.	3.53	0.533			
Grand satisfaction (x̄ = 3.41)					
Overall, the tourist advantages outweigh the expenses to Gökceada as a whole.	3.38	0.782	1.81	11.32	0.656

(continued)

Table 7.3 (continued)

Dimension/Items	$\bar{x}$	Factor loadings	Eigen value	Variances explained (%)	Cronbach's alpha
Overall, I am satisfied with Gökceada's tourism.	3.48	0.748			
Overall, the advantages of tourism outweigh the costs to the local community.	3.42	0.665			
KMO sampling adequacy: 0.801 Barlett Chi-Square: 931.431 p: 0.000 Total variances explained: 52.28%					

to be acceptably reliable as contributions to community (0.644), destination living standards (0.616), and grand satisfaction (0.656).

The overall average of the Residents' Attitudes toward Tourism Development scale was determined to be 3.53. As the averages of the dimensions are examined, above average participation has been observed support for tourism development ($\bar{x} = 3.89$), contributions to community ($\bar{x} = 3.14$), destination living standards ($\bar{x} = 3.55$), and grand satisfaction ($\bar{x} = 3.41$) (Table 7.3).

Independent sample t-test and one-way analysis of variance were used to test the differences of dimensions' mean between demographic groups. As a result of the t-test conducted in terms of gender, it was not determined ($p > 0.05$) that any mean of dimension is different. In other words, it can be said that the attitudes of male and female towards the development of tourism in Gökceada are similar.

An Independent sample t-test was conducted to determine whether the averages of dimensions of Residents' Attitudes toward Tourism Development differed in terms of marital status. Again, it was determined that the dimensions of Residents' Attitudes toward Tourism Development did not differ statistically ($p > 0.05$) in terms of marital status. In other words, married and single participants exhibit similar attitudes to tourism development in Gökceada.

One way analysis of variance was conducted to determine whether the averages of dimensions of Residents' Attitudes toward Tourism Development differed in terms of respondents' age. Again, it was determined that the dimensions of Residents' Attitudes toward Tourism Development did not differ statistically ($p > 0.05$) in terms of age group. From here, it can be concluded that the attitude of tourism to development in Gökceada does not change according to age.

One way analysis of variance was conducted to determine whether the averages of dimensions of Residents' Attitudes toward Tourism Development differed in terms of respondents' professions. Again, it was determined that "Support for tourism development", "Destination living standards", and "grand satisfaction" dimensions did not statistically vary ($p > 0.05$) in terms of professions. On the other hand, it is seen that the average of contributions to community dimension differed statistically in terms of profession. LSD test from post hoc tests was used to find the source of the difference in mean. While private sector and self-employed people are found to have a higher average than retired and housewives, the unemployed and public employees have higher average participation than all other groups (Table 7.4).

Table 7.4 Comparison of contribution to community dimension means in terms of the profession

Factor	Group		n	$\bar{x}$	s.d.	F	p.	Source of difference
Contributions to community	1	Private sector	79	3.08	0.804	2.392	0.038	5 < 1–2 < 3–6
	2	Self-employed	49	3.03	0.823			
	3	Public services	40	3.40	0.921			
	4	Student	37	3.09	0.908			
	5	Retired&Housewife	35	2.90	0.767			
	6	Unemployed	33	3.42	0.754			
		Total	273	3.14	0.840			

Table 7.5 Comparison of contribution to community scale means in terms of education level

Factor	Grup		n	$\bar{x}$	s.d.	F	p.	Source of difference
Contributions to community	1	Primary school	60	3,28	0,797	2940	0,021	4 < 3 < 2 < 1
	2	High school	96	3,15	0,777			
	3	Associate degree	56	3,03	0,803			
	4	Undergraduate	48	2,91	0,842			
	5	Graduate	13	3,67	1305			
		Total	273	3,14	0,840			

One way analysis of variance was conducted to determine whether the averages of dimensions of Residents' Attitudes toward Tourism Development differed in terms of respondents' education. Also, it was determined that "support for tourism development", "destination living standards", and "grand satisfaction" dimensions did not statistically differ ($p > 0.05$) in terms of professions. Moreover, it is seen that the average of contributions to community dimension differed statistically in terms of respondents' education. LSD test from post hoc tests was used to find the source of the difference in mean. The average of attitudes by education level are listed statistically as follows: undergraduate, associate degree, high school, and primary education graduates. It is seen that as education increases, the positive perspective towards the development of tourism in Gökceada decreases (Table 7.5).

One way ANOVA analysis was conducted to determine whether the averages of dimensions of Residents' Attitudes toward Tourism Development differed in terms of respondents' monthly income. Again, it was determined that the dimensions of Residents' Attitudes toward Tourism Development did not differ statistically ($p > 0.05$) in terms of respondents' monthly income. Based on this, it can be concluded that people's attitudes towards the development of tourism in Gökceada do not change according to their income groups.

In order to determine whether the averages of Residents' Attitudes toward Tourism Development dimensions differed in terms of residence duration, one way ANOVA analysis was conducted. The attitude averages of the respondents who resided for different periods were not found statistically different.

An independent samples t-test was conducted in order to determine whether the averages of Residents' Attitudes toward Tourism Development between those who think there is an improvement in Gökceada since the last 5 years and do not. According to the t-test results, there is not a statistical difference between these two groups.

An independent sample t-test was conducted in order to determine whether the averages of Residents' Attitudes toward Tourism Development in terms of respondents' future wishes about tourism in Gökceada. According to the t-test results, there is a statistical difference ($p < 0.05$) between these two groups for all four dimensions. Those who want tourism to develop in Gökceada have a statistically higher average on Support For Tourism Development dimension (4.11 > 3.11), Contributions To Community dimension (3.20 > 2.95), Destination Living Standards dimension (3.76 > 2.95), and Grand Satisfaction dimension (3.53 > 3.07) than those who want tourism to remain the same in Gökceada (Table 7.6).

A regression analysis was conducted for the effect of Residents' Attitudes toward Tourism Development scale dimensions on Grand satisfaction. The established model was found to be statistically significant and explains approximately 21% of grand satisfaction. According to results, Contributions to the community ($\beta = 0.268$) and Destination living standards ($\beta = 0.261$) have statistically positive significant effects on grand satisfaction. On the other hand, Support for the tourism development dimension has no significant effect on grand satisfaction (Table 7.7).

Table 7.6 Residents' attitudes toward tourism development means in terms of respondents' future wishes about tourism in Gökceada

Factors	Would tourism change in Gökceada for the future?	n	$\bar{x}$	s.d.	t	p (two-tailed)
Support for tourism development	Remain the same	72	3.11	0.93	−2.192	0.000
	Increase	201	4.17	0.67		
Contributions to community	Remain the same	72	2.95	0.86	−2.152	0.029
	Increase	201	3.20	0.83		
Destination living standards	Remain the same	72	2.95	0.82	−7.242	0.000
	Increase	201	3.76	0.79		
Grand satisfaction	Remain the same	72	3.07	0.95	−3.491	0.000
	Increase	201	3.53	0.91		

Table 7.7 Regression analysis for the effect of grand satisfaction

Variables	Beta	t value	Sig.
Constant	1.102	3.838	0.000
Support for tourism development	0.087	1.349	0.178
Contributions to community	0.268	4.813	0.000
Destination living standards	0.261	4.052	0.000
$R^2 = 0.212$ $F = 24.097$ $p. = 0.000$			

Conclusion and Discussion

This study aims to identify and examine residents' perceptions of tourism's socio-cultural impacts in Gökceada, Turkey. The study sample consisted of 273 residents of Gökceada. It has been determined that Residents' Attitudes toward Tourism Development scale average is above the average. It can be concluded that the local people living in Gökceada are generally satisfied with the development of tourism in the region.

In general, it was found that the levels of participation in the Residents' Attitudes toward Tourism Development scale did not differ in terms of demographic variables. Some studies support these results (Akdu and Ödemiş, 2018; Korça, 1998) and are conflicted by some studies (Ritchie & Inkari, 2006; Williams & Lawson, 2001; Mason & Cheyne, 2000; Harvey et al., 1995). On the other hand, it has been determined that the level of participation in the scale of contribution of tourism development to the community varies by participants' profession. This result contradicts the research of Korça (1998). This may be due to the higher expectations of tourism-related private sector employees and self-employed participants. The opinion that the tourism season on the island is short and that there is not enough economic benefit may have been considered by the respondents.

Similarly, it has been determined that the contribution to the community dimension differs in terms of the participants' education level. Accordingly, it has been determined that as the education level of the participants' increases, the contribution of tourism to society is perceived less. This result contradicts Haralambopoulos & Pizam, 1996; Kuvan & Akan, 2005; Korça, 1998 and Teye et al., 2002. This may be due to the increase in expectations from the tourism sector as education increases and fewer socio-economic opportunities due to public transportation difficulty with the mainland.

It is seen that the participants who want the development of tourism have more positive responses contributions of tourism to the society, the effect of tourism on living standards, the tourism development until now, and the higher satisfaction with tourism in general. Many studies accepted that perceived personal benefits from tourism are positively associated with an attitude that supports additional tourism development (Lankford & Howard, 1994; Haralambopoulos & Pizam, 1996; Gonzalez et al., 2018; Vargas-Sánchez et al., 2011).

Lastly, it is founded that people who think tourism contributes to the community and increases destination living standards are more satisfied. Logically, individuals are satisfied with tourism development, as it contributes to the society they live in and increases their living standards.

References

Abdool, A. (2002). *Residents' perceptions of tourism: A comparative study of two Caribbean communities*, PhD thesis. Bournemouth University, England.

Akdu, U., & Ödemiş, M. (2018). Examining the impacts of tourism on Gümüşhane residents according to the Doxey index. *Turizm Akademik Dergisi, 5*(2), 33–45.

Alexis, P. (2017). Over-tourism and anti-tourist sentiment: An exploratory analysis and discussion. *"Ovidius" University Annals, 17*(2), 288–293.

Andereck, K. L., & Nyaupane, G. (2011). Exploring the nature of tourism and quality of life perceptions among residents. *Journal of Travel Research, 50*(3), 248–260.

Andereck, K. L., & Vogt, C. A. (2000). The relationship between residents' attitudes towards tourism and tourism development options. *Journal of Travel Research, 39*(1), 27–36.

Andereck, K. L., Valentine, K. M., Knopf, R. C., & Vogt, C. A. (2005). Residents' perception of community tourism impacts. *Annals of Tourism Research, 32*(4), 1056–1076.

Andriotis, K., & Vaughan, R. D. (2003). Urban residents' attitudes toward tourism development: The case of Crete. *Journal of Travel Research, 42*(2), 172–185.

Boğan, E., & Sarıışık, M. (2016). Yerel halkın turizm faaliyetine yönelik görüş ve algılamalarının belirlenmesi üzerine Alanya'da bir araştırma. *Kastamonu Üniversitesi İktisadi ve İdari Bilimler Fakültesi Dergisi, 12*, 325–342.

Brent, R., & Incari, M. (2006). Host community attitudes toward tourism and cultural tourism development: The case of the Lewes District, southern England. *International Journal of Tourism Research, 8*(1), 27–44.

Bresson, G., & Logossah, K. (2011). Crowding-out effects of cruise tourism on stay-over tourism in the Caribbean: Non-parametric panel data evidence. *Tourism Economics, 17*(1), 127–158.

Choi, H., & Murray, I. (2010). Resident attitudes toward sustainable community tourism. *Journal of Sustainable Tourism, 18*(4), 575–594.

Cnt Editors. (2018) 15 Beloved places struggling with overtourism. *CN Traveler*. Retrieved from https://www.cntraveler.com/galleries/2015-06-19/barcelona-bhutan-places-that-limit-tourist-numbers/, Accessed: 26 Oct 2020.

Coldwell, W. (2017). First Venice and Barcelona: Now anti-tourism marches spread across Europe. *The Guardian*. Retrieved from https://www.theguardian.com/travel/2017/aug/10/anti-tourism-marches-spread-across-europevenice-barcelona/, Accessed: 26 Oct 2020.

Duyar, M., & Bayram, M. (2019). Overtourism and tourismphobia: Evolution of host and tourism relationship. *International Journal of Geography and Geography Education, 40*, 347–362.

Express Newspaper. (2017). *WATCH: Shocking moment Spanish anti-tourism activists hit big Brit tourist destination*. Retrieved from https://www.express.co.uk/news/uk/836224/Anti-tourist-protests-violent-British-tourist-destination-attack-Spain-Majorca-Palma. Downloaded: 20/09/2020.

Garcia, F. A., Vazquez, A. B., & Macias, R. C. (2015). Resident's attitudes towards the impacts of tourism. *Tourism Management Perspectives, 13*, 33–40.

Gökçeada Municipality. (2020). *2019 Gökçeada Nüfus İstatistikleri*. Gökçeada: İlçe Nüfus Müdürlüğü. http://www.gökçeada.com/turizm.html. Accesed 21 Sept 2020.

Gonzalez, V. M., Coromina, L., & Galí, N. (2018). Overtourism: Residents' perceptions of tourism impact as an indicator of resident social carrying capacity – Case study of a Spanish heritage town. *Tourism Review, 73*(3), 277–296.

Goodwin, H. (2017, October). *The challenge of overtourism, responsible tourism partnership* (Working Paper 4, pp. 1–19).

Gursoy, D., Jurowski, C., & Uysal, M. (2002). Resident attitudes. *Annals of Tourism Research, 29*(1), 79–105.

Gutiérrez-Taño, D., Garau-Vadell, J., & Díaz-Armas, R. (2019). The influence of knowledge on residents' perceptions of the impacts of overtourism in P2P accommodation rental. *Sustainability, 11*(4), 1043.

Haralambopoulos, N., & Pizam, A. (1996). Perceived impacts of tourism. *Annals of Tourism Research, 23*(3), 503–526.

Harvey, M. J., Hunt, J., & Harris, C. C. (1995). Gender and community tourism dependence level. *Annals of Tourism Research, 22*(2), 349–366.

Inbakaran, R., & Jackson, M. (2006). Resident attitudes inside Victoria's tourism product regions: A cluster analysis. *Journal of Hospitality and Tourism Management, 13*(1), 59–74.

Jacobsen, J. K. S. (2000). Anti-Tourist attitudes. *Annals of Tourism Research, 27*(2), 284–300.

Jacobsen, J. K. S., Iversen, N. M., & Hem, L. E. (2019). Hotspot crowding and over-tourism: Antecedents of destination attractiveness. *Annals of Tourism Research, 76*, 53–66.

Jani, D. (2018). Residents' perception of tourism impacts in Kilimanjaro: An integration of the social exchange theory. *Tourism (Turizam): An International Interdisciplinary Journal, 66*(2), 148–160.

Kahraman, S. (2005). Gökçeada'da göçlerin nüfus gelişimi ve değişimi üzerine etkileri. *Coğrafi Bilimler Dergisi, 3*(2), 39–53.

King, B., Pizam, A., & Milman, A. (1993). Social impacts of tourism: Host perceptions. *Annals of Tourism Research, 20*(4), 650–665.

Kline, R. B. (1998). *Principles and practice of structural equation modeling*. Guilford.

Koens, K., Postma, A., & Papp, B. (2018). Is overtourism overused? Understanding the impact of tourism in a city context. *Sustainability, 10*(12), 1–15.

Korça, P. (1998). Resident perceptions of tourism in a resort town. *Leisure Sciences, 20*(3), 193–212.

Kuščer, K., & Mihalič, T. (2019). Residents' attitudes towards overtourism from the perspective of tourism impacts and cooperation–the case of Ljubljana. *Sustainability, 11*(6), 1823.

Kuvan, Y., & Akan, P. (2005). Residents' attitudes toward general and forest- related impacts of tourism: The case of Belek, Antalya. *Tourism Management, 26*(5), 691–706.

Lankford, S. V., & Howard, D. R. (1994). Developing a tourism impact attitude scale. *Annals of Tourism Research, 21*(1), 121–139.

Martín, J. M., Guaita Martínez, J. M., & Salinas Fernández, J. A. (2018). An analysis of the factors behind the citizen's attitude of rejection towards tourism in a context of overtourism and economic dependence on this activity. *Sustainability, 10*(8), 1–18.

Mason, P., & Cheyne, J. (2000). Residents' attitudes to proposed tourism development. *Annals of Tourism Research, 27*(2), 391–411.

Milano, C., Cheer, J. & Novelli, M. (2018) Overtourism: A growing global problem. *The Conversation*. Retrieved from https://theconversation.com/overtourism-a-growing-global-problem-100029, Downloaded: 26/ 11/2020.

Oklevik, O., Gössling, S., Michael Hall, C., Jens Kristian, S. J., Ivar Petter, G., & McCabe, S. (2019). Overtourism, optimisation, and destination performance indicators: A case study of activities in fjord Norway. *Journal of Sustainable Tourism, 27*(12), 1804–1824.

Öztürk, A. B., Özer, Ö., & Çalışkan, U. (2015). The relationship between local residents' perceptions of tourism and their happiness: A case of Kusadasi, Turkey. *Tourism Review, 70*(3), 232–242.

Panayiotopoulos, A., & Pisano, C. (2019). Overtourism dystopias and socialist utopias: Towards an urban armature for Dubrovnik. *Tourism Planning and Development, 16*(4), 393–410.

Park, D. B., Lee, K. W., Choi, H. S., & Yoon, Y. (2012). Factors influencing social capital in rural tourism communities in South Korea. *Tourism Management, 33*, 1511–1520.

Pearce, P. L. (2018, 11–14 March). Limiting overtourism; the desirable new behaviours of the smart tourist. *The Tourism Intelligence Forum*, Palma, Spain.

Peric, M., Durkin, J., & Wise, N. (2016). Leveraging small-scale sport events: Challenges of organising, delivering and managing sustainable outcomes in rural communities, the case of Gorski Kotar, Croatia. *Sustainability, 8*(12), 1337.

Peter, L. (2017) 'Tourists go home': Leftists resist Spain influx. *BBC*. Retrieved from https://www.bbc.com/news/world-europe-40826257/. Downloaded: 26/11/2020.

Pinke-Sziva, I., Smith, M., Olt, G., & Berezvai, Z. (2019). Overtourism and the night-time economy: A case study of Budapest. *International Journal of Tourism Cities, 5*(1), 1–16.
Pinkster, F. M., & Boterman, W. R. (2017). When the spell is broken: Gentrification, urban tourism and privileged discontent in the Amsterdam canal district. *Cultural Geographies, 24*(3), 457–472.
Ritchie, B. W., & Inkari, M. (2006). Host community attitudes toward tourism and cultural tourism development: The case of the Lewes District, southern England. *International Journal of Tourism Research, 8*(1), 27–44.
Rouleau, J. (2017). Every (nocturnal) tourist leaves a trace: Urban tourism, nighttime landscape, and public places in Ciutat Vella, Barcelona. *Imaginations, 7*(2), 58–71.
Şahin, Ö. (2018, 21–23 December). An analysis of the impacts of overtourism (overcrowding in tourism destinations) on culture and heritage, 4. *International Culture and Civilization Congress*, Mardin, Turkey.
Shani, A., & Pizam, A. (2012). Community participation in tourism planning and development. In M. Uysal, R. R. Purdue, & M. J. Sirgy (Eds.), *Handbook of tourism and quality-of-life research: Enhancing the lives of tourists and residents of host communities* (pp. 547–564). Springer.
Sharma, B., & Dyer, P. (2009). Residents' involvement in tourism and their perceptions of tourism impacts. *Benchmarking: An International Journal, 16*(3), 351–371.
Smith, M. D., & Krannich, R. S. (1998). Tourism dependence and resident attitudes. *Annals of Tourism Research, 25*(4), 783–802.
Sommer, C., & Helbrecht, I. (2017). Seeing like a tourist city: How administrative constructions of conflictive urban tourism shape its future. *Journal of Tourism Futures, 3*(2), 157–170.
Tatoglu, E., Erdal, F., Ozgur, H., & Azakli, S. (2002). Resident attitudes toward tourism impacts: The case of Kusadasi in Turkey. *International Journal of Hospitality and Tourism Administration, 3*(3), 79–100.
Teye, V., Sirakaya, E., & Sönmez, S. F. (2002). Residents' attitudes toward tourism development. *Annals of Tourism Research, 29*(3), 668–688.
United Nation World Tourism Organization. (2018). *Overtourism? Understanding and managing urban tourism growth beyond perceptions*. UNWTO.
Vargas-Sánchez, A., Porras-Bueno, N., & Plaza-Mejía, M. D. L. Á. (2011). Explaining residents' attitudes to tourism: Is a universal model possible? *Annals of Tourism Research, 38*(2), 460–480.
Wang, Y., & Pfister, R. E. (2008). Residents' attitudes toward tourism and perceived personal benefits in a rural community. *Journal of Travel Research, 47*(1), 84–93.
Ward, C., & Berno, T. (2011). Beyond social exchange theory: Attitudes towards tourists. *Annals of Tourism Research, 38*(4), 1556–1569.
Weaver, D. B., & Lawton, L. J. (2001). Resident perceptions in the urban–rural fringe. *Annals of Tourism Research, 28*(2), 439–458.
Weber, F., Stettler, J., Priskin, J., Rosenberg-Taufer, B., Ponnapureddy, S., Fux, S., Camp, M., & Barth, M. (2017). *Tourism destinations under pressure* (Working paper) (pp. 1–214). Institute of Tourism ITW, Lucerne University of Applied Sciences and Arts.
West, S. G., Finch, J. F., & Curran, P. J. (1995). Structural equation models with nonnormal variables: Problems and remedies. In R. H. Hoyle (Ed.), *Structural equation modeling: Concepts, issues and applications* (pp. 56–75). Sage.
Williams, J., & Lawson, R. (2001). Community issues and resident opinions of tourism. *Annals of Tourism Research, 28*(2), 269–290.

Chapter 8
The Role of Community Based Tourism in Avoiding Overtourism

Işıl Arıkan Saltık and Uygar Turgut

Abstract Overtourism is one of the terms that has appeared in the literature recently, meaning the excessive number of tourists and the impacts of their flows on local communities and ecology. The increase in anti-tourism events around the world is a significant indicator of the effects of overtourism on communities. The underlying reason for the problem is mainly based on the decisions which were taken in a hurry, without proper evaluation, consideration of necessary matters, and often without active participation and inclusion of communities. Community development intends to flourish social, economic, environmental and cultural welfare and prosperity of communities, particularly made up of disadvantaged and marginalized people. Community based tourism development is one of the key factors to avoid that senseless anti-tourism sentiment and proceed to bring societies and cultures closer together through developing understanding and contact. Growing guest arrivals inflict an intense pressure on the environment and local community and their culture in a destination, thereby deteriorating both the quality of tourists' experiences and the life quality of local people. Rapid growth in tourism and undesirable condensation disrupt the sustainability of a destination causing local community and other stakeholders no more show any interest in further tourism development and making them feel they suffer since their access to opportunities, comfort, and health etc. is inhibited. It is significant that the technological revolution and intense usage of social media by tourists have been triggers to overtourism. However, COVID-19 Pandemic period may be an opportunity to invest both in technology and the further to help mitigate overtourism after COVID-19 when the crowd come back. In this context, this chapter explains the importance of community development in tourism and indicates considering community frustration to avoid adverse effects of tourism, in particular overtourism, and mentions how technology could be utilized in this period.

I. Arıkan Saltık (✉)
Fethiye Faculty of Business Administration, Muğla Sıtkı Koçman University, Muğla, Turkey
e-mail: isilas@mu.edu.tr

U. Turgut
Graduate School of Social Sciences, Muğla Sıtkı Koçman University, Muğla, Turkey

A. Hassan, A. Sharma (eds.), *Overtourism, Technology Solutions and Decimated Destinations*, https://doi.org/10.1007/978-981-16-2474-2_8

Keywords Community development · Community based tourism · Overtourism

Introduction

Overtourism is the fact that it arises when the amount of tourists arriving a destination surpasses the destination's carrying capacity, consequently, it negatively affects the residents' living conditions (Kim & Kang, 2020). The underlying reason for the problem is mainly based on the decisions taken in a hurry, without proper evaluation, consideration of necessary matters, and often without active participation and inclusion of communities, particularly at the beginning phases of tourism development in the destinations. Hence, communities feel they are inadequately equipped to deal with the swiftly occurring tourism impacts involving environmental, social, and economic issues. Local communities may feel left under the responsibility of major impacts that lower their quality of life and lead the community frustration, which contrasts with the idealized purpose of tourism. Community development is amongst the basic practices to prevent community frustration. Community development includes people's reaching a consensus in a community to launch an action collectively not only for developments such as economic growth and infrastructural improvements, etc. but also for communication and cooperation between people, informal groups and official institutions (Thedori, 2005). Thus, it could be possible for communities to benefit from the achievements of sustainable tourism. Besides, taking advantage of technological applications and tools that have an accelerating power in the process is important in ensuring effective tourism management in the digital world. Hence, it makes sense to examine the community development, community-based tourism approach, and the technological applications that can be used to prevent the emergence of overtourism.

Community and Community Development

Community is defined by a powerful web of social connections which cater the capability to cooperate, collaborate and interact (Stone & Hughes, 2002). Joppe (1996) describes the community as a "sense of shared purpose and common goals" and explains it could be geographical or a community of interest which is built on legacy and cultural values shared among the members of a community. Hence, he stresses that the community described in this way is not necessarily interchangeable with municipality or municipality boundaries. Definition of community plays a critical role to discern community development and its benefits when it arises. According to Hunt & Smith (2006), community refers to four properties, as shown in Fig. 8.1.

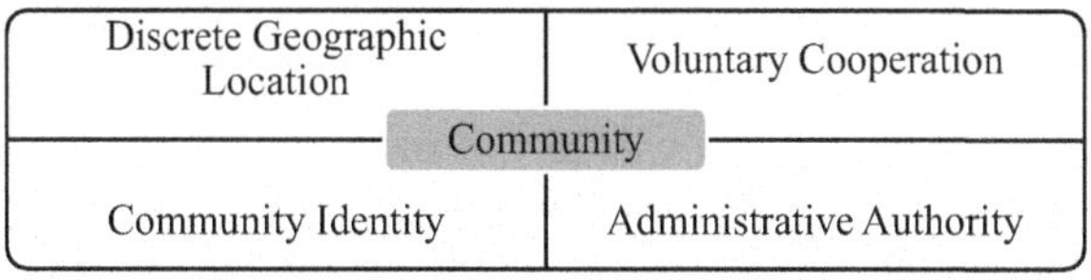

Fig. 8.1 Elements of community. (Source: Illustrated by the authors based on Hunt & Smith, 2006)

Discrete geographic location involves for instance, a spatial territory or residential such as a neighborhood, city or rural town etc. *Community identity*, states net of people and associations and their participation not because of geographic coexistence but because of cultural and historical reasons. The community of interest renders a *Voluntary Cooperation* or partnership of people that might have different perspectives on customs and lifestyle, yet, they share a set of common objectives; or a set of organizations that serve as the interests of a wide group of people shapes a community interest. Finally, *political* or *administrative authority* points out a state authority such as procedures and legal grids of individuals (Hunt & Smith 2006).

One of the facets of community is the belief of shared connection or psychological identification with a community and people. Most of the time people can give the name of the community where they live. People get reliant on a specific area for buying goods and getting benefits for employment, entertainment and coming together with the members of the community (Bhattacharya, 2004). Consequently, people identify themselves with this locality as a community. Although communities are important in terms of many aspects, community development is required and essential for any community to have a voice in society.

Researchers have brought various views to the concept "community development" including varying ideas, notions, approaches, objectives and values. Bhattacharya (1995) defines the objective of community development as the creation of harmony and power. In his definition, harmony means a common identity (come from a place, beliefs or concerns) and a feature for an attitude or standards that discrepancy impacts the members intensely and other possibilities (Bhattacharya, 2004). The UN describes community development as "a process formed to build opportunities of economic and social development for the entire community with an active involvement" (Moser, 1989).

Reid and Van Dreunen (1996, p. 49) state that "community development is a process for empowerment and transformation of individuals and communities". Keller and Mbewe (1991) describes empowerment as a process which enables people to arrange themselves to increase their own autonomy and stress their self-dependent right to choose and regulate capabilities that will help them to challenge and remove their own dependence. Community development means indicating willingness about the degree to which community members are empowered (Colton & Whitney-Squire, 2010).

In a community development process, the focus is particularly on relationships instead of on place. Thus, without healthy communication improved by healing, capacity building chances and robust link to culture and nation, any kind of development involving tourism will presumably not be sustainable (Colton & Whitney-Squire, 2010). Galston & Baehler (1995 as cited in Mair & Reid, 2007) suggest that tourism development includes mindful decisions related to the life style of a community rather than job creation and more research are needed to see the economic, cultural, political, racial, gender-related indications to economic development. Tosun (2005) claims that different forms of local involvement are more likely to remove current power and disproportionate development.

Aref and Redzman (2009) argue that for the success of community development, building the capacity of local communities is an essential factor and it helps communities to improve their capability to take part in the tourism decision-making, thereby ensuring their involvement in decisions. They also state that communities' capacity building assists individuals, communities, and organizations in realizing their unused and unimproved skills, resources, and geographic advantages that help review strengths and opportunities of the capabilities of community capacity in tourism development. The framework developed by Bell (1999 as cited in Colton & Whitney-Squire, 2010) to examine aboriginal community development is respectable for any communities in general. The elements that construct the framework involves *Community Empowerment* (governance, organization structures, policy frameworks etc.), *Community Wellness* (physical, mental and spiritual health; self-identity through traditional culture etc.), *Community Economic Development* (investments, job developments, community economic development strategies etc.) and *Community Learning* (individual and group learning, community as a classroom, skill development and training etc.). He also argues that facilitating community wellness that refers to community's capacity to create, support and continue healthy, stabilized relationships individually and cooperatively assures sustainable development, particularly within aboriginal communities.

Community Development in Tourism: Community-Based Tourism

Globalization of capitalism increased the mobility of people and progress of technologies in transportation and communication that assisted in advancing tourism into one of the world's largest industries in the twentieth century (Choi & Sarikaya, 2006). Although it enables to promote of economic benefits including income, taxes, jobs and hard currency, tourism also causes class, and regional inequalities arouse economic, environmental and social issues (Tosun, 1998). Among the adverse effects of tourism are a rise in the expenses, an inequitable allocation of tourism income, employment with low-skilled workforce and low-wages policy, natural and cultural resource degradation, loss of jobs to outsiders as well as crime and crowded living areas and low degree of empowerment (Lee & Jan, 2019; Tascı et al., 2013). Many scholars state that these unwanted effects of tourism caused the growing interest for the protection and security of natural resources, community health and comfort and economic growth in the long run (Choi & Sarikaya, 2006). Consequently, as a concept, sustainable development became an alternative approach to the conventional economic development of the neo-classical approach.

Alternative approaches to tourism that emerged within sustainable development support tourism that is small-scale, environmentally-friendly, locally-oriented and facilitate locally-guided actions, manage economic advance, environmental and cultural protection, removal of poverty and capacity building particularly in less developed areas (Dangi & Jamal, 2016). Community based tourism (CBT) is amongst the emerging forms of tourism within the scope of sustainability (Reed, 1997).

CBT that develops one of the sustainable tourism forms has gained popularity since almost 1995 and switched the approach for the growth to "bottom-up", attempting to deliver real and extensive community involvement at all stages of the growth. CBT is a type of "local" tourism, backing local service suppliers and producers and emphasizing defining and corresponding local culture and environment (Asker et al., 2010). Dixey (2005) defines CBT owned and managed by communities creates greater community advantage. He also states that communities may own tourism enterprises such as a hotel and outsource the management to a travel agency or they may own public spheres such as national parks. They must be responsible for management in case that there is a goal for greater community advantage. CBT varies from many other types of tourism since it exclusively aims to boost the advantages not for investors who do not actually share responsibilities but primarily for other stakeholders of the society (Dodds et al., 2016). Mearns (2003) explains that for the sustainable development of CBT, all the activities planned must have four equally important aspects. Accordingly, tourism activities (i) must be economically viable-the revenue surpasses the costs, (ii) must be ecologically sustainable-the environment should not be destroyed, (iii) must assure the equitable allocation of cost and advantages among all citizens, (iv) must have the transparent organization, appreciated by all shareholders as representing the concerns of the whole community members and indicating genuine ownership.

CBT addresses to build a more sustainable tourism industry, concentrating on the host communities in terms of planning and supporting tourism development (Salazar, 2012). Many scholars argue that CBT is based on the development of community and its involvement in tourism products' design and development processes (Dodds et al., 2016). The following characteristics are seen as common attributes of CBT (Asker et al., 2010; Hamza & Khalifah, 2009, Tascı et al., 2013):

- CBT is exploited as a means of economic growth for underdeveloped areas,
- The descriptors used for its target groups usually involve remote, rural, poverty-stricken, undeveloped, marginalized, indigenous, economically distressed, poor, ethnic minority and people in small towns,
- CBT aims to contribute to the well-being of local people, especially in rural areas or small settlements and prosperity of their cultural and environmental values,
- CBT includes accommodating visitors within the local community,

- CBT involves sharing benefits equitably. Generally, community interests come before the individual,
- CBT includes the involvement of communities during tourism planning, decision making, development and operations,
- CBT destinations include cultures that are not Westernized and surroundings where the residents' lifestyle, culture and folklore, gastronomic products, materials, clothing and natural environment are parts of CBT outputs,
- CBT involves education and interpretation as part of the activity for tourists,
- CBT generally arranged for a small group of people by small, specialized enterprises owned by local people,
- CBT boosts local and guest awareness of conversation.

Tascı et al. (2013) mention that economical, socio-cultural and environmental benefits of CBT, highlighting that the most important advantage of CBT is the shift in perspective of local people on life and themselves. They also underline that CBT gives hope and willingness for local communities to participate, produce and organize their lives due to the fact that CBT opens up the horizons of the future of the community. Benefits of CBT are summarized below (Dangi & Jamal, 2016; Sebele, 2010):

*CBT provides economic empowerment (economic benefits, generating an increase in the revenue, assisting rural development, capacity building, alleviating poverty, management of internal and external financial resources, training and skills development, vision, goals, strategies, marketing, networking, enabling the supplying of local communities' products and services)

*CBT consolidates psychological empowerment (participation, involvement, protecting local identity, educational activities to identify self needs, contributing to the conservation of the habitat of animals and the natural environment, collaboration, enhanced sense of pride, tourist satisfaction)

*CBT contributes to social empowerment (participation, involvement, community cohesion, interaction among stakeholders, networking, quality of life, respect for local culture and tradition, sense of community, collaboration, tourism resource conversation, consideration the important role of women in development)

*CBT maintains political empowerment (participation, involvement, collaboration, support from local/national government visionary and passionate leaders).

Despite these benefits of CBT, there are also various barriers to successful CBT. Due to these barriers for successful CBT (Asker et al., 2010), benefits expected from CBT may be interrupted, and CBT activities will result in frustration and eventually, failure (See Table 8.1).

In the relevant literature, different examples show improper tourism development without establishing community development at a destination. Locals are stayed outside of the way to make way for tourists and their needs. Thus, they are customarily omitted for senior positions to supervise or manage in hotels and resorts, thereby making the locals feel resented. These negative impacts are the

Table 8.1 CBT barriers and the results

CBT barriers	Frustration/ failure in
Community foundations and organizations (of men/women/youth) are disintegrated and unorganized.	PE & SE & EE
Powerful individuals (usually, males) are in the leading position in decision-making, and the benefits are distributed unequally.	PE & SE & EE & PoE
Land and resource conflicts are common and recurring.	EE
Strictly hierarchical decision making and management structures when CBT is "settled" by an external, in particular, when the source for CBT is provided from international actors and local people perceive the motivations are solely financial.	PE & SE & EE
The absence of real, local decision making or limited decision making is based on insufficient information and the absence of evaluations of local decisions.	PE & SE & EE
Participation decreases while implementing the CB facility.	PE & SE & EE & PoE
Drivers are merely financial.	PE & SE & PoE
Insufficient marketing or inaccurate marketing.	EE
Locals believe that they will earn money without any effort but just by inviting tourists, and there is the absence of planning that affect community and natural landscape adversely.	PE & SE & EE &PoE
Established through external funding mechanisms.	PE & SE & EE &PoE
There is a false assumption that "one form of CBT is convenient for all ventures".	PE & SE
CBT is seen as a fast way to escape from the poverty cycle.	PE & SE
Lack of sense of the uniqueness of 'place' because of not informing visitors about authenticity of local natural and cultural heritage.	PE & SE
Inadequate infrastructure and lack of investment potential.	EE

Source: Adopted by authors from Asker et al. (2010), Sebele (2010), and Dangi and Jamal (2016)
PE Psychological empowerment, *EE* Economic empowerment, *SE* Social empowerment, *PoE* Political empowerment

consequences of local people's neglect being completely kept out of tourism development (Reid & Sindiga, 1999).

Chirenge et al. (2013) reveal that local communities are not involved while policies are established, and decisions are made during the planning and determining the budget of the actions in African countries due to the lack of their participation causing conflicts between the community members and the project team. Kuscer & Mihalic (2019) confirm in their study that the impacts of tourism on life quality, air pollution, traffic and crowding influence irritation of local people and offer a potential overtourism risk in Ljubliana, Slovenia. The survey indicates that an appropriate tourism development that provides cooperation among stakeholders can be used to develop the influence of tourism-related superstructure on community benefit, life quality and hospitality enterprises mitigating the negative perceptions of residents at the destination.

Canavan (2016) in his research in the Isle of Man, British Islands, cites between tourism stakeholders, there are ineffective, fractured, mistrustful and dysfunctional relationships. Therefore, stakeholders feel that they must be included in tourism decision making regarded as a non-transparent process dependent on a small group of unnamed agents. Tourism development should include efforts that are already valued, praised and supported by community members so that local people should not be seen as outsiders yet as partners (Mair & Reid, 2007). Moreover, all channels for dialogue between stakeholders should stay open and activities to be planned should be determined with the inclusion of all sides impartially and without any prejudice.

The Frustration of Community: Overtourism

Since 1950, global tourism has increased from 25 million to 1.4 billion visitors (UNWTO, 2019), witnessing rapid and consistent growth. The number is forecast to reach 1.8 million by 2030 (UNWTO, 2011). The estimated number of domestic tourist arrivals will reach 9.05 billion exceeding the world's population by 2.55 billion (Hall, 2015). This massive growth of tourism has been accompanied by tourists flows to particularly specific destinations, causing crowding and challenges related to carrying capacity (Kim & Kang, 2020), environmental sustainability and the imitation effect, whereby emerging economies imitate the tourist behavior of Western economies (Cappochi et al., 2019). When the number of tourists exceeds a destination's capacity, destinations, particularly urban destinations, suffer from the negative effects of tourism on the long-term residents of these destinations (Oyla et al., 2017). Within relevant literature "overtourism" has not been emerged recently since numerous notions associated with maintaining the numbers of visitors and resources at tourism destinations were relieved almost fifty years ago. Carrying capacity (Kuss & Morgan, 1980; Sinden, 1975; O'Reilly, 1986), the life cycle model by Butler (1980), community antagonism and Doxey's irritation index (Doxey, 1975) are the related concepts of overtourism discussed in early studies.

Nowadays, overtourism issues maintain these long-lasting debates on tourism development. Nonetheless, overtourism issues today arise in much wider kind of destinations and with a greater degree of complicacy (Milano et al., 2019) due to the level of awareness of the possibly harming impacts of the permanent quantitative growth of mass tourism by the wider public (Saltık, 2018; Benner, 2019). According to Goodwin (2019, p.110), overtourism describes "destinations where hosts and guests, locals and visitors, feel that there are too many visitors and that quality of life in the area or the quality of the experience has deteriorated unacceptably." Overtourism refers to the case that the quality of visitors' experiences and the quality of life of citizens are affected extremely adversely due to the tourism impacts on a destination (UNWTO, 2018).

Goodwin (2019) mentions multiple causes of overtourism that the rise of low-cost flights, especially short-haul ones, and cheap coach travel have made the journey affordable for people. Besides, he also remarks that Airbnb and similar platforms have helped more people to find competitive accommodation, thereby enabling them to travel more easily. Cappochi et al. (2019) argue that overtourism is connected to three areas: first, growth of tourism and massification of tourist flows, second, concentration bringing about crowding, carrying capacity and environmental sustainability concerns and third, governance with respect to the resources of the territories, relations with airline companies and technologies. Koens et al. (2018) stress that the aspects such as visitor behavior, timing, concentration, location, experience with local tourism etiquette are as crucial as tourist numbers that help emerge overtourism. Benner (2019) states that technological developments form tourist behavior and could thus cause undesirable results. Social media can lead tourists to the influx to areas that require the fundamental infrastructure, accordingly boosting the adverse effects of overtourism.

Coccossis & Mexa (2004) explain three indicators to assess overtourism in an urban context. The physical and ecological indicator is associated with the aspects such as cultural heritage with a specific area, noise pollution, the tourism infrastructure and the mobility within the defined area. The socio-cultural indicator is related to present tourism flows and satisfaction of tourists and inhabitants, respectively, with their experience in a tourist destination. The political and economic indicator associates tourism income and investment, decision-making process, governance and public expenses allotted to the area. The reasons that include irritation of residents are the emergence of touristic monocultures, the loss of authenticity through touristification of the places in hotspots and a sense of insecurity because of the housing rented to tourists through internet portals (Koens et al., 2018), loss of sense of belonging, making public places privatized, excessive development of cruise tourism and rapid rise in the number of guests visiting seasonally or daily (Cheer et al., 2018).

"Overtoursim" and "tourismphobia" emerged as a consequence of unsustainable mass tourism practices and the destructive exploitation of urban, rural and coastal fields for tourism objectives. The term "Tourismphobia", borrowed by Spanish mass media, defines both the development of the displeasure of the society caused by the burden related to the growth of tourism and discrediting the activities that are results of society-led actions and the local bodies of the society (Milano et al., 2019). When the population of the visitors reaches certain stage that disrupts the life of the community, locals start to grow negative feelings for them (Oyla et al., 2017). Some problems of overtourism are the congestion, the destruction of the resource base, the removal of cultural identity, the annoyance of local people, the threat to a country's heritage and environment and the agitation for local authorities, which in recent years have led to various protest across Europe (Manzin, 2020; Avond et al., 2019).

These issues caused anti-tourist attitudes among residents in Venice, Barcelona, Berlin, Hong Kong, New Orleans and Prague (Colomb & Novy, 2016). Adie et al. (2019) support the authors above and argue that locals living in big cities are faced with the threat most caused by overtourism whereas locals in rural areas perceive overtourism as a threat more than locals in small cities. Goodwin (2019) states that rural areas have the similar pressures as in urban areas with trampling, litter and congestion and there is a pressure on the availability of housing for residents in view of renting for holiday purposes, second home ownership and internal migration by the retirees. Due to the unwanted consequences of overtourism, Iceland is now displaying a greater unwillingness to develop and taking new measures such as increasing prices and restricting visitors' numbers to protect their natural sites regardless of the sector's boosting the island's economy at different points in the past (Avond, et al., 2019).

A more sensible perspective taking into account community members is needed for an appropriate advancement and management (Kuscer & Mihalic, 2019) to identify common problems, their solutions and common values. Throughout this process, collaborating in trust between stakeholders help ensure effective tourism development (Kim & Kang, 2020). Mass tourism destinations should make sustainable tourism planning before any conflict or rage occurs to avoid anti-tourism actions. An effective system should be promoted so as to establish social carrying capacities and perceptions of society towards the negative effects of tourism should be identified through reliable sources. Eventually, it should be decided how social carrying capacities will be evaluated, planned and managed (Schofield, 2011).

UNWTO (2020) considers that export income/revenue from tourism may decline by $ 910 billion to $ 1.2. trillion in 2020. It is expected to cause a broader impact and curtail global GDP by 1.5–2.8%. Taking into account the outbreak of COVID-19 pandemic in which tourism sector has been severely affected, each one of the stakeholders who is a part of the phenomenon, tourism, must participate in the decision to be made on tourism development during the process of planning tourism after 2023.

Although new technology is part of the problem today in that the impacts of the trends and communication on social networks, online tourism sites, YouTube channels and bloggers' pages and mobile applications have caused an over-visit of different places all over the world thus creating a perception of overtourism (Bourliataux-Lajoinie et al., 2019), technology can be used to reduce the effects of overtourism. The impact of COVID-19 has arisen new challenges of how the tourism industry will be able to deal with and get over this pandemic. History has proven that technological advancements and progress result from the crisis (Colombo et al., 2016).

Gretzel et al. (2020) believe that information and technology has the key to recognize the emerging conditions related to a pandemic with respect to how mankind manage to travel along with their daily life and how they predict a better future for e-tourism defined (Neithard & Werther, 2018, p. 1) as "analysis, design, implementation and application of IT/e-commerce solutions in the travel and tourism industry as well as the analysis (of the impact) of the respective technical/economic processes and market structures".

Therefore, during COVID-19, especially for popular destinations, this period may be an opportunity to invest both in technology and the further to help mitigate overtourism after COVID-19 when the crowd come back. Many places have chosen to make use of reality and 360-degree content to inspire and draw travelers' attention over the last several years and during the days with COVID-19, it could be useful for people particularly locked down. Applications used in tourism for fighting overcrowding are widespread in our today's world. Some of these technologies and their purposes and functions are as follows (See Table 8.2).

Zeng et al. (2020) suggest that the progress in artificial intelligence and other technologies have given robotics support in enhancing the characteristics of travel and tourism experiences and the sustainability of them particularly with the advent of COVID-19. They also identify some of the benefits that robotics might provide during and post COVID-19 including; more convenient observing and protection of delicate resources both cultural and natural, systems that help tourists move without causing overtourism, public information systems that identify improper behavior and educate tourists on culturally tolerable behavior when needed, information systems attracting tourists attention to more local suppliers so that the advantages of tourism can be delivered impartially.

Table 8.2 Purposes and functions of technologies in tourism

Purpose	Function
Collecting data for market research	Wi-Fi and Bluetooth tracking allow a comprehensive understanding of visitor numbers to destinations at various times. Guests use smartphones to record travel patterns through via an application. The application also creates pop-up surveys to provide socio-democratic status, knowledge of a destination, and cultural background before tourists visit a destination.
Dispersing tourists	As overcrowding takes place at key "pitch points" at a destination, applications that keep travelers away from the most visited places and allow them to see different sites are a key strategy. These applications send users notifications informing when an attraction is busy and notably offer alternatives, especially less visited places.
Speeding and managing tourist flow	New technology is used to manage the movement of people through key parts of a site/an attraction, helping people move in an organized fashion. Such applications store passengers' data securely and thereby accelerating identification. New facial recognition technology allows travelers to board planes without presenting their passports. Many attractions such as the Louvre Museum in Paris, Hong Kong Disneyland etc. allow visitors to book online in advance so that they will not be exposed to long ques. If attractions and destinations ask all visitors for online bookings, which consequently improves tourist traffic flow (for a better visitor experience), they will help reduce noise pollution, advancing the living conditions for residents.
Increasing interaction with visitors	Augmented and virtual reality technology gives travelers a chance to interact with places. Smart museums use a range of apps allowing visitors to interact with exhibits and providing background information, imagery and video. Moreover, these audio handset tours will be presented using augmented reality handsets and allow visitors to witness the warriors from the past and ancient ruins when they visit a castle or historical places. Besides, people will be able to visit overcrowded destinations via this technology more in the near future.

Source: Adopted by the Authors based on WTTC (2018) and Koh (2020)

Discussion and Conclusion

In addition to the sudden and excessive enhancement in the number of visitors, the fact that the direct or indirect negative effects of tourism are felt more than the positive effects due to the increase of awareness in the society about the damage caused by tourism to the environment has created sensitivity and induced discontent in the society about tourism (Saltık, 2018). Overtourism is one of the terms that has appeared in the literature recently, meaning the excessive number of tourists and the impacts of their flows on local communities and ecology. In recent years, the overstretching of tourism carrying capacity in many European cities, which have been important cultural and tourism destinations, has caused discomfort and anger of the local people. As a result of this sensitivity developed against tourism, local people have gained a negative perspective towards effects of tourism on the community and even tourism has started to be seen as a problem by a large part of the local people (Castela, 2018). Although it is difficult to determine the definite effects of anti-tourism, it is necessary to act with the awareness that the sustainability of tourism is related to the community support for tourism. As one of the basic practices to prevent community frustration, community development becomes prominent in this process. In this way, it will be possible to realize sustainable tourism activities that not only prevent the formation of anti-tourism reactions but also ensure the community development by getting the support of society. It is significant that technological revolution and intense usage of social media by tourists have been triggers to overtourism. However, they could be managed to give mass tourists a taste of their own medicine. In other words, social media and technological applications could be utilized for guiding mass tourists to uncrowded areas and keeping the tourist flow under control in accordance with the carrying capacity of the tourist attraction centers in the destination. Last but not least, the adverse effects of overtourism can be overcome by people's working together for the shared interest and finding solutions for common problems. Consequently, it will be possible to support community development in the destination in the short term and to contribute to sustainable tourism development in the world in the long term.

References

Adie, B. A., Falk, B., & Savioli, M. (2019). Overtourism as perceived threat to cultural heritage in Europe. *Current Issues in Tourism, 23*(14), 1737–1741.

Aref, F., & Redzman, M. B. (2009). Community capacity building for tourism development. *Journal of Human Ecology, 27*(1), 21–25.

Asker, S., Boronyak, L., Naomi, N., & Paddon, M. (2010). *Effective community based tourism: A best practice manual*. Retrieved from: https://www.apec.org/Publications/2010/06/Effective-Community-Based-Tourism-A-Best-Practice-Manual-June-2010. Accessed: 27 Sept 2020.

Avond, G., Bacari, C., Limea, I., Seraphin, H., Gowreesunkar, V., & Mhanna, R. (2019). Overtourism: As a result of the Jaus-faced character of the tourism industry. *Worldwide Hospitality and Tourism Themes, 11*(5), 552–565.

Benner, M. (2019). *From overtourism to overtourism to sustainability: A research agenda for qualitative tourism development in the Adriatic*. Retrieved from: https://mpra.ub.uni-muenchen.de/92213/1/MPRA_paper_92213.pdf. Accessed: 14 Sept 2020.

Bhattacharya, J. (1995). Solidarity and agency: Rethinking community development. *Human Organization, 54*(1), 60–69.

Bhattacharya, J. (2004). Theorizing community development. *Community Development, 34*(2), 5–34.

Bourliateaux-Lajoinie, S., Dosquet, F., & Arriaga, J. L. D.-O. (2019). The dark side of digital technology to overtourism: The case of Barcelona. *Worldwide Hospitality and Tourism Themes, 11*(5), 582–593.

Butler, R. W. (1980). The concept of a tourist area cycles of evolution: Implications for management of resources. *Canadian Geographer, 24*(1), 5–12.

Canavan, B. J. (2016). Tourism stakeholder exclusion and conflict in a small island. *Leisure Studies, 36*(3), 1–17.

Cappochi, A., Vallone, C., Pierotti, M., & Amaduzi, A. (2019). A literature review to assess implications and future perspectives. *Sustainability, 11*(12), 3303.

Castela, A. (2018). Impacts of tourism in an urban community: The case of Alfama. *Athens Journal of Tourism, 5*(2), 133–148.

Cheer, J. M., Milano, C., & Novelli, M. (2018). Tourism and community resilience in the anthropocene: Accentuating temporal overtourism. *Journal of Sustainable Tourism, 27*, 554–572.

Chirenge, L. I., Giliba, R. A., & Musamba, E. B. (2013). Local communities' participation in decision-making process through planning and budgeting in African countries. *Chinese Journal of Population Resources and Environment, 11*(1), 10–16.

Choi, H. C., & Sarıkaya, E. (2006). Sustainability indicators for managing community tourism. *Tourism Management, 27*, 1274–1289.

Coccossis H. & Mexa A. (2004). Tourism carrying capacity: Future issues and policy considerations. In H. Coccossis & A. Mexa (Eds.). *The challenge of tourism carrying capacity assessment: Theory and practice* (pp. 277–288). Aldershot: Ashgate Publishing.

Colomb, C., & Novy, J. (Eds.). (2016). *Protest and resistance in the tourist city*. Retrieved from: https://www.researchgate.net/publication/332718780_Protest_and_Resistance_in_the_Tourist_City. Accessed 14 Sept 2020.

Colombo, M. G., Piva, E., Quas, A., & Rossi-Lamastra, C. (2016). How to high-tech entrepreneurial ventures cope with global crisis: Changes in product innovation and internationalization strategies. *Industry and Innovation, 23*(7), 647–671.

Colton, J. W., & Whitney-Squire, K. (2010). Exploring the relationship between aboriginal tourism and community development. *Leisure/Leosir, 34*(3), 261–278.

Dangi, T. B., & Jamal, T. (2016). An integrated approach to "sustainable community-based tourism". *Sustainability, 8*, 475.

Dixey, L. (2005). *Inventory and analysis of community based tourism in Zambia*. The USAID-Funded PROFIT Programme.

Dodds, R., Ali, A., & Galaksi, K. (2016). Mobilizing knowledge: Key elements of success and barriers in community based tourism. *Current Issue in Tourism, 21*(13), 1547–1568.

Doxey, G. V. (1975). A causation theory of visitor-resident irritants: The impact of tourism. Methodology and research inferences. *The Impact of Tourism, Sixth Annual Conference Proceedings of the Travel Research Association*, pp. 195–198.

Goodwin, H. (2019). *Overtourism, symptoms and treatment*. Retrieved from: https://responsibletourismpartnership.org/wp-content/uploads/2019/06/TWG16-Goodwin.pdf. Accessed 13 Sept 2020.

Gretzel, U., Fuchs, M., Baggio, R., Hoepken, W., Law, R., Neidhardt, J., Pesonen, J., Zanker, M., & Xlang, Z. (2020). e-Tourism beyond COVID-19: A call for transformative research. *Information Technology & Tourism, 22*, 187–203.

Hall, C. M. (2015). On the mobility of tourism mobilities. *Current Issues in Tourism, 18*(1), 7–10.

Hamza, A., & Khalifah, Z. (2009). *Handbook on community-based tourism: How to develop and sustain community based tourism*. APEC Secretariat.
Hunt, J., & Smith, D. (2006). *Building indigenous community governance in Australia: Preliminary research findings* (Center for Aboriginal Economic Policy Research Working Paper No.31). Australian National University.
Joppe, M. (1996). Sustainable community tourism development revisited. *Current Issues, 17*(7), 475–479.
Keller, B., & Mbewe, D. C. (1991). Policy and planning for the empowerment of Zambia's women. *Canadian Journal of Development Studies, 12*(1), 75–88.
Kim, S., & Kang, Y. (2020). Why do residents in an overtourism destination develop anti-tourist attitudes? An exploration of residents' experience through the lens of the community-based tourism. *Asia Pacific Journal of Tourism Research, 25*(8), 858–876.
Koens, K., Postma, A., & Papp, B. (2018). Is overtourism overused? Understanding the impact of tourism in a city context. *Sustainability, 10*(12), 4384.
Koh, E. (2020). The end of overtourism? Opportunities in a post-Covid-19 world. *International Journal of Tourism Cities, 6*(4), 1015–1023.
Kuscer, K., & Mihalic, T. (2019). Residents' attitudes towards overtourism from the perspective of tourism impacts and cooperation-the case of Ljubljana. *Sustainability, 11*, 1823.
Kuss, F. R., & Morgan, J. M., III. (1980). Estimating the physical carrying capacity of recreation areas: a rationale for application of the universal soil loss equation. *Journal of Soil and Water Conservation, 35*(2), 87–89.
Lee, T. H., & Jan, F.-H. (2019). Can community-based tourism contribute to sustainable development? Evidence from residents' perceptions of the sustainability. *Tourism Management, 70*, 368–380.
Mair, H., & Reid, D. H. (2007). Tourism and community development vs. tourism for community development: Conceptualizing planning as power, knowledge and control. *Leisure/Loisir, 31*(2), 403–425.
Manzin, A. (2020). Challenges in sustainable development of Dubrownik as a tourist destination. *Tourism & hospitality industry 2020, congress proceedings*, pp. 162–175.
Mearns, K. (2003). Community-based tourism: The key to empowering the Sankuyo Community in Bostwana. *Africa Insight, 33*(1/2), 29–32.
Milano, C., Novelli, M., & Cheer, J. M. (2019). Overtourism and tourismphobia: A journey through four decades of tourism development, planning and local concerns. *Toursim Planning & Development, 16*(4), 353–357.
Moser, C. O. N. (1989). Community participation in urban projects in the Third World. *Progress in Planning, 32*(part 2), 71–133.
Neithard, J., & Werther, H. (2018). IT and tourism: still a hot topic, but do not forget IT. *Information Technology & Tourism, 20*, 1–7.
O'Reilly, A. M. (1986). Tourism carrying capacity: Concept and issues. *Tourism Management, 7*(4), 254–258.
Oyla, H. T. G., Shahmirzdi, E. K., & Alipour, H. (2017). Pro-tourism and anti-tourism community groups at a world heritage site in Turkey. *Current Issues in Tourism, 22*(7), 763–785.
Reed, M. G. (1997). Power relations and community-based tourism planning. *Annals of Tourism Research, 24*(3), 566–591.
Reid, D. G., & Sindiga, I. (1999). Tourism and community development: An African example. *World Leisure & Recreation, 41*(2), 18–21.
Reid, D. G., & van Dreunen, E. (1996). Leisure as social transformation mechanism in community development practice. *Journal of Applied Recreation Research, 27*(1), 45–65.
Salazar, N. B. (2012). Community-based cultural tourism: Issues, threats and opportunities. *Journal of Sustainable Tourism, 20*(1), 9–22.
Saltık, I. A. (2018). Karşıt Turizm (Anti-Tourism). In O. İçöz & M. Uysal (Eds.), *Turizm Ansiklopedisi-Türkiye: Turizm ve Ağırlama Endüstrisinin Temel Kavramları*. Detay Yayıncılık.

Schofield, P. (2011). City resident attitudes to proposed tourism development and its impacts on the community. *International Journal of Tourism Research, 13*, 218–233.

Sebele, L. S. (2010). Community-based tourism ventures, benefits and challenges: Khama Rhino Sanctuary Trust, Central District, Bostwana. *Tourism Management, 31*, 136–146.

Sinden, J. A. (1975). Carrying capacity as a planning concept for national parks: Available and desirable capacity. *Landscape Planning, 2*, 243–247.

Stone, W., & Hughes, J. (2002). Understanding Community Strengths. *Family Matters, (61)*, 62–67. https://search.informit.org/doi/10.3316/informit.746181059943515. Accessed: 8 Oct 2020.

Tascı, A. D. A., Semrad, K. J., & Yılmaz, S. S. (2013). *Community based tourism finding the equilibrium in COMCEC context: Setting the pathway for the future*. COMCEC Coordination Office.

Thedori, G. L. (2005). Community and community development in resource-based areas: Operational definitions rooted in an interactional perspective. *Society and Natural Resources, 18*, 661–669.

Tosun, C. (1998). Roots of unsustainable tourism development at the local level: the case of Urgup in Turkey. *Tourism Management, 19*(6), 595–610.

Tosun, C. (2005). Stages in the emergence of a participatory tourism development approach in the Developing World. *Geoforum, 36*, 333–352.

UNWTO. (2011). *Tourism towards 2030/Global review*. UNWTO.

UNWTO. (2018). *Overtourism? Understanding and managing urban tourism growth beyond perceptions*. UNWTO.

UNWTO. (2019). *International tourism highlights, edition 2019*. UNWTO.

UNWTO. (2020). *Tourism and Covid-19 – Unprecedented economic impacts*. Retrieved from: https://www.unwto.org/tourism-and-covid-19-unprecedented-economic-impacts. Accessed: 4 Oct 2020.

WTTC. (2018). *How can new technologies help deal with overcrowding?* Retrieved from: https://medium.com/@WTTC/how-can-new-technologies-help-deal-with-overcrowding-dda554fb164b. Accessed; 3 Oct 2020.

Zeng, Z., Chen, P.-J., & Lew, A. A. (2020). From high-touch to high-tech: COVID-19 drives robotics adoption. *Tourism Geographies, 22*(3), 724–734.

Part V
Challenges

Chapter 9
Overtourism: Issues, Risk and Responsibility

Aslı Özge Özgen Çiğdemli

Abstract Tourism has become a sector in which many countries and destinations compete to attract tourists due to its contribution to world revenues. As a result of the competitive environment of tourism and promotion efforts to captivate more and more tourists lead to a human concentration above the carrying capacity in destinations. Density in tourist inflows causes, on the one hand, enhancing destination popularity but on the other hand complaints about crowding both tourists and local communities. The negative effects of tourism like deterioration of the culture, damage to the local language, noise pollution, tensions caused by crowds in daily life, increase in the cost of living, incidents of violence, the widespread use of drugs, pressures due to overcrowding in public spaces are risky issues for both local community and tourists. This situation will adversely affect the quality of life of locals, which is an important stakeholder for the development of tourism. Overtourism, which has emerged as a concept outcome of excessive tourism demand, affects tourist experiences as well as the quality of life. The footprint effect caused by crowding in sensitive areas can damage historical sites, cultural heritage, environmental quality, and scenery. Given these components as the main sources of attraction of tourism, it is obvious that the benefit of the tourist from the destination will decrease. Considering all these factors, the main purpose of this study is to define the concept of over tourism, to reveal the issues related to overtourism as carrying capacity, density and overcrowding, to determine the outputs of overtourism in terms of local people, tourists, environment and socio-cultural aspects. Finally, the measures to be taken to prevent the negative effects of overtourism will be discussed within the framework of marketing, managerial, and tourist's aspects.

Keywords Overtourism · Carrying capacity · Overcrowding · Negative effects of tourism · Local community

---

A. Ö. Özgen Çiğdemli (✉)
Department of Recreation Management, Osmaniye Korkut Ata University, Osmaniye, Turkey

A. Hassan, A. Sharma (eds.), *Overtourism, Technology Solutions and Decimated Destinations*, https://doi.org/10.1007/978-981-16-2474-2_9

Introduction

The mobility of humans and the increase in individual disposable income brought by globalization have led to the development of tourism in terms of the number of visitors and the area where tourism activities are carried out. With this development, the increase in the number of tourists has generated economic, social, environmental, and administrative effects in respect to redesign the accommodation infrastructure in countries attracting more tourists, increasing and managing the capacity of tourism services. Managing tourist mobility has become an urging necessity for tourist attracting countries. On the one hand, tourism investments should be financed, and on the other, regulations should be made to reduce the negative spillovers of tourism. Additionally, mass tourism, being still the main tourism type for the twentieth century, creates a risk factor, affecting the popular tourist destinations and the quality of life for the local people residing in these places.

With developments such as making travel easier, the possibility of benefiting from private housing opportunities, like Airbnb, besides hotels for tourists, the explosion of international tourism movements, the enhancement of knowledge and awareness about destinations due to the viral effect created by tourists visiting the destinations, the structure of traditional shops in destinations to change and become completely serving tourists, tourism movements have been the subject of the world agenda, especially in Europe (HOTREC Hospitality Europe, 2018). Bertocchi et al. (2020) pointed out that decreasing travel costs, facilitating access to information about destinations due to the frequent sharing of tourist experiences as a result of the spread of the internet, becoming popular of cruise tourism, diversification of accommodation supply, and cheaper accommodation with applications such as Airbnb, has increased the tourist flux and intense tourist movements.

The negative effects caused by mass tourism, especially in European destinations, for example, Venice, Barcelona, Dubrovnik, were protested by residents. For example, demonstrations were held in Venice on the grounds that the number of tourists crossing the borders increased rental prices and environmental pollution in the region. In Barcelona, the government has started to take measures to prevent unlicensed rentals through Airbnb, which reveals overtourism (Coldwell, 2017). Mass tourism poses serious dangers in these popular city centers as well as in the rainforests of Galapagos Island and Costa Rica within the framework of the problem of overtourism. As a result of the protests stemming from excessive tourism movements, the concept of "overtourism" has become a popular concept both in scientific research and among tourism stakeholders (Papathanassis, 2017). However, the concept's emergence can be attributed back to the "Irridex Index" developed by Doxey (1975), "Tourist Area Life Cycle" developed by Butler (1980), and "Tourism carrying capacity" by O'Reilly (1986) (Cunha, 2019).

Overtourism triggers the "tourismphobia" known as tourist hatred and rejection, as well as anti-tourism movements that bring together all the antagonists of tourism (Seraphin et al., 2019). The term overtourism, which is sometimes used in place of the word tourismphobia, has an extremely negative meaning and is a concept that emerged as a result of excessive use and consumption, overcrowding, and destination attractiveness becoming extremely popular (Volo, 2020).

In this chapter, the aim is to explain the concept of overtourism and its reasons in detail, to reveal the risks posed by overtourism with local people, tourists, and environmental perspectives, and to explain the responsibilities that should be undertaken in order to minimize the negative effects of overtourism with different perspectives. In addition, the possible effects of the COVID-19 crisis that emerged in 2020 on overtourism and the ability to eliminate the negative effects of ovarian tourism are discussed under a separate heading.

The Concepts Related to Over Tourism

Although overtourism is a concept that has been discussed together with sustainable tourism, where uncontrolled tourism development is criticized, it has become a frequently used concept as of the end of 2010 with the name of overtourism (Namberger et al., 2019). Overtourism is defined as tourism activities that create changes in the lifestyles of residents in the country that receive tourists due to the excessive increase in the seasonal peak of the number of visitors, creates difficulties in terms of accessibility to facilities in the destination area, resulting in damage to infrastructure, scenery and natural resources (Milano et al., 2019). According to Goodwin (2017), overtourism means that the quality of life of residents or experience perceived by visitors or locals deteriorates to unacceptable limits due to a large number of visitors. Seraphin et al. (2018) called overtourism broadly movements in any destination exposed to the negative effects of tourism. Therefore, a maximum limit/threshold should be found in tourism development (Seraphin et al., 2018). In general, the concept of overtourism is defined by, on the one hand, perceived overcrowding, on the other hand, the discomfort and resistance to tourism and protests in residents due to feeling overcrowded (Zemla, 2020).

Overtourism term which is seen as a tourism form where the negative effects of tourism arise as a result of hosting more tourists than the physical, economic and social thresholds that a destination can tolerate, is shaped by mass tourism and tourist influxes (Dodds & Butler, 2019). Density and over-crowding, on the other hand, are closely related to the concepts of carrying capacity, which have social, economic, and environmental dimensions shaped by reactions to these intense human movements. Similarly, Papathanassis (2017) states that the main reason for the anti-tourism and tourist hostility emerging in popular destinations is the exposure to tourist influxes above their carrying capacity rather than the quantitative increase in tourism movements. The main point that the overtourism concept indicates is the tourism destination is damaged socially, economically, culturally, and environmentally as a result of the concentration of visitors. Thus the sense of belonging in the local community is weakened (Seraphin et al., 2018).

Explaining the Term of overtourism Better: Irridex Model, Tourist Destination Life Cycle and Social Exchange Theory

The term overtourism is not a precisely defined concept, but meanwhile, it is a field of study that needs to be developed through empirical studies. Although the concept has a blurry structure yet, the most important foundations of overtourism will become more understandable with tourists, local people, and models explaining the development of destination tourism. One of the best knowns of these models is the Irritation Index model developed by Doxey (1975). According to the model, the excessive increase in the number of visitors to a destination and the development of tourism beyond its borders generates discomfort within the local people in the destination, but also cause negative attitudes towards tourists. As the number of tourists increases and the tourism achieves saturation in the destination, the appreciativeness of the local people turns into antagonism, apathy, and irritation starting from the maturity period to the period when it has started to decline of destination, and this situation prevents the development of tourism (Doxey, 1975). According to Harrill (2004), the Doxey Model is used to explain the change in local people's attitudes towards the evolutionary development of tourism in terms of the social effects. In this respect, when the costs perceived by the local people exceed the expected positive benefits with the increase in the number of tourists and the development of tourism, the unrest in the local population will continue to rise as the tourist-local people mismatch extension.

The second approach explains the attitude changes between the tourist-host society that overtourism is the "Tourist Area Life Cycle" model developed by Butler (1980). According to the model, the tourist area, which has gone through a seven-stage development process, as its tourism infrastructure and superstructure develop, its popularity and visitor numbers improve, the disturbance in the local community will become more pronounced. Butler (1980) announced that the reaction that will emerge because of the balance/imbalance between the duration of visitors' stay at the destination, the socio-cultural and economic characteristics of the tourism destination, and the spatial distribution of economic activities would change. In each phase of the tourism life cycle model, for example, although the impact of in exploration stage tourists on the socio-cultural structure of the local community is almost at the lowest point; the increase in the number of tourists will remain limited due to the insufficiency of the tourism infrastructure, and in stagnation stage where the tourist accommodation capacity increases residents evaluate tourism development negatively due to the increase in the number of tourists, different capacity threshold emerges (Martin & Uysal, 1990).

Another theory used to explain the relations between tourists and local people and adapted to tourism is the "social exchange theory". According to the theory, if the gains as a result of the social, cultural, and economic relations established between the tourist and the locals exceed the costs incurred, the locals are more likely to support tourism development. In theory, it is assumed that both parties, who establish a relationship, will display a rational behavior and compare the benefit-cost elements. The judgments obtained after this comparison will determine the local community's support/opposition to tourism (Ward & Berno, 2011). In addition, support for tourism development is higher in societies economically dependent on the tourism sector than in non-affiliated societies (Andereck et al., 2005).

These theories lead to the studies of local-tourist relations, state that the excessive tourists' numbers alone in the context of overtourism does not cause unrest. The discomfort feelings also occur and differ whether the destinations have reached saturation and whether the sources of income are provided from tourism. In this respect, overtourism includes both the imbalance in the quantitative increase and the unrest or discontent that turns into feelings of hostility, antagonism, and hatred.

Relationship Between Density, (Over) Crowding and Overtourism

Tourism development engenders great potential in terms of both revenues and tourist movements. International tourist arrivals aggregated 1,4 billion in 2018 according to World Tourism Organization (UNWTO), and this number increased by 6% compared to the previous year, and this was the second largest increase since 2010 was recorded. In 2018, 713 million to Europe, 343 million to the Asia Pacific, 217 million to America, 67 million to Africa, and 64 million to the Middle East have traveled for touristic purposes (UNWTO, 2019). While these large tourist influxes specifically affect each destination, it also closely affects the locals who host tourists and the tourists involved in the tourism event. Once, tourists who visit a region affect the social, cultural, and economic activities of the area, the lifestyle of residents, and also the construction of public policies (Brandolini & Mosetti, 2005).

There are three main factors that cause overtourism, and these factors trigger overtourism by generating congestion. One of these factors is a large number of visitors due to the tourism season. Secondly, the negative effects caused by these visitors to the extent that they are disturbing to the local people, and third and lastly, tourism facilities and other supportive products that are designed for visitors to the region and exceed optimal limits are physical effects (UNWTO, Centre of Expertise Leisure, Tourism & Hospitality, NHTV Breda University of Applied Sciences, & NHL Stenden University of Applied Sciences, 2018).

Over tourism is in a close relationship with "density" which is an objective quantitative dimension, and "overcrowding", a subjective dimension that is perceived as an output of density (Wall, 2019). Tourism density is measured by "the number of tourists per km^2" (Manera & Valle, 2018: 4547) of a destination, which is composed of the total of inbound and domestic tourism in a per km^2. While density in tourist inflows cause enhancing destination popularity, it leads to complaints about crowding in terms of both other tourists and local communities. Jacobsen et al. (2019) state that in addition to negative sentiments against the increase in human density, people's socialization and following the interactions brought by socialization can also create positive sentiments that increase the emotions of excitement and fun. This situation varies according to tourist and destination type. In this context, a higher density than the threshold of the acceptable number may negatively affect visitor satisfaction, loyalty, and attitudes (Jacobsen et al., 2019). It is indicated that determining optimal crowding levels in a town or village is quite difficult. Extensive tourism development and visitor crowding together can cause dissatisfaction (Jacobsen et al., 2019; Ryan & Cessford, 2003).

Simancas Cruz and Penarrubia Zaragoza (2019) stated that density alone could not cause discomfort and, in this context, it will create uncertainty when evaluated in terms of the periods before the saturation phase when the number of visitors is high, and there is not yet a feeling of hostility in the local population. In addition, they argued that the real emergence of overtourism is the perception of "overcrowding"that comes out when the accommodation capacity reaches saturation. In this framework, tourism development creates overcrowded, noisy, architecturally flavorless, and place pressures on the infrastructure (Dwyer & Edwards, 2010).

Crowding, which constitutes the subjective and negative perception part of the density of visitors, expresses a cognitive and perceptual judgment that density reveals in the individual from a psychological and behavioral point of view (Shelby & Haberlein, 1984; Wall, 2019). It can be described as the reaction(s) caused by the inhabitants of this region by reason of density (Gürkaynak & LeCompte, 1979). In this respect, crowding can be defined functionally different according to related scientific fields and approaches. According to Mehta (2013), the perception of crowding exceeded optimal levels and experienced in the same environment leads to negative evaluations. It is appointed that overcrowding and congestions create displeasure in a destination for locals (Brunt & Courtney, 1999: 507; Jin & Pearce, 2011). However, Stokols (1972) stated that density alone is not sufficient to have a one-to-one effect on this feeling of disturbance but is a necessary precondition for the formation of overcrowding perception. In addition, the psychological stress felt as a result of crowding differs according to social situations. Social stress is caused by relationships arising from the presence of others in the same environment, and non-social stress differs depending on whether there is an appropriate spatial gap (Stokols, 1972).

The feeling of resentment is associated with the intervention of the personal space of the individuals (Jacobsen et al., 2019). Hall (1966) notes that in this space-home, city, town, or office, individual tries to keep the distance between himself and others to eliminate alienation or improve self-knowledge. Little (1965) emphasized that crowding affects the staff space via individual distance violations. If the deviations in the distance occur and up to the point of threat threshold unexpectedly, the discomfort reactions and negative consequences have begun (Burgoon, 1978). A defense system is engaged.

It would be an inadequate explanation to make sense of overtourism and set a threshold through numbers alone. In this context, if the density and crowding in a region can be managed with the appropriate usage level approach and if it is subjected to a specific classification according to the use or protection purpose, the negative effects of overtourism can be minimized (Wall, 2020).

Carrying Capacity and Overtourism

The growth process of tourism should be managed by taking into consideration the tourism capacity. Tourism capacity refers to the tourist density that the region can absorb before the feelings of hostility are felt by the local people in the destination that accepts tourists and the negative effects of tourism emerge (O'Reilly, 1986). Here, the main point emphasized by the tourism capacity is the number of tourists desired to be accepted in the host country.

Carrying capacity (CC) is also a concept associated with sustainable tourism development. The concept CC is frequently used in the implementation of sustainable tourism indicators (Lopez-Bonilla & Lopez-Bonilla, 2008). Dwyer and Edwards (2010: 25) defines carrying capacity as "*maximum number of people who can use a destination without an unacceptable alteration to the physical environment, an unacceptable decline in the living environment of the community and an unacceptable decline in the visitor experience*". CC in tourism means that the economic growth of tourists is responsible towards the local community and cultural values of the community, consistent with protection and improvement of the local natural environment and with the conservation of the local traditional economic activities (Brandolini & Mosetti, 2005).

CC has dimensions as the physical, economic, and social-psychological (Massiani & Santoro, 2012). Physical carrying capacity refers to a destination's or site's availability of space, the form of the landscape, and the CC of the land (Da Silva 2002). This concept is connected with ecological capacity, which focuses on the demands on the landscape and requires tourism growth levels which not damage the soil, plants, or animals. Overuse of environmental resources concludes in negative effects such as increasing pollution (noise, water, waste discharges), damage of historic sites, deterioration of flora and fauna, scenery, wildlife habitat, etc. (Castellani et al., 2007).

Social-psychological capacity is related to both residents' lifestyles and interests, and their attitudes towards tourism and activities, behavior and attitudes of visitors. This concept accepts that there is a limit based on the tolerance of visitors by residents. Increasing tourist numbers can lead to annoyance and hostility in the host community towards tourists. In terms of visitors, enormous tourist numbers can lead to a decrease quality of their visitation experience (crowding, congestion, etc.) (Dwyer & Edwards, 2010). In this context, the social-psychological CC, which also represented the social dimension of the CC and called the socio-cultural CC that is a part of forms of tourism CC. Psychological CC aims to obtain information about the feelings of both tourists and residents. The social dimension of tourism carrying capacity plays an important role in the sustainable development of tourism destinations (Zhiyong & Sheng, 2009). Given the psychological capacity, the long-term sustainability of tourism can be negatively impacted by any effect between tourists and residents, which may cause anger, discomfort, and anger (Postma & Schmuecker, 2017).

A touristic system is an integrated system consisting of different sub-systems such as ecological (biological and physical), social, cultural, infrastructure and management (institutional and economic) sub-systems. The levels of these different CC may be in conflict; for example, a large number of tourist may be economically desirable, although it is doubtful, as it increases the income of the local community, but from a social and ecological point of view, it can cause permanent injuries if the crime rate increases, etc. The social CC is generally analyzed from the perspective of both residents and visitors. Therefore, decision-makers need to plan considering these CCs carefully (Brandolini & Mosetti, 2005).

The local, socio-economic and physical environment and the image of the touristic product play a key role in determining the CC. While the influx of tourists above the saturation level reduces the quality of the destination, it causes permanent damage in cultural, social, and environmental terms (Inskeep, 1991).

Perceived Risks Caused by Over Tourism

Overtourism causes local people to have some negative effects on the environment in some destinations (Bayram & Arıcı, 2019). Mccool and Lime (2001) state that excessive tourist number in a destination affects factors negatively such as environmental integrity and social life, Opinions about the difficulties caused by these risks of overtourism can affect stakeholders such as residents, tourists, excursionists, destination marketing organizations or local businesses (Aylan & Aylan, 2020). Çetinoğlu and Meydan Uygur (2020) state that overtourism can cause irreversible damage to the natural environment, cultural heritage, social structure, economy, ecological balance, and most importantly, psychologically. In this context, there are various risks or impacts that make a destination prone to overtourism. These risks are handled from a tourist, local community, and environmental point of view in this section.

Local Community Perspective

When the definition of overtourism is examined, overcrowding due to the increase in the number of tourists above the available supply sources affects the general welfare of the local people in destinations, forces community to make permanent changes in their lifestyle, bear the consequences of temporary and seasonal tourism concentration, make difficult to access the opportunities used in daily life (Milano et al., 2018). From this point of view, overtourism is considered as a concept that most concerns the local people. If the tourism sector is taken into consideration, tourists are welcomed by the local people due to provide the direct income increase through foreign exchange expenditure. However, as the number of tourists in the region increases, local people may think that the quality of life is under threat, which may cause local people to be less hospitable towards tourists (Croes et al., 2017). Overtourism creates an obstacle to sustainability with excessive consumption of resources and negatively affects the quality of life of local people (Koens et al., 2018). Similarly, Sputniknews (2019) states that overtourism is a concept that can be undesirable by most of the locals. In addition, overtourism has been considered a major problem for popular tourist destinations in the European continent, and local people in the relevant destination expressed their complaints about overtourism by protesting tourists with banners and even attacking tour buses.

Çetinoğlu and Meydan Uygur (2020) also state that overtourism is a movement that will disrupt all kinds of order of the local people and cause feelings of hatred and grudge towards incoming visitors. In this context, it is stated that when the social CC is exceeded in the destinations, the local people are most affected, and the visitors start not to enjoy coming to their destination. On the other hand, Lino and Marie (2019) state that factors such as overcrowding, traffic congestion, and negative tourist behaviors that have emerged in some destinations due to the size of tourism in recent years may cause unwanted economic, physical, and social effects for the local population. In addition, local people may be concerned about the development of tourism because of increased living costs, noise, loss of employment opportunities, economic leakages, and cultural changes due to overtourism (Çolak et al., 2020). The intensive use of social areas such as communication, public transportation, and hospitals used by the local people may cause the local people to be unable to meet their needs (Kurnaz & İpar, 2020). Due to the decrease in the gains from tourism activities for residents in destinations (Martin et al., 2018); the alienation of local people (Sarantakou & Terkenli, 2019), and the lack of control of local governments, it is revealed that consequences of negative effects such as the security problem in the region, the increase in house prices, noisiness and decrease in quality of life, local people migrate to other places. Yumuk and Altıntaş (2019) found that the local people evaluate rental house prices were high, and infrastructure service was insufficient in the context of quality of life due to overtourism. Such causes show that overtourism has significant risks in destinations, especially for local people.

Tourist Perspective

Tourists visits to a popular destination at certain times and the increase in demand in that destination are among the most important reasons for overtourism (Phi, 2019). In this context, Goodwin (2017) states that due to the increase in demand, overtourism negatively affects the tourist experiences in the destination. A tourist can find himself questioning the satisfaction of his travel when he cannot comfortably visit a destination where he goes with great expectations, waits in the open buffet queue at every meal at his hotel is exposed to noise at the place where he has a coffee break, and cannot find an empty sunbed on the beach (Genç, 2020). These situations are some factors that negatively affect the quality experiences of tourists. Çolak et al. (2020) explained that in disproportionately developing destinations when evaluated in terms of the tourist experience, problems arise due to difficulties in accessing products, queues, and crowds. Jamieson and Jamieson (2019) asserted that the increasing number of tourists due to excessive tourism affects the population balance of the destination, and this is a factor that eliminates the attractiveness of the destination for tourists. Çolak et al. (2020) also affirmed that the tourists who cannot find adequate infrastructure and superstructure facilities in the city destinations, who cannot benefit from the operations and services sufficiently, and who are welcomed by the local elements of the destination they visit, leave the destination unsatisfied manner. Tourists cannot get the satisfaction they expect from their vacations or travels due to the excessive flood of people they encounter in the tourism regions, streets, transportation vehicles, indoor and outdoor touristic areas where they work all year round and on limited annual leave to rest, have fun, and spend good and quality time (Aylan & Aylan, 2020). In this context, the concentration of tourists on a specific destination, activity, or business and their actions cause congestion in these destinations; and so this situation turns out negatively affects the experiences of tourists. Therefore, spreading tourists to different areas or destinations, ensuring that tourists benefit from all tourism businesses, including accommodation in a balanced way, can be considered as a factor that can prevent congestion in crowded places.

Environmental Perspective

While tourism development is perceived positively due to the economic benefit in any destination, negative changes in the physical environment turn the positive perceptions that occur at the beginning into a negative statement gradually (Akova, 2006). Increasing the participation of more people in tourism causes exceeding the CC, environmental problems in the destinations (Altanlar & Kesim, 2011). In this context, when the risks posed by overtourism on the environment are evaluated, environmental pollution (visible increase in water, soil, air, and noise pollution), infrastructure, visual pollution (visual/aesthetic pollution of natural or cultural landscapes), environmental damage (natural, historical and architectural damage to areas) and excessive population density (overcrowding in natural, historical and architectural areas) (Aylan & Aylan, 2020). Yüksel et al. (2020) also include the environmental effects of overtourism; sea pollution, environmental pollution, noise pollution, damage to forest, and habitat degradation.

Genç (2020) affirmed that the negative environmental effects experienced as a result of overtourism are such as destruction, pollution (waste, noise, and visual origin), and congestion in the infrastructure in natural, historical, and cultural areas. In addition, overtourism can also cause waste management in destinations and wear on endangered flora and fauna (Çolak et al., 2020). Similarly, Koens et al. (2018) articulated that overtourism causes more damage to the region than its economic return to the sector and has negative effects such as the sea, air, noise, and environmental pollution, traffic, and crowding. In this context, overtourism leads to negative environmental degradation due to exceeding the number of visitors that can stay in a destination, and these destinations do not want too many tourists to come to their regions. For example, Kuhn (2015) explained that the Chinese government blacklists the individuals who behave in this way due to the environmentally disturbing actions of Chinese tourists in various destinations and restricts the travel rights of these individuals. Morris (2018) indicated that the island of Boracay, Philippines, was closed to tourists until it was restored to its former state due to the damage caused by overtourism, and actions such as smoking, consuming food and beverages, and using plastic bags on the coast are prohibited. According to Sputniknews (2019), due to the environmental degradation and pollution experienced by excessive tourists coming to Fjadrargljufur Canyon in Iceland, the destination managers decided to close the canyon at the peak time of the tourists. Therefore, all these factors show that the most important negative effect of extreme tourism in destinations is environmental degradation, and destination managers are trying to implement various measures.

Responsibilities for Coping with Overtourism

There are various responsibilities for governments, visitors, and all other marketing entities - state or private - to tackle overtourism. Although overtourism is the subject of different interests for each tourism stakeholder, it creates great damage to the physical environment and authentic culture, which are the main attraction elements of overtourism. In addition, the constant consumption of environmental resources due to hedonic habits, the deterioration of the socio-cultural environment, and the uniformization of culture undermine tourism. In this section, responsibilities to prevent the negative consequences of overtourism are handled in the marketing dimension, governments/administrative, and tourist responsibilities.

Marketing Efforts

Encouragement of luxurious consumption habits emerges the desire of the individual to gain a status against the other and have led to reveal the consumer type that is completely oriented to consumption via panache, bandwagon, and Veblen effects. This popular culture takes full advantage of digital media and commodifies all tangible and intangible values. In this context, content created on the web, such as "destinations you need to see before you die", pressures people to visit a region. In addition, the consequences of unhindered tourism demand in certain regions, included in the world cultural heritage list, damage the intangible and tangible assets in the region.

Many destination management organizations (DMOs) have used their resources in promotions as a touristic attraction rather than managing destinations sustainably. The increase in tourism mobility at the international level is called "success" by the destinations, and the level of tourism development is measured by the excess number of visitors. Because of this "fake" approach to success, many DMOs build their marketing activities with the perspective that they build on increasing the number of visitors. However, holding DMOs solely responsible for overtourism would be an insufficient assessment (Dodds & Butler, 2019). Low-cost passenger transportation by airline companies (Dodds & Butler, 2019), price campaigns in hotels, promoting mobility with tourist loans, early payment campaigns, cheap holidays arising from the exchange rate difference between the host destination and tourist generated regions triggered overtourism.

DMOs have also made various arrangements to deal with overtourism. For example, for Venice, it is recommended to remove the Taj Mahal from all promotional materials and to review the Tourism Management Plan (Seraphin et al., 2018). On the other hand, advertising campaigns were carried out with the emphasis on organizing tours to "less visited areas" to reduce the pressure in regions intensified by overtourism, respecting the culture of the local people and making a positive contribution to the economy of the local people with visits to untouched natural areas (Morris & Parry, 2017). In addition, the change caused by tourist mobility accelerates through social media and other media platforms with the widespread use of the internet by many individuals (Peters et al., 2018), the destination becomes popular in a very short time.

In consideration of coping with overtourism, some destinations used the power of "soft measurements" or "soft power" in their promotional activities. For example, in the Cinque Terre region in Italy, with applications that show the current tourist density at the hot points they want to visit, with the imposition of a ban on opening new cafes and restaurants in tourist-intensive areas in Copenhagen, with the arrangement of a specially designed event calendar as a promotional activity in Bruges, in Belgium; with methods such as the daily limit for cruise ship passengers in Dubrovnik in Croatia and Santorini in Greece, and the spread of tourist density in Iceland by diffusion to less crowded places. (Bershidsky, 2019).

For this reason, it is vitally important for both the public and private sectors to conduct marketing studies with a quantity-oriented demand perspective-taking sustainability factors into consideration. While many people can have an idea for a holiday spent in another part of the world, especially with consumer-generated content today; promotional activities and tools need to be carried out in a conscious and planned manner so that the marketed place/culture/cultural heritage does not harm itself at least.

Governmental and Managerial Measurements

Overtourism has emerged as the consequence of policies that tend to expand in volume (Peeters et al., 2018). One of the most important methods of dealing with overtourism movements is the management of tourism development and carrying capacity without exceeding the limits of tourism with a sustainable understanding (Brandolini & Mosetti, 2005; Dwyer & Edwards, 2010). However, it is not easy to measure these boundaries quantitatively. For this reason, official authorities are required to determine the concrete data to calculate the vulnerability levels specific to each destination and to measure these data scientifically and periodically. In addition to measuring these data, to minimize the negative effects caused by tourism, it is necessary to carry out these activities in the presence of a committee of experts who can consult economically, socio-culturally, and environmentally, considering the interests of each stakeholder. Consultation activities should create sufficient time and a suitable environment for tourism providers to properly regulate the market. For this reason, the plans to be made by the official authorities should be arranged in a concrete way (Koh & Fakfare, 2020).

Legislation power is the most important power that governments have with respect to regulate the tourism market. In this context, especially in order to prevent the damage of tourism to the environment, it may be necessary to organize environmental impact reports more effectively, to make spatial restrictions in areas that are vulnerable to tourism movements, to avoid unfair competition and densely clustered in tourism-related business, and to distribute workplaces gathered in a certain region, and legal regulations for waste management. However, according to Seraphin et al. (2018), state measurements such as imposing entry and exit bans (trexit) may have negative and drastic effects on the destination image. These measures are not sustainable and are temporary arrangements that limit certain periods in which tourists pose a serious threat to the destination (Seraphin et al., 2019). Other measures that include the reduction of the negative effects caused by overtourism by the official authorities are the distribution of the visitor from the tourism-intensive points to the points where the concentration is less, the creation of new routes, the rejuvenation of the legal regulations according to the conditions, the elimination of the deficiencies in the city infrastructure, the strengthening of the communication with the stakeholders, the monitoring of the tourism movements (UNWTO, Centre of Expertise Leisure, Tourism & Hospitality, NHTV Breda University of Applied Sciences, & NHL Stenden University of Applied Sciences, 2018).

Separating tourism and recreation areas into specific spectrums according to their sensitivity is one of the measures that can be used by official authorities. For example, in the province of Ontario, it provides visitor management by dividing into sections such as parks, other zones separated within parks, zones reserved for wildlife protection, recreation areas (Wall, 2019). Also, Wall (2020) included that actions and regulations at the administrative level in the context of direct and indirect measures that can prevent the negative consequences of overtourism. Setting entrance and user fees on the basis of direct measures, restricting activities and equipment, imposing permissions and licenses for participation, intensifying and enforcing rules, implementing a certain area for certain activities, assigning more staff who can perform and supervise them. In the context of indirect measures, it is stated that limits should be set in ancillary facilities and accommodation, diversifying accessibility, site design, type and degree of maintenance, providing information and training to improve environmental awareness (Wall, 2020).

Although the regulation of these measures by official authorities has drastic importance, not any measures taken by states and official authorities alone can prevent overtourism movements. A stakeholder participatory approach must be developed for these regulations to be effective. One of the most important duties of the state at this point is to create an environment that will facilitate communication between stakeholders.

Tourists and Potential Tourists

Development in tourism will be short-lived without the planned and local support and participation in tourism development. Many anti-tourist movements have proven how important local people and visitor relations are in determining the attitudes of local people towards tourism development and developing a sustainable tourism development strategy (Eusébio et al., 2018).

Exceeding the tourism capacity may cause negative reactions from the local people in the host country, and from the tourist point of view, the tourist feels uncomfortable or unwanted in the host country due to the feeling that it is uncomfortable (O'Reilly, 1986). Some empirical studies have been demonstrated that (for example, Neuts & Nijkamp, 2012), the density of the number of tourists, interaction with others, other tourist's behavior, level of similarity is directly and positively related to the crowd perception of tourists, and nationality and experience perceptions are negatively related to crowd perception. The crowd perception of the tourist is a factor that can affect the tourist experiences in this respect. As a result of this crowd perception, the experience quality of the tourist may decrease (Goodwin, 2017). Therefore, tourists think that the attractiveness and authenticity of the destination have come to an end due to the excessive commercialization of the destination (O'Reilly, 1986).

One of the most important responsibilities of the tourists is to respect the local culture and to be conscious about protecting the environment. On the other hand, it is one of the methods that can be developed by tour guides to create awareness and consciousness about the features of the destination in order to minimize the negative effects of overtourism. So, a tourist should be responsible manner. In this sense, a responsible tourist should be informed about the environment, culture, security, local problems; respected and appreciated the environment, people, land, laws; tolerant, unbiased and open to celebrating differences, interested and affected by the environment, people, culture (Stanford, 2008). Among the steps that can be taken by responsible tourists; to be able to communicate with the local people in a moderate atmosphere, to behave in accordance with the religious, cultural, and gastronomic values of the destination, not to violate the responsible policies on electricity and water consumption adopted and implemented by tourism enterprises, and to give importance to shopping from local sellers (Zengin & Bayram, 2018). Besides focusing on environmental protection, responsible tourists are also tending to display ethical consumer behavior such as being altruistic and empathic, tends to stick to the rules, adopting a sustainable lifestyle (Weeden, 2011).

For tourism authorities, visitor management is of great importance, especially during peak tourism seasons. In this context, in addition to activities such as managing the number of visitors, spatial distribution, subject to special permissions for certain activities, measures such as the management of tourist behavior is taken (Wall, 2019). However, Papathanassis (2017) states that creating codes of conduct for tourists and license restriction measures in accommodation services are short-term, simple, and superficial measures.

Local authorities should plan to deal with overtourism by considering the natural, socio-cultural and economic dimensions of the density while also taking into account the experiential dimension. At this point, tourists should clearly identify and departmentalize by destination before promotion, and destinations also create alternative tourism potential by diversifying activities for incoming tourists. With the development of technology, negative evaluations about a destination in digital environments and especially on social media can easily and quickly affect other tourists and potential tourists in a negative way. For this reason, planning the planning without damaging the quality of experience can prevent negative image perceptions.

Can COVID-19 Be a Cure for Overtourism Induced Pain?

According to December 2020 data due to Coronavirus (COVID-19) pandemic disease, which emerged in March 2020, 79 million people infected, and approximately 1754 million people died due to the spreading of the virus (WHO, 2020). One of the sectors most economically affected by the disease has been the tourism and travel industry. The International Civil Aviation Organization (ICAO) stated that the number of international passengers decreased by approximately 2900 million people in 2020, and a decrease of approximately 400 billion dollars was realized (ICAO, 2020). Again, the United Nations World Tourism Organization (UNWTO) stated that international tourism mobility lost its capacity by 72% between January–October 2020 (UNWTO, 2020). Apartment rental booking rates through Airbnb sites fell between 41% -96% (Laurent, 2020). This scenario indicated that although the vulnerability of the tourism sector to the crisis is high, it has also shown that the crisis preparedness level is also at low levels. The disease has transformed everyday life, habits, needs and wants institutions, and consumer behaviors (Koh, 2020). Pandemic crisis wrecked and drastic effects on travel and tourism's all sector as hospitality, events, arts and all affiliated sectors (Higgins-Desbiolles, 2020). While in terms of tourism consumption, the luxury wants lost importance, meeting basic needs has gained importance. In the center of the future of tourism, consumption will be safety and security issues.

Tourism hotspots have been able to breathe a little with the prevention of travel and curfews. However, will tourism, responsible consumption, and sustainability be a solution to the new crowd problem after the process is over? In this context, some destinations have taken various measures after the pandemic, such as the government COVID Secure Protocols in Machu Picchu, where social distance will be maintained and strict ticketing policy, such as the negative COVID-19 PCR test application for visitors to Dubrovnik (Street, 2020), To reduce the number of participants for the protection of physical distance for adventure tourists, new health and hygiene protocols, determination of contact tracking points by using digital facilities, to benefit from vehicles that will cause the least damage to nature, and to meet the needs of tourists such as skiing, bicycles, and helmets by WTTC. It tries to prevent overtourism, such as online ticketing, online check-in, and check-out transactions (WTTC, 2020) in order to use belongings, protect physical distance and limit contact points.

Gössling et al. (2021) explained that one of the most important consequences of the crisis would be to implement tighter border policies that will support nationalist movements, and domestic tourism will gain importance. Higgins-Desbiolles (2020) demonstrated another perspective on the future of tourism as; if infectious disease crisis is considered in terms of a system change, more sustainable practices in the new era, the global fight against climate change, opportunities for equality and justice may result in a restriction in the number of visitors. If seen as a means of economic gain, the negative consequences of overtourism will continue. Approaches to the future of tourism will determine this (Higgins-Desbiolles, 2020). However, the COVID-19 crisis will create an "awakening" for some individuals about how important awareness, responsibility, and sustainability are. The crisis has revealed the fact that when the boundaries of natural and wildlife are intervened, "invisible threat-viruses" and other defense organisms that may emerge in a short time can put an end to human life. Therefore, tourists traveling responsibly can at least reduce the old crowd.

Conclusion

Considering the rapid intensification of the tourist influx in depth, tourism generates both harmful and favorable effects to a destination according to cost-benefit analysis. Primarily, if a concentration that exceeds the CC levels and crowding causes negative environmental pressures, economic added value cannot reach a level that can tolerate this effect. Therefore, overtourism should be fought with a sophisticated management approach specifically for each destination (Seraphin et al., 2018). This versatile management approach should be carried out within the framework of a sustainable approach shaped by technological developments and innovation. The sustainable framework should create an environment in which interests do not conflict for both tourists and tourism providers. The most important issues in the implementation of this aim are the fair use of all long-term tangible/intangible tourism resources, all aspects of which are regulated in detail, and all these approaches are determined under a general umbrella within the framework of sustainability.

Managing the change in the destination plays a critical role during the tourism development phase. Especially in the maturity phase, preventing the damage caused by the intensity that exceeds the tourism capacity should be a priority in terms of tourism management (Getz, 1992). In this critical stage, in terms of determining the future position of the destination, instead of producing policies based on preventing only tourist arrivals, the correct management of change and transformation is proportional to the development of tourism should be designed appropriately. The yield management can be applied in this situation, which can be identified as "right tourist", "right price", "right tourism capacity met", "right time", "right destination" "right accommodation type". This value-oriented approach will only be possible with the correct segmentation, correct analysis of tourists, and correct visitor management.

Considering the negative consequences of overtourism in terms of political authorities, tourism suppliers, and especially companies (travel and accommodation sector), they are neglected now and before, especially at the expense of profit. For this reason, making a versatile tourism plan and managing the plan within the scope of public interest is very critical in terms of protecting tourism resources and, most importantly, cultural heritage. Instead of making the same recipe for each destination, specific plans should be created by making separate evaluations according to the characteristics, economic and socio-cultural structure of each destination. In this respect, the meaning attributed to destinations, visitor management processes (Wall, 2019), and tourist perspectives differ from culture to culture, creating a different balance of protection and use. Therefore, the process of dealing with overtourism can be complicated and difficult. The main problem in terms of overtourism is to ensure and manage the interaction between tourism stakeholders. Problems that may arise in this regard are under the responsibility of the tourism sector and most importantly the state. Effective management of tourism development and communication between stakeholders should be carried out using the information infrastructure of both tourism and economics and other related disciplines.

The decrease in tourism movements related to overtourism due to COVID-19 pandemic disease has led to the intensive use of digital visit methods in terms of visiting tourist-intensive areas and sensitive areas. It has shown that virtual tours and virtual reality applications can be used as an alternative travel and travel tool in crisis situations. The rapid spread of the epidemic has greatly affected the tourism sector in accommodation, travel, and especially entertainment.

Although overtourism is not a new subject in scope, it cannot be denied that the concept itself brings a new approach to the literature. It is clear the concept overtourism is evolving in a direction that can be measured and developed analytically, at least with applied studies (Simancas Cruz & Peñarrubia Zaragoza, 2019). Studies in the literature show that empirical studies on overtourism have not yet reached a generalizing level and are still insufficient. In addition, the local people's point of view towards the tourist also varies in areas with high tourism density. Whether the destination has reached the maturity level can also differ depending on whether it is a destination that has just entered the market. Also, the cultural characteristics of societies can determine the attitude of local people towards tourists.

References

Akova, O. (2006). Yerel Halkın Turizmin Etkilerini Algılamalarına ve Tutumlarına Yönelik Bir Araştırma. *Akademik İncelemeler Dergisi, 2*(1), 77–109.

Altanlar, A. ve Kesim, G.A. (2011). Sürdürülebilir Turizm Planlaması İçin Yöre Halkı ve Yerli Turistlerin Davranış ve Beklentilerini Anlamaya Yönelik Bir Araştırma: Akçakoca Örneği. Ankara Üniversitesi Çevrebilimleri Dergisi, 3(2), 1–20.

Andereck, K. L., Valentine, K. M., Knopf, R. C., & Vogt, C. A. (2005). Residents' perceptions of community tourism impacts. *Annals of Tourism Research, 32*, 1056–1076.

Aylan, F. K. ve Aylan, S. (2020). Aşırı Turizm Kavramının Farklı Boyutlarıyla İncelenmesi: Instagram'da Paylaşılan Fotoğraf ve Yorumların İçerik Analizi. Journal of Tourism and Gastronomy Studies, 8 (1), 584–607.

Bayram, M. ve Arıcı, S. (2019). Aşırı Turizm ve Destinasyon Yönetimi: Avrupalı Destinasyonların Aşırı Turizm Politikalarının Değerlendirilmesi. *5. Uluslararası Türk Dünyası Turizm Sempozyumu,* 13–15 Haziran, Taraz, Kazakistan.

Bershidsky, L. (2019). *Europe's top vacation sites need options to fight overcrowding*. Retrieved from: https://www.bloomberg.com/opinion/articles/2019-06-09/europe-s-top-vacation-sites-need-options-to-fight-overcrowding. Accessed 12 Dec 2020.

Bertocchi, D., Camatti, N., Giove, S., & Van der Borg, J. (2020). Venice and over-tourism: Simulating sustainable development scenarios through a tourism carrying capacity model. *Sustainability, 12*(2), 512–527.

Brandolini, S.M.D.A. and Mosetti, R. (2005). *Social carrying capacity of mass tourist sites: theoretical and practical issues about its measurement*. Nota di Lavaro, no:144.2005, Fondazione Eni Enrico Mattei (FEEM) Working Paper, Milano.

Brunt, P., & Courtney, P. (1999). Host perceptions of sociocultural impacts. *Annals of Tourism Research, 26*(3), 493–515.

Burgoon, J. K. (1978). A communication model of personal space violations: Explication and an initial test. *Human Communication Research, 4*(2), 129–142.

Butler, R. W. (1980). The concept of a tourist area cycle of evolution: Implications for management of resources. *The Canadian Geographer/Le Géographe Canadien, 24*(1), 5–12.

Castellani, V., Sala, S., & Pitea, D. (2007). A new method for tourism carrying capacity assessment. *WIT Transactions on Ecology and the Environment, Ecosystems and Sustainable Development VI, 106*, 365–374.

Çetinoğlu, D. & Meydan Uygur, S. (2020). Aşırı turizm sorununa çözüm olarak pazarlamama stratejileri kullanılabilir mi? Pamukkale örneği. *İşletme Araştırmaları Dergisi*, 12(3), 372–388.

Çolak, O., Kiper, V. O. & Batman, O. (2020). Ölçüsüz turizme (over-tourism) dair kavramsal bir yaklaşım. *Akademik Araştırmalar ve Çalışmalar Dergisi*, 12(23), 609–621.

Coldwell, W. (2017). First Venice and Barcelona: now anti-tourism marches spread across Europe. *The Guardian*. Retrieved from: https://static1.squarespace.com/static/53109b11e4b05040160f0a8f/t/59bfcc8b90badeccf7cc1564/1505741995422/First+Venice+and+Barcelona_+now+antitourism+marches+spread+across+Europe+_+Travel+_+The+Guardian.pdf. Accessed 15 Nov 2020.

Croes, R., Rivera, M. A., Semrad, K., & Khalizadeh, J. (2017). *Happiness and tourism: Evidence from Aruba*. Unpublished Manuscript, The Dick Pope Sr. Institute for Tourism Studies. Orlando, Florida.

Cunha, N. V. D. (2019). Public policies and tourist saturation in the favelas of Rio de Janeiro. In C. Milano, J. M. Cheer, & M. Novelli (Eds.), *Over-tourism: Excesses, discontents and measures in travel and tourism* (pp. 152–166). CABI.

Da Silva, C. P. (2002). Beach carrying capacity assessment: How important is it? *Journal of Coastal Research, 36*, 190–197.

Dodds, R., & Butler, R. (2019). The phenomena of over-tourism: A review. *International Journal of Tourism Cities, 5*(4), 519–528.

Doxey, G. V. (1975). A causation theory of visitor-resident irritants: Methodology and research inferences. In *Conference proceedings of the 6th annual travel research association* (pp. 195–198). Travel Research Association.

Dwyer, L., & Edwards, D. C. (2010). Sustainable tourism planning. In J. J. Liburd & D. Edwards (Eds.), *Understanding the sustainable development of tourism* (pp. 20–44). Goodfellow Publishers.

Eusébio, C., Vieira, A.L. & Lima, S. (2018). Place attachment, host–tourist interactions, and residents' attitudes towards tourism development: the case of Boa Vista Island in Cape Verde. *Journal of Sustainable Tourism, 26*(6), 890–909.

Genç, K. (2020). *İstanbul'da Aşırı Turizm: Sivil Toplum Kuruluşları Görüşleri Üzerinden Yorumsayıcı ve Karşılaştırmalı Bir Araştırma (Yayınlanmamış Doktora Tezi)*. Sakarya Uygulamalı Bilimler Üniversitesi.

Getz, D. (1992). Tourism planning and destination life cycle. *Annals of Tourism Research, 19*, 752–770.

Goodwin, H. (2017). The Challenge of Over-tourism. *Responsible Tourism Partnership Working Paper 4*, Cape Town, South Africa, 1–19.

Gössling, S., Scott, D., & Hall, M. (2021). Pandemics, tourism and global change: A rapid assessment of COVID-19. *Journal of Sustainable Tourism, 29*(1), 1–20.

Gürkaynak, M. R., & LeCompte, W. A. (1979). Introduction. In M. R. Gürkaynak & A. W. Lecompte (Eds.), *Human consequences of crowding* (pp. 1–7). Plenum Press.

Hall, E. T. (1966). *The hidden dimension*. Anchor Books.

Harrill, R. (2004). Residents' attitudes toward tourism development: A literature review with ımplications for tourism planning. *Journal of Planning Literature, 18*(1), 1–16.

Higgins-Desbiolles, F. (2020). The "war over tourism": challenges to sustainable tourism in the tourism academy after COVID-19. *Journal of Sustainable Tourism, 29*(4), 551–569.

HOTREC Hospitality Europe. (2018). *Position paper on over-tourism: Putting sustainable tourism on top of the EU policy agenda for the benefits of society*. Retrieved from: https://www.hotrec.eu/wp-content/customer-area/storage/9637095ddab8cf5b3edadc9d61af5fe8/Over-tourism-position-paper1.pdf. Accessed 30 Nov 2020.

ICAO (International Civil Aviation Organization) (2020), *Effects of Novel Coronavirus (COVID-19) on civil aviation: Economic impact analysis, Montreal, Canada*. Available at: https://www.icao.int/sustainability/Documents/Covid19/ICAO_coronavirus_Econ_Impact.pdf. Accessed: 25.12.2020.

Inskeep, E. (1991). *Tourism planning: An integrated and sustainable development approach*. Wiley.

Jacobsen, J. K. S., Iversen, N. M., & Hem, L. E. (2019). Hotspot crowding and over-tourism: Antecedents of destination attractiveness. *Annals of Tourism Research, 76*, 53–66.

Jamieson, W. ve Jamieson, M. (2019). Overtourism management competencies in Asian urban heritage areas. International Journal of Tourism Cities, 5(4), 581–597.

Jin, Q., & Pearce, P. (2011). Tourist perception of crowding and management approaches at tourism sites in Xi'an. *Asia Pacific Journal of Tourism Research, 16*(3), 325–338.

Koens, K., Postma, A., & ve Papp, B. (2018). Is Overtourism overused? Understanding the impact of tourism in a City context. *Sustanability, 10*, 1–15.

Koh, E. (2020). The end of over-tourism? Opportunities in a post-Covid-19 world. *International Journal of Tourism Cities, 6*(4), 1015–1023.

Koh, E., & Fakfare, P. (2020). Overcoming "over-tourism": The closure of Maya Bay. *International Journal of Tourism Cities, 6*(2), 279–296.

Kuhn, A. (2015). *For Chinese tourists behaving badly: A government blacklist*. NPR. Available at: https://www.npr.org/sections/parallels/2015/05/08/405183120/beijing-clamps-down-on-chinese-tourists-behaving-badly. Accessed: 05.12.2020.

Kurnaz, H. A. ve İpar, M. S. (2020). Yavaş Şehir Akyaka'da Aşırı Turizm Olgusunun Esnafın Bakış Açısıyla Değerlendirilmesi. Journal of Hospitality and Tourism Issues, 2(1), 39–55.

Laurent, L. (2020). *Coronavirus gives cities taste of a post-overtourism world*. Bloombergnews, 2 April 2020. Available at: https://www.bnnbloomberg.ca/coronavirus-gives-cities-taste-of-a-post-overtourism-world-1.1416235. Accessed 10.12.2020.

Lino, B. ve Marie, A. (2019). *Has Overtourism reached the Maltese Islands?* Islands and Small States Institute.

Little, K. B. (1965). Personal space. *Journal of Experimental Social Psychology, 1*(3), 237–247.

Lopez-Bonilla, J. M., & Lopez-Bonilla, M. L. (2008). Measuring social carrying capacity: An exploratory study. *Tourismos: An International Multidisciplinary Journal of Tourism, 3*(1), 116–134.

Manera, C., & Valle, E. (2018). Tourist intensity in the world, 1995–2015: Two measurement proposals. *Sustainability, 10*(12), 4546–4566.

Martin, B., & Uysal, M. (1990). An examination of the relationship between carrying capacity and the tourism life cycle: Management and policy implications. *Journal of Environmental Management, 31*, 327–333.

Martin, J. M. M., Martinez, J. M. G. ve Fernandez, J. A. S. (2018). An analysis of the factors behind the Citizen's attitude of rejection towards tourism in a context of Overtourism and economic dependence on this activity. Sustainability, 10, 1–18.

Massiani, J. & Santoro, G. (2012). The relevance of the concept of capacity for the management of a tourist destination: theory and application to tourism management in Venice. *Rivista Italiana di Economia Demografia e Statistica*, LXVI, 141–156.

McCool, S. F. ve Lime, D. W. (2001). Tourism carrying capacity: Tempting fantasy or useful reality? Journal of Sustainable Tourism, 9(5), 372–388.

Mehta, R. (2013). Understanding perceived retail crowding: A critical review and research agenda. *Journal of Retailing and Consumer Service, 20*, 642–649.

Milano, C., Novelli, M. & Cheer, J. M. (2018). *Overtourism a growing global problem*. Retrieved from: https://theconversation.com/overtourism-a-growing-global-problem-100029. Accessed 5 Dec 2020.

Milano, C., Novelli, M. & Cheer, J.M. (2019). Over-tourism and tourismphobia: a journey through four decades of tourism development. Planning and Local Concerns, *Tourism Planning & Development*, pp.1–4.

Morris, H. (2018). Boracay, Philippines reopens to tourists with bans on beach parties, plastic bags and smoking. *Traveller*. Retrieved from: http://www.traveller.com.au/boracay-philippines-reopens-to-tourists-with-bans-on-beach-parties-plastic-bags-and-smoking-h16ov6/. Accessed 5 Dec 2020.

Morris, J., & Parry, T. (2017). Intrepid trips to explore 'least visited' locations. *TTG Media Limited*. Retrieved from: https://www.ttgmedia.com/news. Accessed 12 Dec 2020.

Namberger, P., Jackisch, S., Schmude, J., & Karl, M. (2019). Overcrowding, over-tourism and local level disturbance: How much can Munich handle? *Tourism Planning & Development, 16*(4), 1–21.

Neuts, B., & Nijkamp, P. (2012). Tourist crowding perception and acceptability in cities: An applied modelling study on Bruges. *Annals of Tourism Research, 39*, 2133–2153.

O'Reilly, A. M. (1986). Tourism carrying capacity: Concept and issues. *Tourism Management, 7*(4), 254–258.

Papathanassis, A. (2017). Over-tourism and anti-tourist sentiment: An exploratory analysis and discussion. *"Ovidius" University Annals, Economic Sciences Series, 17*(2), 288–293.
Peeters, P., Gössling, S., Klijs, J., Milano, C., Novelli, M., Dijkmans, C., Eijgelaar, E., Hartman, S., Heslinga, J., Isaac, R., Mitas, O., Moretti, S., Nawijn, J., Papp, B. & Postma, A. (2018). *Research for TRAN Committee - Overtourism: impact and possible policy responses*. Brussels: European Parliament, Policy Department for Structural and Cohesion Policies.
Phi, T. G. (2019). Framing overtourism: A critical news media analysis. *Current Issues in Tourism*, 1–6.
Postma, A., & Schmuecker, D. (2017). Understanding and overcoming negative impacts of tourism in city destinations: Conceptual model and strategic framework. *Journal of Tourism Features, 3*(2), 144–156.
Ryan, C., & Cessford, G. (2003). Developing a visitor satisfaction monitoring methodology: Quality gaps, crowding and some results. *Current Issues in Tourism, 6*(6), 457–507.
Sarantakou, E. & Terkenli, T. (2019). Non-institutionalized forms of tourism accommodation and overtourism impacts on the landscape: The case of Santorini, Greece. *Tourism Planning & Development,* 16(4), 411–433.
Seraphin, H., Sheeran, P., & Pilato, M. (2018). Over-tourism and the fall of Venice as a destination. *Journal of Destination Marketing & Management, 9*, 374–376.
Seraphin, H., Zaman, M., Olver, S., Bourliataux-Lajoinie, S., & Dosquet, F. (2019). Destination branding and over-tourism. *Journal of Hospitality and Tourism Management, 38*, 1–4.
Shelby, B., & Haberlein, T. A. (1984). A conceptual framework for carrying capacity determination. *Leisure Sciences, 6*, 433–451.
Simancas Cruz, M., & Peñarrubia Zaragoza, M. P. (2019). Analysis of the accommodation density in coastal tourism areas of ınsular destinations from the perspective of over-tourism. *Sustainability, 11*(11), 3031.
Sputniknews. (2019). *Hollanda artık turizm tanıtımı yapmayacak: Ülke aşırı kalabalık*. Retrieved from: https://tr.sputniknews.com/yasam/201905081038988329-hollanda-artik-turizm-tanitimi-yapmayacak-ulke-asiri-kalabalik/. Accessed 5 Dec 2020.
Stanford, D. (2008). "Exceptional" visitors: Dimensions of tourist responsibility in the context of New Zealand. *Journal of Sustainable Tourism, 16*(3), 258–275.
Stokols, D. (1972). A social-psychological model of human crowding phenomena. *Journal of the American Planning Association, 38*(2), 72–83.
Street, F. (2020). These destinations were overwhelmed by tourists. Here's how they're doing now. *CNN Travel*. Retrieved from: https://edition.cnn.com/travel/article/overtourism-coronavirus/index.html. Accessed 12 Dec 2020.
UNWTO, Centre of Expertise Leisure, Tourism & Hospitality, NHTV Breda University of Applied Sciences, & NHL Stenden University of Applied Sciences. (2018). *'Overtourism'? – Understanding and Managing Urban Tourism Growth beyond Perceptions, Executive Summary*. UNWTO, Madrid.
UNWTO. (2019). *International tourist arrivals reach 1.4 billion two years ahead of forecasts*. Retrieved from: http://www2.unwto.org/press-release/2019-01-21/international-tourist-arrivals-reach-14-billion-two-years-ahead-forecasts. Accessed 13 May 2020.
UNWTO. (2020). *International tourism and COVID-19*. Retrieved from: https://www.unwto.org/international-tourism-and-covid-19. Accessed 20 Dec 2020.
Volo, S. (2020). Over-tourism: definitions, enablers, impacts and managerial challenges. In H. Seraphin, T. Gladkikh, & T. Vo Thanh (Eds.). *Over-tourism: Causes, implications and solutions* (pp. 11–26). Cham: Palgrave Macmillan, Springer.
Wall, G. (2019). Perspectives on the environment and over-tourism. In R. Dodds & R. W. Butler (Eds.), *Over-tourism: Issues, realities and solutions* (pp. 27–45). Berlin: De Gruyter.
Wall, G. (2020). From carrying capacity to over-tourism: A perspective article. *Tourism Review, 75*(1), 212–215.
Ward, C., & Berno, T. (2011). Beyond social exchange theory. *Annals of Tourism Research, 38*(4), 1556–1569.

Weeden, C. (2011). Responsible tourist motivation: How valuable is the Schwartz value survey? *Journal of Ecotourism, 10*(3), 214–234.

WHO (World Health Organization). (2020). *Coronavirus Disease (COVID-19) Dashboard*. Retrieved from: https://covid19.who.int/. Accessed 27 Nov 2020.

WTTC (World Travel & Tourism Council). (2020). *Leading Global Protocols for the New Normal*. Retrieved from: https://wttc.org/COVID-19/Safe-Travels-Global-Protocols-Stamp. Accessed 27 Dec 2020.

Yüksel, F., Ön Esen, F., Kılıç, B. & Akçay, S. (2020). Paydaşların gözüyle yavaş şehir Akyaka'da aşırı turizm. *Turizm Akademik Dergisi*, 7(1), 257–268.

Yumuk, Y. ve Altıntaş, V. (2019). Bodrum'da Yaşayan Yerel Halkın Kentsel Yaşam Kalitesi Algılarının Ağırı Turizm (Over tourısm) Kapsamında Değerlendirilmesi. *Uşak Üniversitesi Sosyal Bilimler Dergisi, XII. Özel Sayı*, 90–102.

Zemla, M. (2020). Reasons and consequences of over-tourism in contemporary cities knowledge gaps and future research. *Sustainability, 12*, 1729–1749.

Zengin, B., & Bayram, G. E. (2018). Sorumlu turizm türleri geliştirme politikaları. In O. Batman, O. Türkay and Ş. Ulama (Eds.). *Turizmde politika ve stratejiler* (pp.369-389). İstanbul: Değişim Yayınları.

Zhiyong, F., & Sheng, Z. (2009). Research on psychological carrying capacity of tourism destination. *Chinese Journal of Population Resources and Environment, 7*(1), 47–50.

Chapter 10
Is Undertourism an Option After COVID-19?

Anurodh Godha

Abstract The world is challenged by the new coronavirus (COVID-19). The world is now struggling with the COVID-19 epidemic, and also, the epidemic would no wonder influence the future. The COVID-19 epidemic has seriously affected the tourism industry, with drastic impacts on jobs and businesses as without any flu vaccination as well as a limited medical capability to cure the symptoms, the primary disease outbreak strategy is non-pharmaceutical interventions (NPI). However, tourism has also been regarded as just a resistant sector, one of the largest yet fast-growing industries. The tourism sector is much more impacted by every internal or external disruption or depression compared to other professions. In this sense, COVID-19's recent epidemic triggered nationwide anxiety. The world's largest disturbance in the world economy since the Second World War is due to unexpected worldwide travel bans and homestay orders. With global travel restrictions affecting more than 90% of the population and broad prohibitions on public gatherings and group mobility, the tourist industry largely stopped in March 2020 (Gössling et al., J Sustain Tour 29: 1–20, 2020). The COVID-19 disease outbreak downturn may provide an unusual and extremely valuable opportunity for a good long-term reconsideration and re-establishment of the tourist industry. However, "leading" approaches to tourism alone will not provide adequate capacity to allow such a reset. Instead: such a perception provides a framework of the tourist industry, predicated on the needs and privileges of regional individuals and communities, to reshape and realign the tourist industry (Higgins-Desbiolles, Tour Geogr 22: 1–14, 2020). COVID-19 is universally acknowledged as a challenge or maybe even a travel & leisure colossal gamble. It has acted as a basis for serious discussion in the "tourism academy", comparing the effect of COVID-19 on the tourist industry as regards issues, risks, and responsibilities. It describes that sector supporters for fast recovery are opposed

A. Godha (✉)
Vardhman Mahaveer Open University, Kota, Rajasthan, India

A. Hassan, A. Sharma (eds.), *Overtourism, Technology Solutions and Decimated Destinations*, https://doi.org/10.1007/978-981-16-2474-2_10

to much more socially responsible, sustainable, and responsible attempts to transform the tourist industry. The fight involves the significant role of tourist industry intellectuals and tourist industry academics. The results of such a discussion possess implications for disciplinary development, the tourist industry in the future of tourism.

Keywords COVID-19 · Tourism industry · Issues · Tourism risks · Responsibility

Introduction

The Corona Virus Pandemic (COVID-19) originated in Wuhan, China, has rapidly spread to different nations, with several global reports. 56,342 positive cases were reported in India on 8 May 2020 (Kumar et al., 2020a, b). India has over 1.34 billion inhabitants, this same world's second-biggest, and will struggle with both the influence of the transmitting amongst the populace of serious Coronavirus 2 debilitating breathing disorder. In handling the disease process, multiple strategies would have been important, such as computer modeling, quantitative study, and statistical analysis to regulate expansion as well as the fast formation of a new diagnosis. The Indian Ministry of Health and Family Welfare has made it conscious of current news but has asked to take steps to manage COVID-19 expansion. These same Central and State governments take a range of programs to achieve these goals and formulate various wartime guidelines. Besides that, the Indian government has locked up the virus for 55 days across the country, which began on 25 March 2020 (Moriarty, 2020). This epidemic is intrinsically tied to the national economy since it has severely hampered industry, as the world is presently careful to engage in business in the region concerned. In keeping with the public authorities' current guidance, India is preparing for the pandemic of the COVID-19, and trying to avoid or undermining particular crisis measures will also have incredibly severe consequences. The positive COVID-19 cases were reported in all neighboring Indian nations. The government of India has introduced requisite as well as rigid actions to protect and against the deadly virus, such as the establishment of health checkpoints among country boundaries, to test if people come to this country (Kumar et al., 2020a, b). With global travel restrictions impacting more than 90% of the population and broad prohibitions on public gatherings and group mobility, the tourist industry largely stopped in March 2020 (Gössling et al., 2020). Tourism, a major and rapidly developing sector, has also been viewed as a disadvantaged sector. Explicit or implicit crises or pressures are concerning its competitors hit (i.e., tourism industry). Throughout this respect, worldwide chaos has been triggered by the recent epidemic of Corona (COVID-19) (Priyadarshini et al., 2020). One of the other industries, which have been overlooked in this rehabilitation package, is a tourist attraction. An on-going epidemic, tourism constraints, and nationwide lockdown had also managed to bring the whole tourism industry to a complete halt, and as opposed to many other areas, the tourist industry, in particular tourism, would then take years to recover. This has real consequences for nations like Uttarakhand, Rajasthan, Kerala, Himachal Pradesh, Goa, Sikkim, and other north-eastern countries that largely rely on tourism as a significant source of state income. Instead: such a perception provides a framework of the tourist industry, predicated on the needs and privileges of regional individuals and communities, to reshape and realign the tourism industry (Higgins-Desbiolles, 2020).

There are 8.75 workers employed throughout the Indian tourism industry (12.75% of the total workforce in 2018–2019), including healthcare professionals, travel agencies, hotel chains, guesthouse service providers, driving companies, guidance, business owners, laborers, as well as other service providers. Each segment is also strongly linked and reversed to many other sectors, including agriculture, transport, traffic handling, and FMCG (Anwer et al., 2019). Disturbances will make many people unemployed in the tourism industry. Higher running expenses and no traps have already put pressure mostly on the food and hospitality sector. The strategy affiliation FAITH has approximated the loss to the industry of Rs 10 lakh crore because of COVID-19by the associations of tourism & hospitality industry. The influx of foreign tourists will also affect a drastic decrease in foreign exchange revenue near Rs 2.10, 981 crores in Q1/Q3 2019 (Bagchi, 2020). The tourism industry needs to be addressed by India. Experts: The most thrilling disease outbreak of modern years has approximated that the effects of this deadly virus are severe. 200 nations around the world have already fallen apart. Strategic objectives to minimizing the global downturn are useful to integrate (Beirman, 2020). Throughout this scenario, travel anxiety and a need to ensure the social distance have crumbled in the tourist industry as the country's third-largest foreign-exchange employee. Visa restrictions were imposed, and the international airport shutdown, nations and towns were cured and closed. Citizens were requested to return to one's nations to provide special flight agreements in several nations (Ranasinghe et al., 2020).

It is ideal for all the tourist trade to create a common hygiene and sanitation requirement for their clients. The sector should also take advantage of these opportunities to adopt environmentally friendly waste disposal practices daily.

It is obvious that the tourist industry sector in India wants to push its comment-COVID-19 recovery and renewal. The government is considering an instant management plan and rehabilitation package that focuses not only on tourism as it is among the most impacted, but because it is undoubtedly the world's largest industry and revenue for several small and medium-sized enterprises. The issues, risks, and responsibilities of COVID-19 scenarios are covered in this chapter.

Is Undertourism an Option After COVID-19?

In reality, fear is the most important thing (the worried of uncertainty). Fear is extended as fear for an uncertain future, fear of transmission, fear of infectious disease, worry about contacting people, fear of touching, fear of sickness and families/friends infecting, fear of illness and very far from proper medical treatment.

effect on air travel. In the very first quarter of 2020, the COVID-19 pandemic resulted in a 22% decrease in foreign tourists, which shows current stats from the UNOT. The UN specialized agency says that, when contrasted with 2019, the disaster can result in a yearly decrease of from 60% to 80%. This jeopardizes thousands and thousands of lives and families and threatens to reverse development towards achieving sustainable development (SDGs) (UNWTO, 2020).

Because of the response of the epidemic throughout particular nations, the landscape of tourist attractions can be severely compromised. When a nation reacted inadequately to COVID-19 and seems to have a delayed reaction, travelers can be unfavorable and also have serious concerns about visits following control of the disease outbreak. Tourism destinations officials must encourage and convince international visitors when the COVID-19 has been controlled. The place is secure to boost tourism (Research and Markets, 2020).

Going to reopen and maintaining its healing of tour operators in a safe, appealing and economical manner for visitors: this same path to redemption in the tourism sector is complex and arduous. Authorities sometimes intervene to support the industry (Durbin, 2020). As an attractive commodity for visitors, cleaning, and sanitation guidelines are at the top of the priority list. During which hotel owners have prevented cleaning changes from becoming noticeable all day long, the frequency of cleaning needs to be increased now, but this prominent for visitors trying to seek security. Hand sanitizers must remain in effect in all customer touch points in rooms and in toiletries (e.g., check-in desks). Its a moment to a system and developing strategies presently. Start preparing anywhere you can and make sure that all facets of your guests, employees, and businesses are very well secured. Hotel owners would then definitely help to successfully reopen with an iterative methodology to changing economic situations and the capacity to keep up with the changing guest business requirements factors.

Governments and the private industry must examine unique and modern financing measures: use the sectoral deals to start preparing the industry for a new traveling period and help raise the population throughout the UK-by sharing data, developing products, and decreasing the actual result deficit (VisitBritain, 2020).

Securing the Long Term Health of the Visitors and the Need of the Industry

Accelerate and expand tourist industry zones when the result of the administration's continuing ambition to "level up" the weaker areas of the country as suggested in the Sector Deal, in assistance of the impact of tourism to recovery.

- Tourism Data Center is helping to fund to develop resources to facilitate recovery, in particular for SMEs.

- The tourism industry must use the downturn to enhance productivity and growth and enhance the country portfolio as a target.
- Promotion of tourists on vacation at residence with such a big domestic marketing strategy.
- The fully automated extension of global economic visas makes it much easier for them all to defer trips owing to a disease outbreak.
- For using reboot to lockdown-forced consumer preferences to take into account broader issues such as achieving sustainable development and consumer preferences.

The tourism industry should develop income accumulation systems for investment property owners, particularly SMEs, to handle differential financial losses (Margaux et al., 2020).

Involvement of Risk in the Tourism Industry COVID-19 Era

Throughout times of the COVID-19 disease outbreak, the tourist industry company has created thorough procedures to implement all kinds of tourist corporations or facilities.

Greater Risk Individuals

Individuals older than 65 may not have been required to visit hotels, guesthouses, and B&Bs. People approximately 60 and 65 are advised not to attend hotels, lodges, and bed and breakfasts (Government of India, Ministry of Health and Family Welfare, 2020). It is suggested that a person with a greater risk factor may not commute and visit hotels, lodges, and B&Bs. Self-catering institutions can be operated by private or employed vehicle for high-risk sections of the visitor. High-risk personnel will be assigned to fewer posts, minimal risk areas, increased PPEs (e.g. visors and work at home as probable).

Rising Numbers of Mass Transit Personnel Give a Bigger Risk of Infection

Facilities will provide on-site staff as far as probable. Premises in which employees also could not live are motivated to use personal vehicles for employees. All establishments would also decrease the use of mass transit much further than possible.

Transmitting Risk Limited by Transportation Among Both Districts and Provinces

To make the journey from and to an accommodation establishment, evidence of reservation should also be held along with a journey planner or instructions. Consumers just avoid buying fuel and other articles from a tank depot or not allowed on-road retail operations. Only persons going to visit personal vehicle accommodation or car rental facilities can cross provincial borders.

Capacity to Detect Contacts in Case of a Positive COVID-19 Test

All designed to operate companies receive and maintain guest/visitor/passenger/customer information additional information on latest and scheduled trips. Required to meet details will be shared, and all employee information on all shifts, vehicle drivers etc., will be carefully recorded.

The arrival of COVID-19 was seen as a turning point. As far as the tourist industry is concerned, the unimaginable has started happening. Whereas the sector has spent initially so long struggling to tourist industry issues, unexpectedly, we are being propelled into the destruction of the world (Higgins-Desbiolles, 2020). Ever since its inception in early 2020, the quickly growing COVID-19 disease outbreak (also known as Corona) has caused worldwide chaos and destruction. Experts, as well as the most thrilling disease outbreak in modern years, that already has 200 countries all over the world, had also estimated the impact of this fatal virus seriously. While many societies have suffered differing duration locks, the economic implications of the virus have also been catastrophic. The international tourism sector has had a great effect, not to consider the millions of destinations across the globe, because our hyper-mobile society has usually stopped (Ioannides & Gyimothy, 2020). Throughout this context, the worry of travel & tourism as that of the nation's third-biggest foreign currency owner has completely collapsed with a need to keep the public distance. During the last days, visa restrictions were imposed and the airport terminal closed, blockade and locking of the cities and countries were made, and citizens were invited to return to their nations, where specific flight provisions were provided in several countries. The situation is difficult now because the whole world is locked down or socially distant. People are frightened to reserve anything until the discomfort from this infectious disease reduces. Are indeed key players in the tourism industry (e.g. airlines, cruise companies, transnational hotel chains) suffering major damage estimated at € 400 billion Nicolas, (2020)? The epidemic of COVID-19 was one of the most dramatic and heart-breaking disease outbreaks of the modern world. Saving a life is presently the highest priority. Even so, we must also finally address the recovery process for the tourist sector in the existence of constant restrictions on international and domestic travel.

It is perceived risk was its starting point for assessing the effects of a recession on the tourist industry. Because of the national emergency, the individual's behavior and preference are significantly influenced by negative risk. Risk perception was described as the probability of bad outcomes (Borko et al., 2020). On this basis, the perspective of tourist industry risk was seen as a personal opinion by customer's response to bad tourism outcomes, which usually results from either the imbalance in tourist industry safety issues or the qualitative perceptual perspective that visitors have. It will significantly raise in a short period scope of the present study of abrupt crises. Personality traits have very distinct risk factors to the very same form of journey. As per, elderly experienced people in travel were less aware of possible hazards like health hazards, terrorist acts, or natural catastrophes (Zhu & Deng, 2020).

Tourism is an important source of jobs worldwide. The job market has certain distinctive characteristics. The sector is labor-intensive. Females have a high percentage of employment. 54% are considerably higher than in other industries and young people, indicating that the industry is often seen as inclusive (International Monetary Fund, 2020). Even so, women are much more likely than most to be business owners in the context of hospitality, but most females have low skill levels in the tourism sector, making them more susceptible to crises. Also, there are indirect negative jobs in infrastructure and public development, in addition to providing travelers with food and beverages and souvenirs. Besides, numerous workers in tour operators, air carriers, boats, hotels, restaurants, shopping centers, and tourist destinations have close communication with visitors. Also, the financial system is being net losses by jobless capital and labor. Because of the spectacular compression in the tourism sector, hotels and tour agents can also face economic crashes in the coming years, as India's number of visits declines (Gossling et al., 2020). The entire inbound, outbound, national, MICE tourism sector would be impacted. The way hospitality reservations are canceled, in same sense tour operators The MICE segment is among the most important segments for tour operators postpone packages offered for sale for inbound, outbound and domestic sectors also, and yet due to such a recession, all types of events, meetings, meetings have been canceled or postponed. This would be no wrong to say that the economy will also move toward more recession if the government and its stakeholders would not take an early step, which would harm the country's GDP rates and, as a developing economy, India would be a challenge shortly (Kumar, 2020). The pandemic will last for approximately two years, according to experts. But the fear of travel and social distance protocols will not disappear soon after the pandemic subsidies. COVID-19 is going to get seriously damaged by the hotel industry, and it is generally agreed that the industry will not be the same long after travel restrictions and lockdowns are lifted. However, hoteliers cannot afford to follow the existing operating model. Hotels must adopt COVID-19 survival strategies. This calls for the hotel service to be repackaged to make it more attractive in the COVID-19 era.

Major Responsibilities of the Tourism Industry in the Era of COVID-19

The tourist trade is linked to a variety of several other business activities such as Travelodge, society, education, finance, farming, medicine, transport and travel, building and buildings, foodservice, and likewise. Safeguarding the whole sector would be a massive investment in nations' economic future of growth. Consequently, if we are to concentrate mostly on the area's long-term progress via the tourist industry, we have to safeguard only those tourist industry providers. One of the powerful founders of economic economics and social systems is the international tourist industry and hotel management. The tourism sector is also a major revenue-generating sector for several and other countries around the world. (Ranasinghe et al., 2020).

Cleaning and disinfecting as a portion of consumption must be stressed and appropriately supported. Guidelines for cleaning will likely shift, and PPE is needed with most tourism activities.

1. Current health safety standards could enable firms to partially renovate their property. Hard floors, air handling structures, and more can are becoming conventional in new construction design elements in a brief period. For instance, hotel front desks can be fitted with counter shields.
2. Reduced occupancy rates for hotels and restaurants could even be desired positioning' safety perceptions. Unless the government, even so, enforces boundaries below its maximum capacity, it could endanger the sustainable development of industry and thus raise the prices.
3. A significant change from daily encounters to innovations to restrict interaction could even take place. Instances of this would be the wider use of keeping look-in kiosks and pay-outs via debit or credit card.
4. Services could restrict access by requiring only a corrected number of attendees to meet. Hotel room service can be substituted for eateries or breakfast bars. Visitor activities could be restricted to those which take place in the open-air contexts only.
5. Hotel rooms can regulate room inventories directly instead of having to rely on private entities.
6. A few other hotels and other tourist industry firms can mean making with public institutions supply their services for those impacted by quarantine in periods of the higher epidemic.
7. Automated systems, robots, and artificial intelligence can contribute to long-term cost reductions, liquidity and sustainability improvement, and social distance preservation. Heavy reliance on a small number of market segments can also be decreased by international markets. The offshoring of certain service providers through suitably adaptable agreements can also reduce risk. The purchase of health coverage could even assist in the forehead of sudden and dramatic risks.

8. Confidence in journeys and interpretation of risks would then directly impact the easily promote the business sector. The tourist industry is likely to come back to post-depression patterns in the long term. Travelers can, even so, but higher benefits on sanitation when choosing their lodging.

Other Responsibilities

Safeguard of Staff Members

Numerous individuals were challenged by the COVID-19 economic recession and have changed everyday life in unparalleled ways. Business as normal is not even a choice for organizations. Those who can begin by designing and delivering a workplace support system that complies with one of the most progressive regulations and triggers new policies can help tourists. A few other major companies measure their attempts against each other to define this same best ideas and support levels for their employees. Among the fascinating designs, we have experienced supplying the management team with clear and concise terms about how to deal with COVID-19 (in line with WHO, CDC, and other Guidelines for health agencies) whilst also offering the others with individual freedom so that they are capable of dealing with any challenging scenario rapidly. This independence is integrated with two-way communication to allow members to experience themselves when they feel threatened and monitor compliance with new regulations (Charumilind et al., 2020).

Cost-Cutting

The downward trend in hotel requirements resulting in a reduction in REVPAR means that hotels are required to take cost reduction initiatives. To reduce or remove economic loss, the team must determine unneeded facilities and regions with significant reductions in demand over this time and mark them for cost-cutting.

Orderliness

We do not have the opportunity and cannot even afford the path; we used it to continue living the pre-COVID-19. COVID-19 created neither a new nor a normal. The new standard comprises social distance and well-being guidelines, which must be conformed to obsessively. Hygiene, safety standards, and operational requirements are included in this. Visitors also will pay more attention to health and protection than to the quality of the service. The luxury apartment specification contains steps to minimize the spread of the disease. To enhance the basic and moral accountability of members and employers' health and security, hotels should try to reach the need to decrease extra expenses. The development and managing quarantine camps for commuters from high-risk locations, thermal detectors at the airport terminal, the protocol to identify the system of healthcare professionals and result in recognized containment potential individuals, curfew and lock the entire region as high-risk zones, low-risk areas, the distribution of essential materials to citizens across the nation, and the strengthen the nearby healthcare sector for effectively managing the medical facilities need to be ensured.

Conclusion

Today, the completely extraordinary public health problem is becoming a disastrous financial recession for the whole world and, therefore, for the tourism industry, hotels and events. The recovering must gradually compare the healing of other industries, the fight against the pandemic, and the recovering of financial outcomes. The practical procedure should therefore be increased the amount in response to changing circumstances. The security of employees and guests is thus a severe worry for the service industry. At the very same time, it is vital to start powerful marketing strategies and campaigns respectively domestically and abroad to regain financial loss and re-establish the good perception of customers. Additionally, promoting "Flexi-services" would be another strategic tool for attracting potential clients. Therefore, looking to offer flexible prices, renewal policies, adaptable working conditions, etc. The industry of tourism is completely dedicated to bringing folks with their well-being first. An international movement is essential to ensure that the industry can significantly contribute to COVID-19 containment. The reaction to the tourist industry needs to be monitored, compatible, and appropriate to it and premised on health threats, risk analysis, and accountability, as set out in the general guidelines and standards of the WHO (WHO, 2020).

References

Abi Younes, G. A., Ayoubi, C., Ballester, O., Cristelli, G., de Rassenfosse, G., Foray, D., & Zhou, L. (2020). COVID-19: Insights from innovation economists. *Science and Public Policy*, 1–13.

Anwer, M. E., Sahoo, B. K., & Mohapatra, S. (2019). Spatio-temporal variations in agricultural diversification in India. *Journal of Agribusiness in Developing and Emerging Economies, 9*(5), 476–502.

Bagchi, B. (2020). *Urban Utopias: Memory, rights, and speculation*. Jadavpur University Press.

Baratti, L. (2020). How technology will reshape the hotel industry post-COVID-19 pandemic travel technology. *Travel Pulse*. Retrieved from: https://www.travelpulse.com/news/travel-technology/how-technology-will-reshape-the-hotel-industry-post-covid-19-pandemic.html. Accessed: 12 Dec 2020.

Beirman, D. (2020). *Restoring tourism destinations in crisis: A strategic marketing approach.* Routledge.

Borko, S., Geerts, W., & Wang, H. (2020). *The travel industry turned upside down: Insights, analysis and actions for travel executives* (pp. 9–87). McKinsey & Comapany.

Charumilind, S., Craven, M., Lamb, J., & Wilson, M. (2020). Preventing future waves of COVID-19. *McKinsey & Company.* Retrieved from: https://www.mckinsey.com/business-functions/risk/our-insights/covid-19-implications-for-business. Accessed: 1 Dec 2020.

Durbin, D- A. (2020). Battered global tourism industry makes reopening plans. *Ksat.com.* Retrieved from: https://www.ksat.com/business/2020/05/03/battered-global-tourism-industry-makes-reopening-plans/. Accessed: 1 Dec 2020.

Gössling, S., Scott, D., & Hall, C. M. (2020). Pandemics, tourism and global change: A rapid assessment of COVID-19. *Journal of Sustainable Tourism, 29*, 1–20.

Government of India, Ministry of Health and Family Welfare. (2020). *SOP on preventive measures in Hotels and Other Hospitality Units to contain spread of COVID-19.* Retrieved from: https://www.mohfw.gov.in/pdf/5SoPstobefollowedinHotelsandotherunits.pdf. Accessed: 19 Dec 2020.

Higgins-Desbiolles, F. (2020). Socialising tourism for social and ecological justice after COVID-19. *Tourism Geographies, 22*, 1–14.

International Monetary Fund. (2020). *Policy Responses to COVID-19.* Retrieved from: https://www.imf.org/en/Topics/imf-and-covid19/Policy-Responses-to-COVID-19. Accessed: 17 Dec 2020.

Ioannides, D., & Gyimóthy, S. (2020). The COVID-19 crisis as an opportunity for escaping the unsustainable global tourism path. *Tourism Geographies, 22*, 624–632.

Kumar, V. (2020). Indian tourism industry and COVID-19: Present scenario. *Journal of Tourism and Hospitality Education, 10*, 179–185.

Kumar, S. U., Kumar, D. T., Christopher B. P., & Priya, D. C. G. (2020a). The rise and impact of COVID-19 in India coronavirus disease (COVID-19): Pathophysiology, epidemiology, clinical management and public health response. *Frontiers in Medicine.* https://doi.org/10.3389/fmed.2020.00250.

Kumar, S. U., Kumar, D. T., Christopher, B. P., & Doss, C. (2020b). The rise and impact of COVID-19 in India. *Frontiers in Medicine, 7*(250), 1–7.

Margaux, C., Steve, S., & Jackey, Y. (2020). Reimagining the $9 trillion tourism economy—What will it take? *McKinsey & Company.* Retrieved from: https://www.mckinsey.com/industries/travel-logistics-and-transport-infrastructure/our-insights/reimagining-the-9-trillion-tourism-economy-what-will-it-take. Accessed: 15 Dec 2020.

Moriarty, L. F. (2020). Public health responses to COVID-19 outbreaks on cruise ships—Worldwide. *MMWR. Morbidity and Mortality Weekly Report, 69.*

Priyadarshini, I., Mohanty, P., Kumar, R., Son, L. H., Chau, H. T. M., Nhu, V. H., & Bui, T. D. (2020). Analysis of outbreak and global impacts of the COVID-19. *Healthcare, 8*(2), 148.

Ranasinghe, R., Damunupola, A., Wijesundara, S., Karunarathna, C., Nawarathna, D., Gamage, S., & Idroos, A. A. (2020). *Tourism after Corona: Impacts of Covid 19 Pandemic and Way Forward for Tourism, Hotel and Mice Industry in Sri Lanka.* Retrieved from: https://papers.ssrn.com/sol3/papers.cfm?abstract_id=3587170. Accessed: 12 Dec 2020.

Research and Markets. (2020). COVID-19 impact on tourism industry: Regional case studies on strategies of destinations to engage travellers. *PR Newswire.* Retrieved from: https://www.prnewswire.com/news-releases/covid-19-impact-on-tourism-industry-regional-case-studies-on-strategies-of-destinations-to-engage-travellers-301060157.html. Accessed: 3 Dec 2020.

UNWTO. (2020). *International tourist numbers could fall 60–80% in 2020.* Retrieved from: https://www.unwto.org/news/covid-19-international-tourist-numbers-could-fall-60-80-in-2020. Accessed: 1 Dec 2020.

VisitBritain. (2020). *Helping the tourism industry recover from COVID-19.* Retrieved from: https://www.visitbritain.org/helping-tourism-industry-recover-covid-19. Accessed: 21 Dec 2020.

World Health Organization. (2020). *Considerations for quarantine of contacts of COVID-19 cases: interim guidance, 19 August 2020*. WHO.

Zhu, H., & Deng, F. (2020). How to influence rural tourism intention by risk knowledge during COVID-19 containment in China: Mediating role of risk perception and attitude. *International Journal of Environmental Research and Public Health, 17*(10), 3514.

Chapter 11
Overtourism in Cox's Bazar in the Pre, During and Post COVID-19 Period

Rashed Al Karim and Azizul Hassan

Abstract Since the outbreak of COVID-19, overtourism was one of the most buzzing terms in many of the world's popular tourism destinations. Overtourism was featured as a problem for decades, and evidently, protests of the residents against the visitors were common. Simultaneously, the tourists reported frustrating holiday experiences owing to overcrowding, longer queues, and socio-environmental degradation of the destinations. Cox's Bazar that is Bangladesh's one of the most popular tourist destinations during the peak season, is not an exception. Thus, this research aims to figure out the causes and repercussions of overtourism in Cox's Bazar before COVID-19. The study then analyses the overall scenario of overtourism in Cox's Bazar during and after the post-COVID-19. This study is carried out relying on both primary and secondary data sources. Besides in-depth interviews of the locals, tourists, industry stakeholders, journal articles, organizational and government publications, website materials, and other sources are used as sources of data and information generation for pre, after, and post COVID-19 overtourism scenario in Cox's Bazar. Based on findings, the chapter suggests guidelines to monitor and mitigate the detrimental influences of overtourism in Cox's Bazar. As expected, this study can support tourism destinations to tackle possible issues generated from overtourism and guide the authorities for the more beneficial tourism industry.

Keywords Overtourism · COVID-19 · Tourist destinations · Hospitality · Cox's Bazar

---

R. Al Karim (✉)
School of Business, East Delta University, Chattogram, Bangladesh

A. Hassan
Tourism Consultants Network, The Tourism Society, London, UK

A. Hassan, A. Sharma (eds.), *Overtourism, Technology Solutions and Decimated Destinations*, https://doi.org/10.1007/978-981-16-2474-2_11

Introduction

Bangladesh is a densely inhabited nations in the world, yet it also has an abundance of natural beauty. As a center of tropical climate and old civilization, this country has many resources and tourist destinations to attract tourists. Due to its distinctive geographical location, the country is blessed with amusing natural beauty such as the world's longest natural sea beach and largest mangrove forest, hill tracts areas, tea estates, islands, and many archaeological resources etc., and as a result, tourism in Bangladesh is always considered as one of the most promising industries in the world (Das & Chakraborty, 2012). There are many remarkable specific areas or locations in Bangladesh, such as Chittagong, Cox's Bazar, Rangamati, Bandarban, Sylhet, Khulna, Patuakhali, where majority of the domestic as well as the international tourists frequently visit. Some major tourism destinations are available in Bangladesh, which are divided into three regions the east (seven destinations), the middle (seven destinations), and the west (nine destinations). Most tourists have visited the world-famous Cox's Bazar frequently among all tourist destinations in Bangladesh. It is known as the country's beach resort district with the world's longest sea-beach. It is said to be one of the most popular tourist sites in the country with up to 30,000 visitors per day (Alam, 2018). In spite of this, this kind of 'overtourism' has a significant effect on the popular tourist destination. Since, though a large proportion of domestic tourists come to Cox's Bazar, there is still a total lack of hotel lodging, facilities, food, security, culture, nature conservation, and so on, this area is now under great threat due to overtourism. At the moment, some unplanned growth in Cox's Bazar are resulted because of overtourism. As the development boom in this region continues, the entire landscape is transforming. Several hundreds of apartment buildings, motels, high-rise hotels, cottages, and restaurants have already sprouted up around the city, particularly along the beaches. There are several shops on the main beach area selling clothing, toys, souvenirs, and fast food. Also, people leave plastic, polythene, and other non-degradable substances on the beach that damage the sea and the beach. Taking into account the above-mentioned issues, the key objective of this chapter is to depict the consequences of overtourism of Cox's Bazar before, during, and after COVID-19 pandemic.

Defining Overtourism and Its Impacts

The tourism industry is the world's one of the fastest developing industries over the last few decades. Generally, this is one of the primary drivers of socioeconomic development due to its contribution to employment, export sales, and infrastructural growth (Peeters et al., 2018). In recent years, the term "overtourism" has got a lot of attention across the world alongside this exceptional development (Phi, 2020). Overtourism is certainly not "new" within tourism literature, as different ideas relating to the management of the number of visitors and tourism destinations' service standards (e.g., carrying capacity–Sinden, 1975; Kuss & Morgan, 1980, community antagonism and Doxey's irritation index – Doxey, 1975) were addressed as early as the 1970s (Phi, 2020). A number of scholars have described "Overtourism" as a dynamic phenomenon (i.e., Bellini et al., 2017; Milano, 2018; Postma, 2013) that powerfully impacts a location's liveability, as well as the interactions of inhabitants, tourists, and multiple stakeholders who are directly or indirectly concerned with or impacted by tourism. According to Goodwin (2017), "Overtourism" refers to situations in which hosts or tourists, residents or visitors realise that there are too many people and that the region's or visitor's quality of life has deteriorated unsatisfactorily. Responsible Tourism, on the other hand, focuses on using tourism to create healthier places to live and better places to visit (Cheung & Li, 2019). Overtourism has been encountered in many destinations, including Venice, Paris, California, Berlin, Hong Kong, and Rio de Janeiro are obvious examples. Protests against tourists have occurred in a number of these places, with some involving low levels of danger and a few major incidents of violence (Cheung & Li, 2019; Colomb & Novy, 2016). Overtourism has a direct influence not only on people, tourist attractions, destinations, and local infrastructure, but also on tourists. Also, tourists can be the potential losers when anti-tourist sentiment increases that does not only predominate substandard service, but also evolves latent hatred into outright violence (Waller, 2011). In 2017, the World Travel & Tourism Council collaborated with McKinsey & Company for producing a report on the impact of overtourism, which identified the challenges linked with overtourism, including degraded tourist experiences, alienated local residents, overburdened infrastructure, environmental damage, and threats to culture and heritage (Forbes, 2018).

Furthermore, according to a World Travel Monitor survey conducted in 2017 on 29,000 international travellers from 24 Asian European, American countries that found about 25% of all international tourists felt that their destination was overcrowded, affecting the quality of their outbound trip and negatively impacting their holiday experience (Peeters et al., 2018; IPK International, 2018). Into the bargain, Milano (2017, 2018) poses a variety of elements and sources of overtourism irritation on the basis of qualitative analysis in three EU cities though field research in Barcelona, and desk research in Venice and Berlin comprise:

- Congestion in city centres' public areas
- Public space privatization.
- Real estate prices rise.
- Increase in the number of cruise ships and the high number of cruise passengers within a short period of time.
- Residents' purchasing power loss.
- Unbalanced population of residents versus visitors.
- Gentrification of the commercial sector.
- Deterioration of the environment, including trash, noise, water quality, and air quality concerns.

Cox's Bazar – At a Glimpse

Cox's Bazar, whose ancient name was Palongkee and which was also known as Panowa, is one of Southeast Asia's most ancient tourism sites and is renowned as Bangladesh's tourism hub for its potential tourism activities. It lies 159 kilometres to the south of Chattogram. It is bordered on the north by Chattogram district, on the east by Bandarban district and Myanmar, and on the west and south by the Bay of Bengal. It is located in the southeast of the nation and is connected to Dhaka and Chittagong by air and road. The Cox's Bazar municipality covers an area of 32.90 Square Km. with 12 wards and has a population of 167,477 (Government of Bangladesh, n.d.). Tourism is the primary source of Cox's Bazar's economy. This type of hospitality and customer service firm employs a large number of employees. For their livelihoods, a variety of individuals participate in fishing and harvesting fisheries and seafood goods. Cox's Bazar is a conventional community with economic and socio-cultural indices that are far below the national average, including literacy (55.7%) (Alam, 2018). Inani, Himchhari, Ramu, Sonadia, Teknaf, Moheshkhali, Saint Martins Island and Kutubdia Island are other desirable tourist places in this district, in addition to the longest beach area. This district is home to a variety of hotels, guest houses, cottages, restaurants, motels, tour guides, and tour guiding activities. So Cox's Bazar, with its distinctive characteristics, transport facilities, services, hospitality, and tourism amenities, is regarded as a more profitable tourism destination for domestic and international visitors than other sites in Bangladesh (Bhuiyan et al., 2020; Bhattacharjee et al., 2018).

According to Daily Industry (2020), Cox's Bazar boasts for the world's lengthiest natural beach, stretching for more than 100 kilometres (62.5 miles). Along the sea shore, at least 500 hotels have been built, including 30 star hotels, furnished flats, world-class resorts, and restaurants. Domestic tourists increased from 1.00 crore a year ago to 1.50 crore in 2019, according to various travel companies, local hotels, motels, and transportation facilities. During peak season, all tourists spend an average of BDTk 25,000, implying that 1.50 crore visitors spent a total of BDTk 37,500 crores in Cox's Bazar each year, creating 1.2 million jobs**. Following** Bhuiyan et al. (2020), Cox's Bazar was placed 77th out of 440 natural wonders in the latest Seven Natural Wonders of the world competition in 2007 and 2008. Cox's Bazar, with its strategic centre location in relation to the Indian subcontinent and the neighbouring South Asian nations of Myanmar and Thailand, has tremendous potential for growth as a major international tourist destination. Surfing, boating, swimming, fishing, and monitoring are the most common tourism activities (MoFA 2014). Given the town's tourist potential, BPC is planning to construct an integrated tourism and entertainment village on a 130-acre beachfront property in Cox's Bazar. The integrated expansion will include a beach resort, entertainment area, golf course, and places for other tourism activities. This city has not yet been able to boost its desire to lure tourists, considering the enticing promise that the destination holds. This is mainly due to rural settlement, excessive unplanned construction of stores, hotels, motels, luxury apartments and residences for time-share, and other administrative irregularities that overlap.

Overtourism in Cox's Bazar (Pre COVID-19 Pandemic)

Apart from maximizing the economic benefits of tourism, several prior studies, newspapers, and secondary sources identified overtourism as a big problem for any tourist destination like Cox's Bazar because of generating several adverse impacts on society and the environment (Zemla, 2020). Overtourism adds to increased water and energy consumption, water, soil, and air pollution, waste and litter in tourism sites in terms of environmental consequences. In terms of social consequences, overtourism may wreak havoc on local residents by lowering their quality of life, typically as a result of increased prices (for example, for short-term rentals or foods) (consequently causing housing shortages). In consequence, dealing with overtourism must now be a priority. Based on various secondary sources, newspapers, and semi-structured interviews with the concerned person, the following the potential threads of overtourism in the Cox's Bazar:

Landscape Changes Due to Construction Boom

Due to the development of infrastructure, buildings, and services for the large number of tourists, the whole scenery of Cox's Bazar has changed. Construction across the beach continually affects the morphology of the beach by increasing construction activities such as dredging, sand removal, mangrove habitat degradation, and seagrass meadows to create open beaches (Sayeda, 2017). In the main beach area, hundreds of businesses offer souvenirs, clothes, toys, and fast food items. As the development boom begins, restaurants and hotels are springing up in practically all parts of the town and along the beachfront. Environmentalists believe that unless the illegally constructed structures around the main beach area are taken out soon, this region will hardly be recover, whilst its beauty can be lost forever (British Broadcasting Corporation, 2012).

Traffic Congestion

Overtourism has made major changes to the Cox's Bazar town's urban structure. In the town center, there are several tourist hotels, motels, and luxury apartments. The potential pressure on the center was also high. The entry point to Cox's Bazar town was packed with inter-highway buses run from 6 pm to 11 pm every weekend. Therefore, heavy vehicles at the point of entry and foot traffic in the beach area are typical in this area (Sayeda, 2017).

Loss of Marine Environment

There is also an influence on the marine environment with the growing number of tourists. Tourist leisure practices such as reckless motor riding on the beach, swimming, and fishing, for instance, by contacting corals, churning up sediments, and dropping anchors, have severely damaged corals in the coastal area. Huge numbers of visitors on the beaches have damaged and disrupted breeding locations for endangered sea turtles. Consumption of seafood is another tourist attraction in the Cox's Bazar area. However, it inevitably puts pressure on the stocks of local fish and often contributes to overfishing. Collecting of shells, corals, and other marine souvenirs -whether by specific visitors or by locals selling them to tourists can have harmful influences on the natural ecology (Bhuiyan et al. 2020; Sayeda, 2017).

Increase in the Amount of Solid Waste

Overtourism has also resulted in a huge buildup of solid trash that is not differentiated. Every day, the growing number of tourists generates a massive amount of garbage. Tourists dump hundreds of plastic bottles and empty boxes into the beach area. As a result, tourism and urbanization have direct and indirect consequences on solid waste. Inevitably, it creates a major environmental problem of disposal and pollution (Framba, 2020).

Topsoil Depletion

The difficulties are not limited to the beach; the surrounding hills also face the threat of overtourism in Cox's Bazar. Hillside trees are cut indiscriminately, and the land is being cleared to make room for the development of infrastructure, dwellings, and facilities for local and foreign tourists in Cox's Bazar. As a result, several landslides occur during the monsoon season in this vacation area, killing many people. As mud banks fell in severe rain, burying homes, dozens of inhabitants were murdered. In less than four years, this was the second significant landslide (British Broadcasting Corporation, 2012).

The Increased Amount of Water Consumption

The imbalance in water consumption between visitors and locals is becoming increasingly problematic. Although no studies on water usage in Cox's Bazar town have been done, it is obvious that excessive water use is a reality to lessen the need for a large number of tourists. Many of Cox's Bazar's hotels and tourist attractions rely on groundwater as their water source. The withdrawal of excessive water has a detrimental effect that can contribute to land subsidence. The visitor attraction facilities face extreme water shortages in the dry season, as the water level falls down. In addition, due to overtourism in Cox's Bazar, overconsumption of groundwater rises by visitors, and daily habitat creates salinity issues in the shallow aquifer (Ahmed, 2010; Sayeda, 2017).

Socio Cultural Impact

Cox's Bazar is facing many socio-cultural disputes due to overtourism. For instance, social unrest is increasing by harnessing the advantage of overtourism, and, therefore, the crime rate is suddenly increasing. Also, the lack of local culture and cultural identity has been defined in Cox's Bazar as another deleterious socio-cultural concern of overtourism (Sayeda, 2017). Locals in Cox's Bazar share some characteristics of a strong culture, ideas, and values. Tourism development, on the other hand, is expanding the usage of such tourism goods as alcoholic beverages, sex tourism, and westernisation that overtly challenges local culture, ideas, and beliefs (Hossin, 2020).

Safety and Security Concern

Cox's Bazar is a well-known tourist destination, but the protection and security infrastructure is not yet in operation. It sometimes causes more difficulties for visitors as a result of abduction and hijacking. Tourists are not permitted to enter tourist areas after midnight, particularly Cox's Bazar sea beach, for fear of being attacked by a hijacker or gang. The Bangladesh government established the Cox Bazar Tourist Police Team for the Tourist Zone in 2013. They aim to serve both locals and visitors from other countries with safety and protection, but it is not enough due to the number of inbound tourists. That is why foreign visitors argue about living in Cox's Bazar for a longer period of time. This concern results in the loss of the country's enormous foreign currencies. Furthermore, due to a lack of sufficient protection, international visitors seldom stay on the beach after dark to experience the unique natural beauty (after sunset) without fear (Alam, 2018). The Government of Bangladesh is attempting to ensure the safety of visitors, local or international tourists. However, as we are currently a heavily populated country, some safety concerns still worry the Government.

Overtourism in Cox's Bazar (During COVID-19 Pandemic)

Bangladesh is one of the most afflicted nations in the globe as a result of the COVID-19 pandemic. The country has a budget imbalance. The Financial Express (2020) mentioned that according to the Tour Operators Association of Bangladesh (TOAB), the tourism industry might lose up to BDTk 60 billion in 2020 because of the COVID-19 pandemic; also, the Pacific Asia Travel Association (PATA) projected that COVID-19 has stumble upon a loss of around BDTk 97 billion in the Bangladesh's tourism industry.

As a result, during the COVID-19 pandemic, all tourism-related institutions (i.e., airlines, transportation networks, hospitality workers, cruise ships, bus companies, and lodging facilities such as hotels, motels, and homestays, as well as providers of many other tourism service such as travel agents or tour operators) suffered massive financial losses. Following Cox's Bazar's governmental designation as a COVID-19 red zone, day-to-day BDTk 25 to 30 crore transactions are no longer happening, according to news sources (The Daily Star, 2020). However, the hotels must pay wages to their employees as well as upkeep fees. In addition, tourism in Cox's Bazar has experienced a precipitous drop in domestic and international visitor visits since being designated a red zone by the government, with the impact of COVID-19 weighing heavily on the local economy. There are no more coastal eateries, and hotel guests no longer go for paragliding, build sandcastles, play volleyball or sail. There is nowhere to watch local fishermen sailing out to sea in their wooden fishing boats. Against this context, tour operators have unavoidably claimed that the unexpected breakout of COVID-19 has plunged Bangladesh's tourism industry into its worst-ever emergent problem, and that the industry need an immediate recovery incentive plan. Hotels in Cox's Bazar have said that their sole source of revenue at the moment is from development and relief workers who work with and stay in their hotels with Rohingya refugees, although the number of overseas development workers has also reduced owing to the quick changeover of COVID-19 in the region (The Daily Star, 2020).

The government banned public meetings at different tourist destinations in Cox's Bazar, Chattogram, Sylhet, and Patuakhali on March 18, 2020, in order to restrict the spread of coronavirus among persons. The district administration of Cox's Bazar imposed restrictions on tourists' access into the city and any sorts of beach gatherings. Following the country's coronavirus pandemic, over 2000 food outlets, around 470 hotels and motels, tourism-based business houses, the Burmese market, and thousands of industry employees have been virtually inactive since March 18, 2020. According to a statement issued by the Secretary-General of the Cox's Bazar Hotel

Motel Guest House Owners Group, Cox's Bazar was devoid of tourists due to COVID-19 restrictions. In the previous four months, several connected firms have also closed, resulting in huge losses. Many people in the tourist business have been laid off, and no one knows when the pandemic will end. In the previous four months, hotel and motel proprietors have slipped into a debt trap in order to pay labour and energy bills. According to a UNB news article, the majority of persons working in tourism today rely on various types of loans to fulfil their monthly household requirements spending (The Independent, 2020).

Businesses of the tourism industry in the seaside resort of Cox's Bazar are looking for a solid recovery after Eid-ul-Adha eyes the next winter chance to stay alive by entertaining guests from home and abroad in the world's tourism industry hub. According to the President of the Cox's Bazar Tour Operators Association, reopening businesses after Eid-ul-Adha would give them an opportunity to survive, but it can be difficult for loss recovery that they have incurred already. Cox's Bazar has returned to visitors after a five-month hiatus owing to the COVID-19 epidemic, with tourism-related attractions reopening on a limited basis on August 17th, subject to the maintenance of hygiene measures. Several precautions have been taken to protect visitors' safety and health, including:

- Tourist police special rescue squad
- Control team for eve-teasing
- Drinking-water zone
- Medical team for immediate medical care
- A special tourist police unit to keep tourists from being bothered
- To combat COVID-19 infections, awareness programmes and mobile courts are now in operation (The Business Standard, 2020).

According to the Cox's Bazar Hotel-Guest House Owners' Association, there are about 450 hotels, resorts, motels, and cottages in Cox's Bazar, hosting millions of guests. Thousands of people go to Cox's Bazar throughout the weekly vacations. Upscale hotels are doing well, with 100% bookings (The Business Standard, 2020). According to the Deputy Commissioner of Cox's Bazar, tourism is a flourishing industry, and the busy tourist season is just around the horizon. At the same time, the second wave of COVID-19 has arrived. We must strictly adhere to cleanliness requirements while also doing all possible to develop the tourism industry. The majority of tourism industry experts in Cox's Bazar have asked the Government of Bangladesh for the following amenities to help them rehabilitate and accommodate tourists:

- Tax breaks for at least two to three years to allow the industry to recover. This includes tax breaks for resort and hotel operators, as well as exemptions from double taxation in the food and beverage sectors)
- Exemptions from bridge and ferry tolls for cars transporting tourists
- Interest-free deferred bank loans for investments in tourism parks, hotels, resorts, picnic spaces, and other recreational facilities (The Daily Star, 2020).

Domestic tourism should be given more attention as part of the growth phase in the long term. Given the length of this health issue, Standard Operating Procedures (SOPs) in accordance with Ministry of Health recommendations might be beneficial to the hotel business. Local visitors may be persuaded to return to Cox's Bazar if they perceive that safety steps are being taken. The government and other stakeholders can develop and implement various initiatives to boost domestic tourism in Bangladesh, which may aid in the recovery of COVID-19 pandemic loss due to overtourism.

Overtourism in Cox's Bazar (Post COVID-19 Pandemic)

Cox's Bazar was once again crowded with people after a protracted five-month halt owing to the COVID-19 epidemic, as tourism-related attractions were formally reopened on a modest scale on 17 August 2020, by preserving sanitary measures. Sea beaches, tourism spots, hotels, motels, bars, souvenir shops, and other theme parks in Cox's Bazar Municipality district have been re-opened on a small scale for visitors. According to tourist experts, the months of February to April are the best time to visit the country's main beaches. This time, more visitors have opted to visit Saint Martins Island along with other islands. In reality, these post-winter months bring domestic tourists to almost every spot of the trip. As domestic tourism has returned, the time has come to help reduce Cox's Bazar's effect on employment and other related businesses. However, true recovery can only be possible if both domestic and foreign tourism returns together and needs local coordination and evidence-based strategies for safe lifting on travel bans. The tourism industry's viability is in jeopardy unless it receives consistent government backing. Although authorities have made significant measures to mitigate the tourist impact, limit employment losses, and start the recovery process in 2021 and beyond, more has to be done, and in a more coordinated manner. The primary strategic aims for enhancing tourist arrivals in Cox's Bazar are as follows:

- For restoring the confidence of travelers
- For supporting tourism businesses for surviving
- For promoting domestic tourism and for supporting international tourism's safe return
- For providing clear information to businesses and travelers and uncertainty limitation
- For evolving response measures for maintaining capacity in the tourism industry of Cox's Bazar and to address support gaps
- Increasing intra- and inter-national cooperation
- For building more sustainable and resilient tourism (OECD, 2020).

Furthermore, the post-COVID-19 pandemic overtourism in Cox's Bazar will be restarted in order to improve the general financial health of this tourist city, which is being driven by the following factors:

- Travel accessibility and affordability
- Traditional policy focused on volume
- Increased international arrivals
- Urbanization pressure
- Gentrification and increasing property prices in city centres and new communities
- Increase in the number of unlicensed tourist lodgings
- Concentration of large tourist groups (source: adapted from Peeters et al. 2018)

The Role of Social Media and Digital Platforms in Reviving Tourism in Cox's Bazar After the COVID-19 Pandemic

Cox's Bazar has experienced a significant increase in visitors in recent years. During the peak season, all hotels, motels, and guest homes are totally filled by millions of tourists visiting Cox's Bazar, and with some guests even spending the night inside the car due to a lack of hotel seats. According to general perception, tourism has brought about significant change in this region. Local regions profit from tourism, and their economies are quite strong in comparison to other backward places. The local economy and other participants, such as developers, travel operators, hoteliers, and others, gain from the socioeconomic element. Over the last few years, the tourist industry has continued to strengthen its position as a minor driver of national economic development.

However, the impact of COVID-19 coronavirus is dreaded to blur the accomplishment. The Government and other stakeholders should also introduce and implement different policies to encourage domestic tourism in Cox's Bazar and other tourism destinations in Bangladesh that can help ensure the recovery of the loss of COVID-19 pandemic. BPC will initiate new domestic tourism initiatives through digital channels, social media, television, and radio commercials. Several projects, such as organizing promotional competitions, developing a new platform/blog for sharing tourism experiences using online resources, can also support domestic tourism businesses in Cox's Bazar (The Financial Express, 2020). Using both visual and print media, BPC will spotlight lesser-known attractions near Cox's Bazar district's rural and urban areas. With the use of the Internet, the media communication network has become more effective and comprehensive in today's world. It can encourage tourism by accurately predicting a destination like Cox's Bazar.

Consequently, contemporary digital media is connected to all the latest technology to sell tourism goods in various future markets. We need to draw significant numbers of domestic and international visitors to recover this industry's loss during the COVID-19 pandemic for earning the proper revenue from tourism-related operations in and around Cox's Bazar and many other tourist locations. The famous tourism-related websites need to promote all the tourist places in Cox's Bazar properly. A partnership should exist between the media and the proprietors of well-known hotels, restaurants, stores, and tourism service providers in Cox's Bazar, in addition to other tourism sites around the country. BPC has a website that should be well-designed, tourist-friendly, and provide all required information. Moreover, there are few local websites at the moment that provide visitors with information on the location and other service deals (Daily Sun. 2016). To conclude, to recover the adverse impact of COVID-19 pandemic and to resurrect the tourism industry of Cox's Bazar, the Government must focus on social media and other digital platforms to promote the tourism of Cox's Bazar digitally and to reach future tourists around the world for sustaining the industry.

Policy Recommendations

The achievement of tourism development with long-term sustainability in Cox's Bazar relies on implementing key policy initiatives, as below:

- To implement tourism-friendly policies in order for making human settlements and cities more inclusive, resilient, harmless, and long-lasting
- To employ information technology for promotional purposes since recent developments and improvements in information and communication technology (ICT) have had a transformative impact on the tourism industry. The Internet can play a leading role in promoting any country's tourism industry by encouraging and sharing information. Tourism authorities in Bangladesh must examine the successful use of ICT in order to promote this business
- For defining the acceptable levels of tourism impact on Cox's Bazar city over a collaborative process by involving all relevant stakeholders in order to make it easier for local communities to create, to benefit joint city experiences for visitors and residents, and for aiding in communication with residents
- To secure the advantages of tourism to Cox's Bazar's local communities, for example, by increasing the number of people working in tourism and incorporating residents in the creation of tourism experiences
- To promote monitoring and evidence-based decisions and planning of Cox's Bazar tourism's key issues as mobility, carrying capacity, cultural and natural resources management and, and attitudes of local residents for tourism
- For investing in technology, innovation, and partnerships for promoting Cox's Bazar as a smart city and for making the best available technology for addressing accessibility, innovation, and sustainability
- For promoting innovative experiences and goods that enable the City of Cox's Bazar for diversifying demand in space and time and, in accordance with its long-term vision and strategy, attract the necessary visitor groups
- To develop and promote tourist experiences using social media and other e-platforms in the Cox's Bazar city to benefit both tourists and residents
- To improve communication facilities, comfortable lodgings, and other infrastructure arrangements and services of the hotels, resorts, historical landmarks, and sightseeing

- To make aware tourists travel out of peak season for their pleasant travelling and staying experience and lesser expense and lesser crowed
- To produce films and documentaries on Cox's Bazar tourism attractions to be widely distributed to attract more and more local and foreign tourists.

Conclusion

Indeed, it is unfortunate that Bangladesh's tourism industry has undergone such unexpected adversity, particularly because, in recent years, owing to the COVID-19 pandemic, it has been seen as a, albeit minimal, the engine of national economic growth. According to credible sources, newspaper reports, and research outcomes, the number of visitor arrivals in the last few months has declined by 70%. Domestic tourists contribute for 70–80% of total tourism in Cox's Bazar, and with the quick development of local COVID-19 events, this figure is expected to rise. The 50% discounts would also not attract them there. Many millions of tourism and travel jobs are at danger as a result of the worldwide health crisis. This is a catastrophic future for the business, which must experience a spectacular comeback while not losing all hope since the industry is robust. Since a result of the coronavirus epidemic, the tourism industry in Cox's Bazar requires more resilience via successful budget allocations, as nearly all business industries confront a slew of complications. To resolve this problem, the whole tourism industry and politicians must go through a comprehensive preparatory procedure before resuming operations. As a result, the government's adequate planning and policy guidelines are required. Bangladesh currently has limited tourism facilities for domestic and international visitors. Because in both the public and private industries, there is less spending in this field than demand. As a result, Bangladesh's performance in the tourism industry is marginal in the international context. However, Bangladesh's potential in the tourism industry is very bright. The tourism industry in Bangladesh will be one of the key sources of jobs, investment, and foreign exchange earnings if it addresses current problems in the tourism industry and creates an acceptable tourism climate.

References

Ahmed, S. U. (2010). *Assessment of groundwater resources and impacts of tourism on groundwater of Cox's Bazar town, Bangladesh*. Master's thesis. Dhaka: University of Dhaka.

Alam, J. (2018). Problems and prospects of tourism industry in Bangladesh: A case of Cox's bazar tourist spots. *International Journal of Science and Business, 2*(4), 568–579.

Bellini, N., Go, F. M., & Pasquinelli, C. (2017). Urban tourism and city development: Notes for an integrated policy agenda. In N. Bellini & C. Pasquinelli (Eds.), *Tourism in the City* (pp. 333–339). Springer.

Bhattacharjee, A., Polas, M. R. H., & Rahman, M. L. (2018). Challenges and prospects of tourism in Cox's Bazar: An empirical study. *Journal of Business and Technology, 8*(1), 63–82.

Bhuiyan, A. H., Darda, A., Habib, W., & Hossain, B. (2020). *Marine tourism for sustainable development in Cox's Bazar, Bangladesh*. Retrieved from: https://www.adb.org/publications/marine-tourismsustainable-development-cox-bazar-bangladesh. Accessed 23 Dec 2020.

British Broadcasting Corporation (BBC). (2012). *Bangladesh's Cox's Bazar: A paradise being lost?* Retrieved from: https://www.bbc.com/news/world-asia-19340259. Accessed 23 Dec 2020.

Cheung, K. S., & Li, L. H. (2019). Understanding visitor–resident relations in overtourism: Developing resilience for sustainable tourism. *Journal of Sustainable Tourism, 27*(8), 1197–1216.

Colomb, C., & Novy, J. (Eds.). (2016). *Protest and resistance in the tourist city*. Routledge.

Daily Industry. (2020). *Cox's Bazar fetches Tk 37,500 crore business from 1.5 Crore Local Tourists*. Retrieved from: http://www.dailyindustry.news/coxs-bazar-fetches-tk-37500-cr-business-1-5-cr-local-tourists. Accessed 4 Jan 2021.

Daily Sun. (2016). *Role of media in promoting tourism*. Retrieved from: https://www.daily-sun.com/arcprint/details/185861/Role-of-media-in-promoting-tourism/2016-11-22. Accessed 16 Jan 2021.

Das, R. K., & Chakraborty, J. (2012). An evaluative study on tourism in Bangladesh. *Developing Country Studies, 2(1)*, 17–27.

Doxey, G. (1975). A causation theory of visitor–resident irritants: The impact of tourism. In *Proceedings of TTRA: The 6th annual conference* (pp. 195–198). San Diego: TTRA.

Forbes. (2018). *How is overtourism impacting travel to popular destinations?* Retrieved from: https://www.forbes.com/?sh=6187774c2254. Accessed 11 Jan 2021.

Framba, A. (2020). *Overtourism: Causes, consequences and solutions*. Ecobnb.

Goodwin, H. (2017). *The challenge of overtourism*. Retrieved from: https://haroldgoodwin.info/wp-content/uploads/2020/08/rtpwp4overtourism012017.pdf. Accessed 19 Jan 2021.

Government of Bangladesh. (n.d.). *Coxsbazar district*. Retrieved from: www.coxsbazar.gov.bd. Accessed 11 Jan 2021.

Hossin, M. Z. (2020). Tourism development and socio-cultural Impact in Cox's Bazar, Bangladesh: Conflicts or Co: exists? *IOSR Journal of Business and Management, 22*(1), 23–33.

IPK International. (2018). *ITB world travel trends report 2017/2018. Boom or bust? Where is tourism heading?* Retrieved from: https://www.itb-berlin.com/itb-berlin/downloads-alle-sprachen/presse/world-travel-trends-report-2017-2018.pdf. Accessed 1 Jan 2021.

Kuss, F. R., & Morgan, J. M., III. (1980). Estimating the physical carrying capacity of recreation areas: A rationale for application of the universal soil loss equation. *Journal of Soil and Water Conservation, 35*(2), 87–89.

McKinsey & Company & World Travel & Tourism Council (WTTC). (2017). *Coping with success: Managing overcrowding in tourism destination*. Retrieved from: https://www.mckinsey.com/industries/travel-logistics-and-infrastructure/our-insights/coping-with-success-managing-overcrowding-in-tourism-destinations. Accessed 7 Jan 2021.

Milano, C. (2017). *Over-tourism and Tourism-phobia: Global trends and local contexts*. Ostelea School.

Milano, C. (2018). Overtourism, malestar social y turismofobia. Un debate controvertido. *PASOS: Revista de Turismo y Patrimonio Cultural, 6*(3), 551–564.

Ministry of Foreign Affairs (MoFA). (2014). *Press release: Press statement of the Hon'ble Foreign Minister on the verdict of the Arbitral Tribunal/PCA*. MoFA.

Organisation for Economic Co-operation and Development (OECD). (2020). *OECD economic outlook, volume 2020 issue 2: Preliminary version*. OECD Publishing.

Peeters, P., Gössling, S., Klijs, J., Milano, C., Novelli, M., Dijkmans, C., Eijgelaar, E., Hartman, S., Heslinga, J., Isaac, R., Mitas, O., Moretti, S., Nawijn, J., Papp, B., & Postma, A. (2018). *Research for TRAN committee – Overtourism: Impact and possible policy responses*. European Parliament Policy Department for Structural and Cohesion Policies.

Phi, G. T. (2020). Framing overtourism: A critical news media analysis. *Current Issues in Tourism, 23*(17), 2093–2097.

Postma, A. (2013). *'When the tourists flew in': Critical encounters in the development of tourism.* PhD thesis. Groningen: Groningen University.
Sayeda, T. (2017). The effects of mass tourism: An evaluative study on Cox's Bazar, Bangladesh. *IOSR Journal of Humanities and Social Science (IOSR-JHSS), 22*(5), 31–36.
Sinden, J. A. (1975). Carrying capacity as a planning concept for national parks: Available or desirable capacity? *Landscape Planning, 2*, 243–247.
The Business Standard. (2020). *Cox's Bazar bustling with tourists again.* Retrieved from: https://tbsnews.net/bangladesh/coxs-bazar-bustling-tourists-again-158644. Accessed: 21 Jan 2021.
The Daily Star. (2020). *Can tourism in Cox's Bazar survive the Covid-19 effect?* Retrieved from: https://www.thedailystar.net/opinion/news/can-tourism-coxs-bazar-survive-the-covid-19-effect-1930397. Accessed 27 Jan 2021.
The Financial Express. (2020). *COVID-19 and domestic tourism development in Bangladesh.* Retrieved from: https://thefinancialexpress.com.bd/views/covid-19-and-domestic-tourism-development-in-bangladesh-1601134828. Accessed 10 Jan 2021.
The Independent. (2020). *Reopening plan brings hope for Cox's Bazar tourism.* Retrieved from: http://www.theindependentbd.com/post/250930. Accessed: 5 Jan 2021.
Waller, I. (2011). Less law, more order. *Acta Criminologica: Southern African Journal of Criminology, 24*(1), I–IV.
Żemła, M. (2020). Reasons and consequences of overtourism in contemporary cities—Knowledge gaps and future research. *Sustainability, 12*(5), 1729.

Part VI
Possible Solutions

Chapter 12
Harnessing Information and Communication Technology Led Interventions for Mapping Overtourism: Prospects and Retrospect

Parag Shukla and Sofia Devi Shamurailatpam

Abstract This chapter gives an insight into the issues of overtourism confronted by countries that can be mitigated through the use of Information and Community Technology (ICT) mechanisms to sustain the tourism industry. The fundamental proposition is that the adverse impacts of the influx of excessive tourism in destinations can be tackled through technology-led ambidextrous management policy such as smart technology, use of mobile apps to detect the tourists' data, analytics for crowd management, censoring tools to observe the behavior of the tourists, tourist cards, eco-friendly apps to conserve bio-diversity, and beach monitoring systems for good governance as well as for restoring peace and harmony of the host country. A model is conceptualized to establish the nature of the association between overtourism and the diffusion of technological innovation as well as diverse aspects of the economy. Our conceptual model generalizes the significance of technology and its feasible solution in mitigating the problems of overtourism.

Keywords Overtourism · Information and communication technology · Good governance · Sustainable development

Introduction

The domain of sustainable development calls for balancing of economic, socio-cultural, ecological, and environmental dimensions that are inter-linked in bestowing well-being to the society. In the best practice, sustainable development in its core construct describes the pace of development in such a way without harming the

---

P. Shukla (✉) · S. D. Shamurailatpam
The Maharaja Sayajirao University of Baroda, Vadodara, Gujarat, India

A. Hassan, A. Sharma (eds.), *Overtourism, Technology Solutions and Decimated Destinations*, https://doi.org/10.1007/978-981-16-2474-2_12

resources being available for the future generation. However, the concept of sustainability becomes much alert and alarming as nations undergo continuous phases in the processes of economic development wherein beyond a threshold limit, negative and adverse impacts set in, bringing the untoward state to the society. One such aspect is the concept of "Overtourism" in the tourism sector. The concept of overtourism is a new phenomenon though countries experienced and debated the adverse impacts of human activities on different aspects of quality of life and state of welfare across time and space. Generally, there is no explicit definition of the term "Overtourism", but it is the excess of demand and supply of tourism services beyond the carrying capacity of the economy, producing negative externalities, thereby impacting the residents, tourists, and population at large. Overtourism describes the situation in which the impact of tourism, at certain times and in certain locations, exceeds physical, ecological, social, economic, psychological, and/or political capacity thresholds (TRAN, 2018). Dealing with this new construct is significant for restoring the attributes of the destination, thereby balancing peace and harmony, socio-cultural values, the eco-friendly environment through minimizing the threats of negative externalities towards a futuristic sustainable development.

The relevant research questions to probe are: (a) What factors determine the optimal size of a tourist destination? (b) What is the carrying capacity of tourism services in a given destination? (c) Will there be "one-size-fits all" solutions across destinations? and (d) What role does technology play in overtaking the issues of "Overtourism"?

To answer these questions, this chapter introduces and highlights the role of technology in situating the issues of overtourism faced by residents, tourists, and the population at large in destination tourism.

Some Theoretical Underpinnings

Although the use of the term "sustainability" is widely adopted in both the disciplines of science and society, less attention is given to the theories of sustainable development (Enders & Remig, 2015). Indeed, given the implicit classical theories of economic development to variants of neo-classical theories, how does the sustainable development framework differ from such conventional models? To specify, the sustainable models take into account both the economic and ecological aspects of the resources and something more (Kadekodi, 2001). In other words, while simulating the framework of sustainability, the dimensions of variables that flow continuously in economic development in terms of efficiency, equity, regenerative, and resilience mechanisms are incorporated invalidating such models. Therefore, one can postulate the proposition that the elements of sustainability models contain the confluence of diverse economic characteristics that directly or indirectly impact the quality of life and well-being of society. Naturally, there cannot be a static sustainability model that fits all state of affairs prevalent in different economies across the globe. Hence a priori or necessary condition calls for taking account of those factors that are detrimental to the cause-effect analysis pertaining to a given destination/site.

One such factor or variable that is indispensable of impact assessment or performance evaluation in sustainability is the elements of by-products of Information and Communication Technology (ICT). In his seminal work, Romer (1986) developed the theory of endogenous technological change, in which technology and innovations as non-rival products that give rise to increasing return to scale. Accordingly, technological progress is achieved by way of investment in R&D, discovery, invention, and innovation, and, hence it increases the human capital. To borrow the significance of technology and innovation in explicit models of economic growth to sustainable economic development, ICT has a crucial role in determining various dimensions and forms of overtourism in destinations. This overarching theory holds significance in conceptual grounding and framework in this research study.

Literature on Overtourism: An Overview

Though the concept of overtourism is imperfectly formed and at a nascent stage, literature on the issues have been bringing forth by academicians and policymakers. It must be addressed in an appropriate scientific manner (Capocchi et al., 2019; Koens et al., 2018; Dodds and Butler, 2019; Gretzel, 2019). In general, the term overtourism is a complex phenomenon, and hence policy design to mitigate such influx of negative externalities calls for taking account of the carrying capacity of the destination in question. In such perspectives study by Camatti et al. (2020) provides an in-depth analysis through the case study of Dubrovnik. By conceptualizing a digital response system for real-time intervention, the undesirable effects of overtourism can be mitigated. Most of the case studies established so far in major worst cities that experienced consequences of overtourism revealed lack of management of excess crowd, deteriorating the quality of life, socio-psychological troubles, touristification, and gentrification among others (TRAN, 2018). By incorporating the socio-psychological' and 'socio-political' capacities, a model framework to assist monitoring, diagnosing, and influencing the risks of any undesirable tourism situation is formulated that encourages more sustainable tourism strategies, policies, and their more effective implementation (Mihalic, 2020). The circumstances of overtourism are not a static phenomenon given the differences in institutional, legal, and geo-political situations perceived in the particular destination.

The pertinent question of measuring overtourism through a common yardstick is a major area of concern. To exemplify in Big Sur, a popular scenic route and major tourist attraction in California, the exploratory structured interviews uses four different major themes to assess overtourism; pre-existing conditions; newly developed conditions; impacts; and responses; determining the various factors involved in the emergence of overtourism at a destination, as perceived by its local stakeholders (Atzori, 2020). Technology and digital platforms such as the internet and web technologies have a significant role towards overtourism from peer-to-peer networking and services, so it is the same as technology as a platform to mitigate overtourism (Hojeghan & Esfangareh, 2011). It is imperative to learn that with the emergence of Information and Communication Technology the world can be better place to live in. Smart-City Lens such as stakeholders, ICT platforms, and social capital can be used to conceptualize a framework of how human-technology interactions reframe overtourism knowledge gaps to orient policy design in urban areas (Pasquinelli & Trunfio, 2020).

No doubt, the technology-led approach instruments are highly suggestive to fight against overtourism starting from data collection of the number of tourists to regulation of the excess tourists to different destinations available to reduce overcrowding and expand the local business within the city in different locations.

Ambidextrous management policies such as demarketing and diversification of tourist spots are strategies adopted to curb overtourism. Branding plays a significant role in overtourism as it contributes to encourage visitation and repeat visitors to a destination (Seraphin et al., 2019; Bianchi & Pike, 2011; Park & Petrick, 2005). To address the issues of overtourism, the locals as volatile groups prove to be resilience in the destination through an ambidextrous approach, which is the balance between exploitation and exploration and identified four archetypes of locals regarding their attitudes towards tourists in an overtourism context: victims, peaceful activists, vandals, and resilient locals (Seraphin et al., 2020). These diverse aspects of overtourism form our current study's basic theoretical framework to frame feasible solutions towards sustainable tourism.

Taxonomy of Overtourism

Defining Overtourism

There is no concrete definition of the term "overtourism". By its traits, overtourism is a phenomenon that is associated with an excessive number of tourists in a destination during a given time frame that generates negative externalities concerning the quality of life and access to resources to the residents, tourists, and the destination environment. Several cities such as Barcelona, Amsterdam, Venice, Milan, Budapest, Bucharest, Reykjavik, Moscow, Athens, and Nantes have reported the worst cities of overtourism in Europe (Statista, 2020). According to the recent report of European Parliament Committee on Transport and Tourism (TRAN) (2018: 22), the term overtourism is defined as: "Overtourism describes the situation in which the impact of tourism, at certain times and in certain locations, exceeds physical, ecological, social, economic, psychological, and/or political capacity thresholds".

Indeed, this definition describes the negative effects of overtourism on different dimensions of human capital and physical capital assets of a particular location or destination during a given time period. One of the widely used definitions of overtourism is provided by Goodwin (2017: 1) as "Destinations where hosts or guests, locals or visitors, feel that there are too many visitors and that the quality of life in the area or the quality of the experience has deteriorated unacceptably". This definition reveals the bad experience faced by tourists from overcrowding in a destination and to the residents the type of quality of life that is deteriorating in the region.

In a similar way, the UNWTO (2018: 4) defines overtourism as: "The impact of tourism on a destination, or parts thereof, that excessively influences perceived quality of life of citizens and/or quality of visitor experiences in a negative way".

In other words, one common feature of each definition of overtourism quoted herewith is the proliferation of negative externalities in the quality of life of residents specifically and tourists generally in a given destination.

In fact, it is associated with the permissible size of traffic, the optimal level of socio-psychological capacity (Russo, 2002) along with the behavior of visitors, the length of their stay, the volume of tourists is equally important (Lindberg et al., 1997). Historically, it would seem that changes in consumerism and technology have mainly fueled the phenomenon of overtourism (Gowreesunkar & Thanh, 2020), and connectivity has not only boosted overtourism but also created a new breed of tourists, which is explicated by a spectrum of emancipation and sophistication, the by-products of technology (Gowreesunkar, 2019). Therefore, measuring the negative effects on the physical environment as well as the socio-psychological impacts is the need of the hour to mitigate such kind of damaging influx in a given site or destinations. In view of the significance of the tourism industry that contributes to micro levels and macroeconomic aspects of the economy on the one side and also the current influx of excessive overcrowding in destinations on the other side of the coin, stringent management policies need to be strategized to restore the peace and harmony of destination.

Enablers or Drivers of Overtourism

Specifically, when we talk about enablers or drivers of overtourism, it is peculiar to a given location – in urban cities, beach and coastal areas, heritage and pilgrimage sites, islands to attractive in rural areas, to mention a few. Most of the tourism industry reports relate either to the positive contribution of the boom in the tourism sector or the negative impacts of worst destinations due to overtourism experienced by major cities of the world. Undoubtedly, the effects of overtourism in rural populations are emerging as many tourist sites and forums are developed in the context of emerging high demand for tourism over the decades. Other things remaining the same, we report common factors that lead to overtourism:

- The increasing number of tourists – the antecedents of overtourism is the number of tourist arrivals in a particular destination during a given time frame (detailed number of tourist arrivals is reported in Table 12.1).
- Affordable Travel – owing to the increasing competition in the destinations and the emergence of new tourist forums and businesses, travels become affordable.
- Emerging Tourist Groups and Forums – more tourist destinations were being institutionalized in the form of new forums and tourist groups as demand far

Table 12.1 International tourist arrivals by (Sub) region

Regions/World	In million				% change	
	2010	2017	2018	2019	18/17	19/18
World	952 (100)	1333 (100)	1408 (100)	1462 (100)	5.7	3.8
Advanced economies	515 (54.1)	732 (54.9)	761 (54.0)	777 (53.1)	4.0	2.0
Emerging economies	437 (45.9)	601 (45.1)	647 (46.0)	686 (46.9)	7.7	6.0
By UNWTO regions						
Europe	487.0 (51.2)	676.6 (50.8)	715.9 (50.8)	744.3 (50.9)	5.8	4.0
Asia and the Pacific	208.2 (21.9)	324.1 (24.3)	347.7 (24.7)	360.6 (24.7)	7.3	3.7
Americas	150.3 (15.8)	210.9 (15.8)	215.9 (15.3)	220.2 (15.1)	2.3	2.0
Africa	50.4 (5.3)	63.3 (4.7)	68.8 (4.9)	73.2 (5.0)	8.8	6.4
Middle East	56.1 (5.9)	57.7 (4.3)	60.1 (4.3)	64.2 (4.4)	4.3	6.8

Note: Figures in the parentheses denote percentage share

exceeds existing destinations backed up by lower and affordable prices due to revision from airlines/cruises. This is also clear from Table 12.1, where a constant rise in the number of tourist arrivals from 952 million in 2010 to 1462 million in 2019, requiring the set up of new organizations to meet the excess demand.

- The dominance of growth focus mind-set – World Bank database indicated that as of 2018, the share of international tourism receipts as a percentage of total exports for the world stood 6.57%, with potential growth in the sector for foreign currency earnings economic growth. This is another factor that contributes to more expansion of the sector with its positive contribution to countries' economic growth.
- Competition for space, amenities, and services – automatically increasing the number of tourist arrivals, resultant foreign exchange earnings, drives to competitions, creating new amenities and services in destinations, and creating overcrowding and unmanageable of the influx.
- Media and information – access to social networking sites also increased with the advent of technology as a platform in marketing and promotions of the sites, thereby making aware of varied sites available for a particular destination or destinations.
- Lack of control over tourists' arrivals – the sudden influx in destinations created the overcapacity of the location to manage the crowd.
- The fragmented base of stakeholders – as the tourism sector is well managed systematically, then the overcrowding will be managed and can divert towards positive contribution to Gross National Income. These are some of the general factors that generate overtourism in a particular destination. The strategic man-

agement of the tourists is essential for the sustainable tourism sector, contributing to the growth of the economy.

Adverse Impacts of Overtourism

Looking at the adverse impacts of overtourism it can be grouped under different dimensions of the economy, which are as follows:

- Reduction in economic welfare: Overcrowding, excess use of resources, rise in tax in the particular sites/destinations lead to a reduction in the welfare of the community.
- Deterioration in socio-cultural values: Long staying of tourists in particular destinations influx the degradation in socio-cultural values attached to a particular location, if not controlled timely illegal and undesirable services in the form of prostitution, betting, gambling, etc. will rise, thereby deteriorating the traditional values.
- Environmental and ecological degradation: Too much use of and access to the resources leads to an imbalance in the ecological environment, difficult to regenerate if it is not contained on time.
- Loss of destination attractiveness: Given the degradation in socio-cultural values, environmental degradation, loss of peace and harmony of the place, gradually the attributes of the destination will fade away and loss their value and attractiveness.
- Excessive tourism carrying capacity: Overcrowding leads to excessiveness in terms of the number of tourist arrivals as well as the corresponding increase in the access to resources and assets belonging to the destination, leading to beyond the carrying capacity of tourism.

The Framework shown in model 1 elaborates the various drivers of overtourism, attributes, and diverse impacts on the economy. (Fig. 12.1)

A Comprehensive Framework of Use of ICT in Situating Overtourism

Decades over the tourism industry dynamics have experienced countries with continuous growth and turned into a significant determinant for socio-economic progress. This calls for strategic management best practices to seize the potential opportunities the industry can establish for the communities across the globe. Ironically the current issue of overtourism is a major concern of the countries, the outcome of excessive demand thereby dispositioning the equilibrium state. It is believed that while new technologies contribute to the problem, peer-to-peer apps

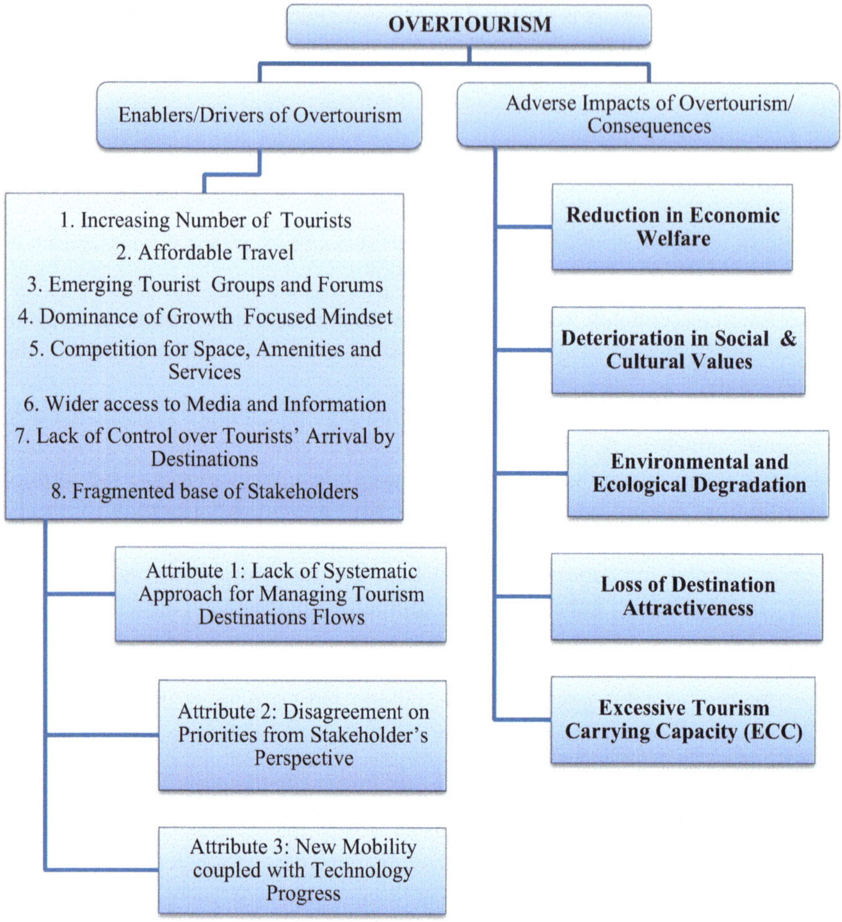

Fig. 12.1 Enablers, impacts, and attributes of overtourism. (Source: UNWTO, 2020)

and social media can offer solutions to this excessive influx to locations. A conceptual model of how technology-led instruments can be used to rationalize against such massive flows of visitors, other things remaining the same, is reported in Framework 2 by the authors.

The starting point of overcoming the major junks of overtourism is collecting data relating to volumes and other qualitative information of the tourists. The use of ICT can get information and data about the number of tourist arrivals in the location. Accordingly, it can evaluate the capacity of tourists that can be absorbed, which will help further detect issues associated with touristification. This kind of information makes it easy to implement the strategy of gentrification in a particular destination that can help reduce the free rider problem that aligns with low and affordable prices of traveling.

Countries can come up with a tech approach to these troubles of overtourism like the use of censoring technologies, interactive web platforms to connect the destinations across, issue of tourist cards, mobile applications, communication through the social media platforms, eco-friendly transportation to restore the ambiance of the location, beach monitoring system, etc. are among others that destination cities can adopt to mitigate the influx of crowd and reap the economies generated through diversification and proper allocation of the services. These instruments can help travelers change their habits of choosing a particular sites or destination on the one hand, and destinations can efficiently manage overcrowding. The role of the local communities has a big role to play on the other.

Recently, in some familiar destinations, such instruments of technology are implemented, and significant results are experienced in the cities to overcome the problems of overtourism. These are namely: (a) "Play London With Mr. Bean", an application used in traveling the London city to help people discover lesser-known parts of the city and surroundings to promote local business; (b) "Amsterdam City Card", used to guide the tourists in Amsterdam about the city and team is able to store data on the card's chip that can study the behavior of the tourist masses, help to distribute the flows more sustainably among the most familiar tech approach to fight against overtourism. As a part of E-governance in India, a software known as "PATHIK" is developed for the security and safety of the tourists, which help in the registration of guests to update the information at the central server that the city police officials can access to check the profile of guests and visitors, identify missing people if any and detect crime. Conceptualizing the substantial role of technology in depositing the problems of overtourism, the need of the hour is to adopt such technology by countries given the institutional and legal framework in the region. (Fig. 12.2)

Conceptualizing the Model

In this research study, we have conceptualized a model of ICT led mechanism to situate overtourism experienced by major destinations. The underlying assumption is that technology as a platform has a significant role in mitigating most of the issues of overtourism, such as crowd management through the help of proper data records of the number of visitors in a location, which helps in limiting the influx of tourist flows; accommodation and arrivals, thereby bringing in control the carrying capacity of tourism.

Moreover, having proper data about the tourists also enhance monitoring systems and strategies like touristification and gentrification be applied easily to divert or fix the excess capacity of tourist. The term touristification has gained a lot of popularity under the platform of social media promoting and availability of new destinations with the expansion of tourism businesses in a given territory to absorb the rising demand from tourists. In other words, this strategy implies a positive response towards the tourism sector by the expansion of more outlets of tourist

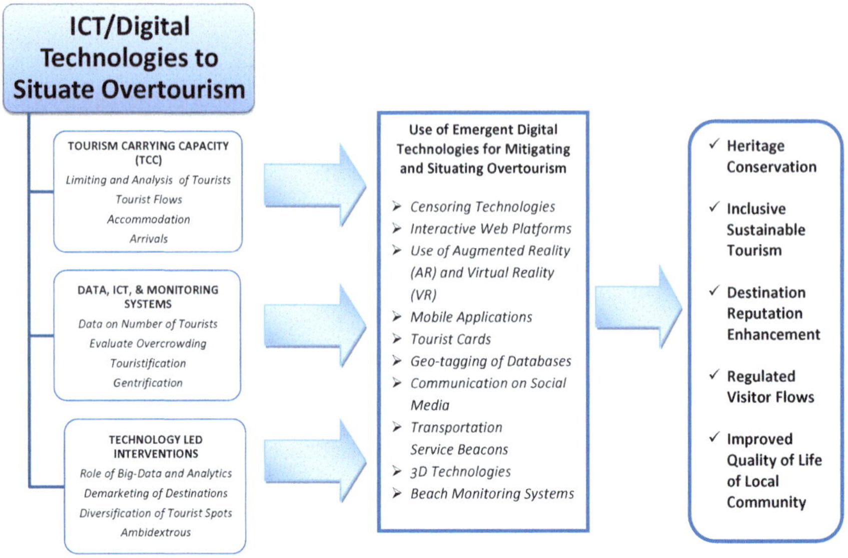

Fig. 12.2 A model strategizing the operationalization of Information and Communication Technology (ICT) and digital technologies to mitigate overtourism. (Source: Compiled by authors)

recreation and leisure in a territory. Nevertheless, there may be the question of "too big to fail" proverb if the absorbing capacity exceeds the carrying capacity of tourism. Gentrification is another strategy focused on socio-spatial consequences of overtourism, specifically in urban areas. The replacement of lower-income population groups by middle and higher income individuals, resulting in higher living and housing costs, is a dislocation of the original inhabitants from the particular location. Generally, it is a negative direction towards overtourism, given its characteristics as temporary to discourage the group of people enjoying the leisure or beauty of a location. The strategy of gentrification is well suited or suggested in the context of urban conflicts relates to space, especially described for niche population to curb excessive entries with affordable costs for living and rentals (for example, housing for tourist use, which is known as "VUT" by the Spanish acronym, marketed through the platforms like Airbnb).

Technology-led transformation through the use of big data and artificial intelligence, demarketing diversification of the tourist spots through ambidextrous strategy is another aspect of curbing overtourism. These are ingenious and elegant ways to divert the excess influx of tourists from one concentrated site to another of lesser-known spots in the same locality to control the masses.

Our conceptualized model highlighted the importance of technology as a platform by using diverse by-products of technology, namely tourist cards, mobile applications, beach monitoring systems, censoring technologies, and transportation services beacon. These are technological interfaces that help to curtail overtourism in popular destinations facing a heavy influx of tourists. The overall aspect is to restore the carrying capacity of tourism so far as peace and harmony, socio-cultural values, ecological balance in the destinations. Together, it will help in the conservation of valuable heritage sites, the collective reputation of the destination, direction, and control of tourists' flows, improvement in the quality of life of the residents and tourists together, which enables the sustainability in the tourism sector followed by economic development.

Managerial Implications and Key Discussions

Having discussed the significance of technology in overcoming the issues of overtourism faced by the countries, the following are some of the strategies that need to be initiated at the earliest:

- As overtourism is in its nascent stage of the policy cycle, it will be effective to implement immediately so that adverse consequences of spill-over-effects are perceived.
- From our conceptualized model also, it can be generalized that technology has a significant role in overcoming the issues overtourism.
- Giant multinational technology companies such as Google, Apple, IBM, Amazon, Microsoft, Intel, etc., can invest in Corporate Social Responsibility (CSR) spending towards sustainable development programs for curbing overtourism, building an image of good-will, and standards.
- Much of the efforts are taken by private tour operators to promote tourism. There are other perspectives that can be incorporated in the tourist packages namely, segmentation, niche tourism and even demarketing policy to wait until the destination in question gains its carrying capacity to manage the resources.

In contrast to other sorts of tourist sites, historic cities are first and foremost neighborhoods where people live, according to the current research study. The impacts of tertiarization of the heritage city's neighborhoods are exacerbated by tourist overpopulation. This occurrence has a cascade of repercussions that are transforming the environment. Traditional or local business is being pushed aside by leisure and tourism-related activities. Increased rivalry for occupying public space is driven by the enormous rise of tourist homes commercialised via online portals, and the housing market is influenced by growing costs. The application of technology in management should be used in conjunction with a social approach to the problem. The use of new technology raises regulatory issues, such as privacy protection, which may be addressed by community-led participation and collaboration. To participate in meaningful conversations with prominent organisations as well as local bodies, a strategy of preparing perception surveys and workshops with people might be used.

The current study hypothesises that the use of new technologies to manage congestion in tourist sites is still in the early stages of development and is being considered in the majority of situations. The findings of this study may be used to provide a set of actionable guidelines for the tourist destination's governance structure. The visitor segmentation or length of stay data is of considerable importance in gaining a better understanding of the tourist experience that may be enhanced via the use of ICT.

Conclusion and Recommendations

In this chapter, the role and significance of technology in overcoming the issues of overtourism faced by the countries is examined based on a conceptual framework. UNWTO (2020) predicts that in the coming future also there are large potentials for growth in the tourism sector, and hence sustainable management of this sector is significant in view of the rising issues of overtourism experienced by major cities of the world.

The negative and ill-effects of overtourism can well be tackled through ambidextrous management strategies like demarketing an overcrowded destination, censoring technologies, interactive web platforms to connect the destinations across, issue of tourist cards, mobile applications, communication through the social media platforms, eco-friendly transportation to restore the ambiance of the location, and beach monitoring system.

It is fair to say that rising visitor populations in a variety of places hasten the effects of people. We discovered that the spatial and time concentration of visitors appear to be key variables in many of the impacts and their size based on the findings of the research articles evaluated in this research study. Overtourism has had a negative influence on the ecosystem as well as the local population. The study highlights only a handful of the numerous harmful consequences. Increased water consumption, trash, and emissions are all environmental consequences of overtourism; displacement of residents, housing shortages, and increased rents and commodities costs are all social consequences. The knowledge obtained about tourism's harmful effects may motivate people to modify their behavior and travel habits. Individual traveler actions, if recorded with the assistance of technology, can help to mitigate and rescue harmful effects, and are thus vital to be recognised. Individuals can travel and discover the world through tourism and travel. For that reason alone, it should be performed in a way that ensures future generations have the chance to do so as well. To ensure the continuation of rational travel patterns, tourism should not be impoverishing its surroundings.

References

Atzori, R. (2020). Destination stakeholders' perceptions of overtourism impacts, causes, and responses: The case of Big Sur, California. *Journal of Destination Marketing & Management, 17*, 100440.

Bianchi, C., & Pike, S. (2011). Antecedents of attitudinal destination loyalty in long-haul market: Australia's brand equity among Chilean consumers. *Journal of Travel & Tourism Marketing, 28*(7), 736–750.

Camatti, N., Baertschi, D., Caric, H., & Borg, J. V. D. (2020). A digital response system to mitigate overtourism. The case of Dubrovnik. *Journal of Travel & Tourism Marketing, 37*(8–9), 87–901.

Capocchi, A., Vallone, C., Amaduzzi, A., & Pierotti, M. (2019). Is "overtourism" a new issue in tourism development or just a new term for an already known phenomenon? *Current Issues in Tourism, 23*(18), 2235–2239.

Dodds, R., & Butler, R. W. (2019). Introduction. In R. Dodds & R. W. Butler (Eds.), *Overtourism: Issues, realities and solutions* (pp. 1–21). De Gruyter.

Enders, J. C., & Remig, M. (2015). Theories of sustainable development – An introduction. In J. C. Enders & M. Remig (Eds.), *Theories of sustainable development* (pp. 1–5). Routledge.

European Parliament's Committee on Transport and Tourism (TRAN). (2018). *Research for TRAN committee – Overtourism: impact and possible policy responses*. Retrieved from: https://www.europarl.europa.eu/RegData/etudes/STUD/2018/629184/IPOL_STU(2018)629184_EN.pdf. Accessed: 2 Oct 2020.

Goodwin, H. (2017). *The challenge of overtourism*. Retrieved from: https://haroldgoodwin.info/pubs/RTP%27WP4Overtourism01%272017.pdf. Accessed: 3 Oct 2020.

Gowreesunkar, V. G. (2019). African union (AU) agenda 2063 and tourism development in Africa: Contribution, contradiction and implications. *International Journal of Tourism Cities, 5*(2), 288–300.

Gowreesunkar, V. G., & Thanh, T. V. (2020). Between overtourism and under-tourism: Impacts, implications, and probable solutions. In H. Seraphin, T. Gladkikh & T. V. Thanh (Eds.), *Overtourism – Causes, implications and solutions* (pp. 45–68). Palgrave Macmillan.

Gretzel, U. (2019). The role of social media in creating and addressing overtourism. In R. Dodds & R. W. Butler (Eds.), *Overtourism: Issues, realities and solutions* (pp. 263–276). Walter de Gruyter GmbH.

Hojeghan, S. B., & Esfangareh, A. N. (2011). Digital economy and tourism impacts, influences and challenges. *Procedia - Social and Behavioral Sciences, 19*, 308–316.

Kadekodi, G. K. (2001). Environment and development. In R. N. Bhattacharya (Ed.), *Environmental economics – An Indian perspective* (pp. 162–217). Oxford University Press.

Koens, K., Postma, A., & Papp, B. (2018). Is overtourism overused? Understanding the impact of tourism in a city context. *Sustainability, 10*(12), 4384.

Lindberg, K., McCool, S. F., & Stankey, G. H. (1997). Rethinking carrying capacity. *Annals of Tourism Research, 24*(2), 461–465.

Mihalic, T. (2020). Conceptualising overtourism: A sustainability approach. *Annals of Tourism Research,* 84, 103025.

Park, S. Y., & Petrick, J. F. (2005). Destinations' perspectives of branding. *Annals of Tourism Research, 33*(1), 262–265.

Pasquinelli, C., & Trunfio, M. (2020). Reframing urban overtourism through the Smart-City Lens. *Cities, 102*. 102729.

Romer, P. M. (1986). Increasing returns and long-run growth. *Journal of Political Economy, 94*(5), 1002–1037.

Russo, A. P. (2002). The "vicious circle" of tourism development in heritage cities. *Annals of Tourism Research, 29*(1), 165–182.

Seraphin, H., Zaman, M., Olver, S., Lajoinie, S. B., & Dosquet, F. (2019). Destination branding and overtourism. *Journal of Hospitality and Tourism Management, 38*, 1–4.

Seraphin, H., Ivanov, S., Dosquet, F., & Lajoinie, S. B. (2020). Archetypes of locals in destinations victim of overtourism. *Journal of Hospitality and Tourism Management, 43*(2), 283–288.

Statista. (2020). *Ranking of the worst cities for over-tourism in Europe in 2017**. Retrieved from: https://www.statista.com/statistics/778687/overtourism-worst-european-cities/. Accessed: 4 Oct 2020.

UNWTO. (2018). *'Óvertourism'? – Understanding and managing urban tourism growth beyond perceptions, executive summary.* UNWTO.
UNWTO. (2020). *UNWTO World Tourism Barometer*. Retrieved from: https://www.e-unwto.org/loi/wtobarometereng. Accessed: 4 Nov 2020.

Chapter 13
Alternative Tourism as a Solution of Overtourism

Yesim Koba

Abstract Since the twentieth century, package tours that consisted 'sea, sun and sand' trio became fashionable. Over time, mass tourism caused rapid and unplanned use of natural, cultural, and environmental resources. This has damaged the natural, cultural and environmental resources in a way that is not possible to renew and reaches a level that is dangerous for future generations. The existence of the phenomenon of tourism continues to depend largely on the quality of the environment (Kozak, Akoğlan Kozak, & Kozak General tourism. Detay Publishing, 2014). Lankford and Howard (Ann Tour Res 21:121–139, 1994) and Butler (Prog Tour & Hosp Res 2:283–293, 1996) describe the harmful effects of overcrowding, noise, damage to physical assets, and reductions in tourist satisfaction. According to Doğan, mass tourism reduces communication between tourists and local people (Doğan, Socio-cultural foundations of tourism. Ofis Ofset Publications, Izmir, 1987, 123). Such negative effects can be reduced by developing customized holiday addressing the special interests of the consumers. Therefore, in addition to the sea-sand-sun trio, alternative tourism types that consist of special interest trips have gained importance. The damage caused by mass tourism on natural, cultural, and environmental resources and the casualization over time has created changes in consumers' travel (holiday) expectations. As consumers started to opt for personalized, high-quality products and services, tourism product providers started to create alternatives that could meet different expectations. Also, the local people living in the regions where tourism is developing want to get a share from the tourism revenues, which has spread the tourism activity to the whole year, and it has started to increase the diversity of tourism products. Sustainable tourism has gained importance in order to minimize the negative effects of tourism. Alternative tourism is seen as an ideal way to eliminate the problems posed by mass tourism and to ensure sustainability. This chapter focuses on alternative tourism as a solution to the aforementioned problems of mass tourism.

Keywords Alternative tourism · Overtourism · Sustainable tourism · Sustainability

---

Y. Koba (✉)
University of Dokuz Eylül, İzmir, Turkey

A. Hassan, A. Sharma (eds.), *Overtourism, Technology Solutions and Decimated Destinations*, https://doi.org/10.1007/978-981-16-2474-2_13

Introduction

The increase in the welfare level with industrialization and the development of technology has brought both negative and positive consequences. The industrial revolution along with technological developments, increased people's income and leisure time and made transportation facilities much more attainable, which led to an increase in the number of people traveling around the world. People who are traveling for tourism look for facilities providing their accommodation, food, and entertainment. This demand has also triggered an increase in tourism supply. However, such demand in tourism supply in a destination causes pollution, consumption, and deterioration of the natural condition of the destination. All these negative results highlighted the importance of sustainability for alternative solutions.

The damage caused by overtourism to natural, cultural and environmental resources have created changes in the travel expectations of consumers. As the customers opted for customized products and services, tourism product providers have started to create alternatives that can meet different expectations. Moreover, the demand of locals to get a share from the tourism revenues and the effort of spreading the tourism activity to the entire year have increased the diversity in tourism products. In this section, the concepts of overtourism and alternative tourism are discussed, and the scope and characteristics of alternative tourism are explained. In order to ensure sustainability in tourism, the chapter offers alternative tourism suggestions to replace overtourism.

Overtourism

The increase in the number of travelers and interest in tourism caused some destinations to be overcrowded from the early 1970s. Especially after World War II, a large number of tourists from Northwest Europe began to visit the Mediterranean Countries for their climate. A tremendous transformation began in the Mediterranean Coasts, which had not been well developed until then (Bramwell, 2004, Obrador Pons et al., 2009). These districts have become overcrowded and experienced the negative effects of tourism. They became exposed to overtourism. Although overtourism seems like a simple concept, it is difficult to define (Robinson & Noveli, 2005). It can be defined as standardized tourism products at relatively low prices available to a wide customer group (Poon, 1993; Vanhove, 1997). These products generally include transportation, accommodation, food, entertainment, and touristic places in a package, and they are offered to customers at low prices by using mass marketing tools (Robinson & Noveli, 2005).

The crowd is perceived as both an indicator of the popularity and reputation of the tourism destination (Petr, 2009). Also, it is perceived as negativity that attracts the reaction of tourists and local people (Dowling, 2006; Jacobsen, 2000; Krebs et al., 2007). In destinations where tourism has begun to dominate local economies and cities, "crowding" creates negative effects and overshadows the positive effects of tourism (Oklevik et al., 2019). Concepts such as "overtourism and tourismphobia" emerge in destinations where tourist density has started to increase. UNWTO (2018) defines overtourism as the perceived negative impact on the quality of life of citizens and/or visitor quality in a destination. On the other hand, Goodwin (2017) argues that overtourism destinations are the ones that receive a huge amount of visitors as well as residents and therefore offer a low quality of life and holiday experience.

The concept of overtourism in the literature has been addressed by focusing on crowding, local people and tourist/tourism relationship. Doxey worked on the perceived human density and its consequences in his work called "Irritation Index" (Doxey, cited in 1975, Akis et al., 1996). O'Reilly focussed on "Tourism Carrying Capacity", and mentioned that people responsible for the management of destinations should take planning of the cities and their capacity into consideration (O'Reilly, 1986). Walter touched upon the concepts of physical carrying capacity and perceived carrying capacity. He stated that when these capacities are exceeded, the experience of tourists and the satisfaction level of the local people will decrease (Walter, 1982). "The Consumption of Tourism" by Urry showed that the "consumption of tourist attractions, buildings, and monuments by looking/seeing" formed the basis of television broadcasts, magazines, and newspapers (Urry, 1990).

Considering the number of visitors, France, USA, Spain, China, and Italy are the leading destinations that are negatively affected by overtourism. Cities such as Paris, Amsterdam, Venice, Barcelona, islands in Thailand, and Kyoto can be given as examples of cities negatively affected by heavy tourist influx. In the list published by UNESCO, the city of Vienna (Austria), Everglades National Park (America), Abu Mena (Egypt), and the former maritime trade city Liverpool (England) are among some destinations exposed to overtourism with some risks (UNESCO, n.d.).

The number of people visiting New Zealand grows by 4.6% each year, and it is estimated to reach 5.1 million people in 2024 (Insch, 2020). New Zealand is one of the destinations with an increasing number of visitors as a result of city branding for the last ten years (MacDonald, 2011; Yeoman & McMahon-Beattie, 2011) and the globally attracted Lord of the Rings movie trilogy. Especially cities that attract tourists, such as Queenstown, Akaroa, and Tarras, have started to see negative effects such as the privatization of public areas, the destruction of historic and natural areas, and environmental pollution (Insch, 2020). Sheivachman stated that if tourism is not managed sustainably within the framework of "overtourism", it will inevitably have negative consequences (Sheivachman, 2016).

The development of overtourism increases the popularity of destinations. Therefore the negative effects of tourism will also increase in over-capacity destinations. In addition, while the large investors benefit from overtourism, local people work for low wages in the tourism sector (Bramwell, 2004). Destinations have adopted alternative tourism in order to ensure sustainability and local development by preventing overcapacity. Alternative tourism is considered as a solution to the problems created by overtourism.

Alternative Tourism

The alternative is an originally French word and means 'an extra option other than what one has." For consumers, the concept of alternative is becoming increasingly important due to the differentiation of human needs and the emergence of different products for each need. Having more than one product for the same purpose means having an alternative to the other. For example, the existence of more than one restaurant serving the same product group is related to the high number of alternatives. Also, different product groups that are not in the same product group and have different purposes may be alternatives to each other. In this case, alternative products can direct consumers to different options in the same product group as well as to different product groups.

This section focuses on creating alternatives for tourists by providing a variety of tourism products and services to mitigate the negative effects of overtourism. Thus, the natural and cultural attractions of the destination can be reserved. Also, such resources can be protected and consumed, or enjoyed within the principles of sustainability.

Alternative tourism mostly refers to special interest tourism but also offers special travel experiences to people interpreted as niche tourism (Noveli, 2005; Morwforth & Munt, 2016; Conway & Tims, 2010, Isaac, 2010). While alternative tourism mostly expresses the way travel is organized, special interest tourism points to the motivations and specific motives that determine the travel (Spilanis & Vayanni, 2004). Alternative tourism has fewer negative effects on destinations and emerged to protect and improve the destinations (Smith & Eadington, 1992; Oriade & Evans, 2011).

The concept of alternative tourism is used with tourism types such as "soft tourism", "conservative tourism", "green tourism", "nature oriented tourism", "slow tourism "eco-tourism", "sustainable tourism", "progressive tourism". Hacıoğlu and Avcıkurt (2008) refer to alternative tourism as diversification of tourism products. Kozak and Bahçe (2012) define alternative tourism as special interest tourism. The literature shows that there are many definitions of alternative tourism. Some of these are as follows:

- The Ministry of Culture and Tourism defines alternative tourism as "a type of tourism that aims to give priority to social and ecological harmony, the cooperation of local and foreign entrepreneurs and the use of local materials in development" (Bahçe, 2013: p.10).
- Alternative tourism has been created by combining new touristic products to reduce the negative effects of traditional, classical overtourism and city tourism (Hacıoğlu & Avcıkurt, 2008).

Alternative tourism is changing the base and commercial perspective of the tourism industry in order to eliminate the negative consequences of overtourism. It has different characteristics than traditional tourism and focuses on the specific needs of tourists, such as natural attractions.

- Alternative tourism has been created in order to reduce the negative effects of traditional and classical overtourism, and city tourism spread to coastal areas such as sea-sand-sun holidays. It differs from traditional tourism, and special interests or natural attractions are prioritized (Britton, 1977).
- Alternative tourism is respectful to the natural and cultural environment and aims to protect and preserve such natural attractions (Bahçe, 2013: p.11).
- Mieczkowski (1995) classifies alternative tourism as cultural, educational, scientific, adventure, rural tourism.

Drawing from these definitions, alternative tourism reveals different attractions besides the classical tourism attractions in a destination. It emphasizes a type of tourism that is sensitive to the natural and social environment and adopts the understanding of sustainable tourism. The emerge of alternative tourism is not to compete with overtourism, but to create new options by reducing the negative effects of overtourism.

Kozak and Bahçe (2012) listed the factors that led to the emergence of alternative tourism as follows;

- The environmental impact of overtourism is increasing day by day, and it creates irreversible results in terms of environment and social aspects.
- As a result of changes in consumer preferences, tourists start to search for products specific to new tourism types, and the numbers of these consumers are gradually increasing.
- Consumers demand products and services that allow them to be more individual but social and to discover the world.
- As a result of unplanned constructions on coastal lines, such as in big cities, consumers from big cities start to escape to smaller destinations,
- Increasing environmental awareness gained advocacy about importance of the environment
- The demand for tourism is increasing day by day, and the regions outside the coastline see tourism as a tool of economic development,
- Organizations such as United Nations, World Tourism Organization, and UNESCO encourage alternative tourism types
- Important actors in the world economy such as the World Tourism Bank and the European Union are financially supporting the small entrepreneurship policies

The environmental and social damages of overtourism have been noted. Additionally, it was necessary to develop alternative tourism types for the reasons outlined above. Alternative tourism aims to prevent unconscious consumption.

What Are the Features of Alternative Tourism?

Bahçe (2013) explained the characteristics of alternative tourism with the following statements;

The understanding of alternative tourism is aimed at protecting the natural, environmental, cultural and social structure. Therefore, the number of people participating in alternative tourism types is generally limited.

Alternative tourism supports local investors and gives priority to regional employment and development. Small entrepreneurs are supported through alternative tourism.

Large tour operators do not prefer engaging with alternative tourism due to economic costs because participation is limited with a low number of people, and the sector gives importance to local architecture. Instead of large tour operators, national and regional travel agencies serve the alternative tourism market.

Local elements come to play a significant role in the products offered in alternative tourism. Alternative tourism attaches importance to the life quality of tourists and local people.

Growth in alternative tourism takes place slowly and in the long run. Accordingly, the capacity is not exceeded, and the supply increases in parallel with the development of tourism.

Alternative tourism takes place actively and ensures positive communication between tourists and locals. Thus, the prejudices that may arise in the locals against tourists are eliminated. The active interaction of tourists and local people creates a positive attitude towards tourists in the local population.

Although alternative tourism is more expensive for tourists, it offers more customized services and products.

What Are the Differences Between Alternative Tourism and Overtourism?

When we compare alternative tourism and overtourism, we see that there are many differences between them in terms of both supply and demand. We can list these differences as follows (Wearing & Neil, 2009; Bahçe, 2013).

Unlike overtourism, where consumption takes place as a group, alternative tourism is carried out individually.

In overtourism, accommodation is generally carried out in large-scale accommodation facilities such as 5-star hotels and holiday villages at international standards. In contrast, in alternative tourism, medium-sized and small hotels, boutique hotels that reflect the local architecture provide the accommodation services. Particularly, local and cultural features are prioritized in accommodation businesses in alternative tourism.

While overtourism travels are carried out mainly by charter flights, alternative tourism mostly depends on the use of cars, trains, coaches, or sea transportation and scheduled flights.

While travel in overtourism is more economical, consumers are willing to pay more in alternative tourism.

While overtourism is generally carried out in a limited space and passively, alternative tourism is carried out actively in open spaces, and it is very much integrated with nature and local people. Since the accommodation in overtourism is generally in the form of a full board or "all inclusive" system, the economic contribution of consumers to the region and local people is gradually decreasing. With alternative tourism, consumers get in touch with local producers and both communicate with local people and economically contribute to the region.

As the package tours prepared within the scope of overtourism consist of the trio of sea-sand-sun, they are carried out intensively in a short period of time in summer. As a result, the demand during a certain season cannot meet the supply in the region, and there is chaos caused by overcapacity in these regions. With alternative tourism types, tourism spreads to different seasons of the year, making it an important step in eliminating tourism's seasonal nature. Thus, by spreading to different seasons of the year and different regions in our country, the seasonal characteristics of tourism are partially eliminated, and the chaos arising from overcapacity is prevented.

Overtourism may cause natural, environmental, and cultural damage because of the high demand during a certain period in one region. Since alternative tourism is being driven by ecological and cultural sustainability, it partially prevents such negative impacts.

Since overtourism usually focuses on the coasts, infrastructure and superstructure investments in these regions are supported by both public and private sectors. Since alternative tourism can be done in different regions of the country, infrastructure and superstructure investments take place in these regions.

Tourism businesses that offer products and services for overtourism are large-scale enterprises and comply with international standards. Businesses that offer products and services in alternative tourism are small-scale and local entrepreneurs. While overtourism is concentrated in the coastal regions during summer, alternative tourism can take place in all regions and in all seasons.

Although there is a general architectural understanding that destroys local architecture in overtourism, alternative tourism aims to protect the original and local architecture.

While overtourism is dependent on large international tour operators managed by large capitals, development is fast, and consumption is one-sided. On the other hand, alternative tourism is under the control of small-scale national tour operators and travel agencies, this development has a slow and sustainable characteristic.

Expected Changes in Mass and Alternative Tourism After the COVID-19 Pandemic

The COVID-19 pandemic affected all sectors, including the tourism sector. The tourism sector is experiencing economic, social, and psychological challenges. Extraordinary circumstances will lead to changes in the tourism sector in the short, medium, and long term. All these circumstances trigger the question of "What will happen to people's understanding of holidays after the pandemic?". Even though an answer to this question cannot be given, it is possible to make some predictions. It is predicted that it will affect the destination and accommodation choice of individuals while they are deciding their holiday.

The pandemic process started the shift from mass tourism to alternative tourism (Hall et al., 2020). People prefer to travel in their vehicles instead of public transport. Individuals plan short-term and close-range travels instead of long-term trips (Li et al., 2020). In the upcoming period, short-term travels will be more common, and businesses that fully comply with the security measures regarding cleaning and hygiene during holidays will be preferred (Wachyuni & Kusumaningrum, 2020).

Individuals who avoid the mass tourism destinations will turn to niche markets in destinations where alternative tourism is developed. Since cleanliness and hygiene will take the first place in the selection of holiday destinations, quiet places away from the crowd will be preferred. The anxiety created by the pandemic causes international travel to be perceived as riskier, and this will cause a decrease in the number of foreign tourists in destinations (Zenker & Kock, 2020). Rather than accommodating large resorts in popular destinations, smaller businesses with a few rooms will be preferred in lesser-known destinations. In addition, nature tourism will come to the fore, and travel in a tent or caravan will become more popular.

With the pandemic, tourism businesses will be involved in technological innovation with the help of artificial intelligence and robotic technologies. Virtual reality, barcode and data matrix applications, e-tickets, contactless fire measuring devices, contactless pos machines, and contactless doors that can be opened in common areas have become widespread. Tourism enterprises have to benefit from the opportunities offered by technology in order to protect the health of their guests and employees and to build trust.

Cases

In order to ensure continuity in the regional and national tourism sector, to achieve internal dynamism, and to get more shares from international tourism, historical, natural, and cultural values should be managed and sustained correctly. Under unfavorable conditions where tourism is not a tool but a goal, an appropriate management model cannot be established, and tourism regions may face the risk of losing their attractiveness gradually.

A holiday concept, "themed tourism"has started to emerge. For example, culture and health tourism shape the holiday understanding of many people. Those who want to meditate, seek mysticism, the dream of having massage and skincare during their holiday, or want to learn the architectural details or traditional culture of a city can enjoy the new tourism market that provides such services.

Niche markets can be created with the development of special interest tourism and nature-based tourism. For example, a tourist who is interested in wine tourism will want to see the vineyards where grapes are grown and cared for, but also want to experience different lifestyles by experiencing the harvest, and will want to learn about the production stages of wine, visit wine cellars, and enjoy wine tasting. Thus, products and services will be created for different time periods, and a contribution will be made to the sustainability of the destination.

In recent years, with the increasing interest in special interest tourism, there have been changes in accommodation types. The importance of small-scale tourism has increased, especially due to the pandemic. Boutique hotels, which generally do not exceed 15–20 rooms, have become very common and popular in the world. The destinations started to work under different policies and carry out comprehensive studies on the protection of cultural and natural values.

One of them is the sustainable tourism policy scheme developed by Aurinkomatkat-Suntours, one of the most important tour operators in Finland. The aim here is to minimize the negative impact of tourism on the cultural and natural structure while maximizing its positive effect.

In order to fulfill this responsibility, the United Nations Environment Program has collaborated with the United Nations Educational, Scientific and Cultural Organization and the World Tourism Organization. They have decided to prioritize issues such as protecting natural and cultural heritage, cooperating with local people, encouraging tourists to respect local culture, protecting designated areas, plants, and animals, and cooperating with relevant units by complying with local, national, and international laws.

Some Suggestions

The preservation of a region's cultural and natural heritage is closely linked with their increasing attractiveness in the region and thus with the improvement of the economy. Tourists often want to see regions with different cultural and natural structures. Therefore, "quality and continuous improvement in tourism" should be adopted to introduce new methods for the development of alternative tourism types.

It is important to properly manage alternative tourism to mitigate the damage caused by overtourism on destinations. In order to support sustainable tourism and ensure its development, alternative tourism should be encouraged as an activity for individuals and small groups; rather than, as crowded group. Thus, instead of large tour operators, small businesses that support regional employment will take place in the alternative tourism market. For this, small entrepreneurs should be supported by local governments.

For the development of alternative tourism, local architecture should be prioritized, local products should be promoted to protect the cultural values of the region. Thus, the development of positive communication between local people and tourists can be supported. Since the development of alternative tourism is not as fast as overtourism, the development will be slow. In the long term, this will prevent the capacity of the region from expanding.

Isaac, R. K. (2010). Alternative tourism: New forms of tourism in Bethlehem for the Palestinian tourism industry. *Current Issues in Tourism, 13*(1), 21–36.
Jacobsen, J. K. S. (2000). Anti-tourist attitudes. *Annals of Tourism Research, 27*, 284–300.
Kozak, M. A., & Bahçe, A. S. (2012). *Special interest tourism* (2nd ed.). Detay Publishing.
Kozak, N., Akoğlan Kozak, M., & Kozak, M. (2014). *General tourism*. Detay Publishing.
Krebs, A., Petr, C., & Surbled, C. (2007). La gestion de l'hyper fréquentation du patrimoine: D'une problématique grandissante à ses réponses indifférenciées et segmentées. In *Proceedings of the 9th international conference on arts & cultural management*. University of Valencia.
Lane, B. (2009). Thirty years of sustainable tourism: Drivers, progress, problems—And the future. In S. Gossling, C. M. Hall, & D. B. Weaver (Eds.), *Sustainable tourism futures: Perspectives on systems, restructuring and innovations* (pp. 19–32). Routledge.
Lankford, S. V., & Howard, D. R. (1994). Developing a tourism impact attitude scale. *Annals of Tourism Research, 21*(1), 121–139.
Li, J., Nguyen, T. H. H., & Coca-Stefaniak, J. A. (2020). Coronavirus impacts on post-pandemic planned travel behaviours. *Annals of Tourism Research*. https://doi.org/10.1016/j.annals.2020.102964
MacDonald, K. (2011). *Expression and emotion: Cultural diplomacy and nation branding in New Zealand (masters of international relations)*. Victoria University of Wellington.
Marson, D. (2011). From mass tourism to niche tourism. In P. Robinson, S. Heitman, & P. Dieke (Eds.), *Research themes for tourism* (pp. 1–11). CABI.
Mieczkowski, Z. (1995). *Envrionmental issues of tourism and recreation*. University Press of America.
Morwforth, M., & Munt, I. (2016). *Tourism and sustainability: Development, globalisation and new tourism in the Third World*. Routledge.
Noveli, M., & Benson, A. (2005). Niche tourism: A way forward to sustainability? In M. Noveli (Ed.), *Niche tourism: Contemporary issues, trends and cases* (pp. 247–250). Elsevier.
O'Reilly, A. M. (1986). Tourism carrying capacity. *Tourism Management, 7*(4), 254–258.
Obrador Pons, P., Crang, M., & Travlou, P. (2009). Introduction: Taking Mediterranean tourists seriously. In P. Obrador Pons, M. Crang & P. Travlou (Eds.), *Cultures of mass tourism: Doing the mediterranean in the age of banal mobilities* (pp. 1–20). Ashgate.
Oriade, A., & Evans, M. (2011). Sustainable and alternative tourism. In P. Robinson, S. Heitmann & P, Dieke (Eds.), *Research themes for tourism* (pp. 69–86). CABI International.
Oklevik, O., Gössling, S., Michael Hall, C., Jens Kristian, S. J., Ivar Petter, G., & McCabe, S. (2019). Overtourism, optimisation, and destination performance indicators: A case study of activities in Fjord Norway. *Journal of Sustainable Tourism, 27*, 1–21.
Peeters, P. (2012). A clear path towards sustainable mass tourism? Rejoinder to the paper "organic, incremental and induced path to sustainable mass tourism convergence" by David Weaver. *Tourism Management, 33*, 1038–1041.
Petr, C. (2009). Fame is not always a positive asset for heritage equity! *Journal of Travel & Tourism Marketing, 26*(1), 1–18.
Poon, A. (1993). *Tourism, technology and competitive strategy*. CABI.
Robinson, M., & Noveli, M. (2005). Niche tourism: An introduction. In M. Noveli (Ed.), *Niche tourism: Contemporary issues, trends and cases* (pp. 1–11). Elsevier Butterworth-Heinemann.
Romeril, M. (1985). Tourism and the environment: Towards a symbiotic relationship. *International Journal of Environmental Studies, 25*, 215–218.
Sheivachman, A. (2016, 16 August). Iceland and the trials of 21st century tourism. *Skift*. Retrieved from: https://skift.com/iceland-tourism/. Accessed: 23 Dec 2020.
Smith, V. L., & Eadington, W. R. (Eds.). (1992). *Tourism alternatives: Potentials and problems in the development of tourism*. University of Pennsylvania Press.
Spilanis, I., & Vayanni, H. (2004). Sustainable tourism: Utopia or necessity? The role of new forms of tourism in the Aegean Islands. In B. Bramwell (Ed.), *Coastal mass tourism: Diversification and sustainable development in Southern Europe* (pp. 269–291). Channel View Publications.

UNESCO. (n.d.). *World Heritage in Danger*. Retrieved from: https://whc.unesco.org/en/158/. Accessed: 1 Aug 2020.

UNWTO. (2018). *'Overtourism'? – Understanding and managing urban tourism growth beyond perceptions: Executive summary.* Retrieved from: https://www.e-unwto.org/doi/pdf/10.18111/9789284420070/. Accessed: 23 December, 2020.

Urry, J. (1990). The 'consumption' of tourism. *Sociology, 24*(1), 23–35.

Vanhove, N. (1997). Mass tourism: Benefits and costs. In S. Wahab & J. J. Pigram (Eds.), *Tourism, development and growth: The challenge of sustainability* (pp. 46–71). Routledge.

Wachyuni, S. S., & Kusumaningrum, D. A. (2020). The effect of COVID-19 pandemic: How are the future tourist behavior? *Journal of Education, Society and Behavioural Science, 33*(4), 67–76.

Walter, J. A. (1982). Social limits to tourism. *Leisure Studies, 1*(3), 295–304.

Wearing, S., & Neil, J. (2009). *Ecotourism: Impacts, potentials and possibilities* (2nd ed.). Elsevier, Butterworth-Heinemann.

Yeoman, I., & McMahon-Beattie, U. (2011). The future challenge. In N. Morgan, A. Pritchard & R. Pride (Eds.), *Destination brands: Managing place reputation* (3rd ed., pp. 169–182). Butterworth-Heinemann.

Zenker, S., & Kock, F. (2020). The coronavirus pandemic: A critical discussion of a tourism research agenda. *Tourism Management*. https://doi.org/10.1016/j.tourman.2020.10416

Chapter 14
Smart Tourism: An Alternative Approach to Clench Overtourism

Muhammad Farooq Akhtar, Sehar Saeed, and Mehreen Waheed

Abstract Tourism is one of the main sources of leisure, exploration, and social gathering. Many factors contribute to this industry's development, such as the growing global economy, emergence of the middle class in developing economies, technological advancement, new business models, low traveling costs, and relaxation in visa policies. Pakistan is blessed with the unique geographical location and endless beauty of nature with natural diversity, covering the range of seashore to high Himalayan Mountains. People visited Pakistan to amaze with endless scenes of nature and celebrating traditional rituals all over the world. Considering the situation of coronavirus, this uncertainty raises multiple questions in the mind of people all around the world. In such a moment of collective confusion, all world economies' economic structures focused on a huge radical paradigm shift. So, there is a need to think holistically about the problem of overtourism and COVID-19. We can provide solid solutions to Government, tourist agencies, tourists, local businesses, health care, infrastructure, etc. The purpose of the study is to provide a viable solution to tackle overtourism and COVID-19 in a holistic way. Smart tourism is a solution to handle multi-faceted problems of this industry. There are few reservations of tourists on their agencies while selecting the destination such as trust, services, help in case of emergency and reputation. Smart tourism is a better option to enhance all stakeholders such as government, tourists, and tourist agencies. It opens the door for the business community to think about a new paradigm in which technology-led businesses are more effective than traditional. This chapter is beneficial for scholars, marketers, tourists and society as a whole.

Keywords Overtourism · COVID-19 · Smart tourism · Pakistan

M. F. Akhtar (✉)
Othman Yeop Abdullah Graduate School of Business, Universiti Utara Malaysia, Kuala Lumpur, Malaysia

S. Saeed
Faculty of Business and Management, Universiti Teknologi MARA, Shah Alam, Malaysia

M. Waheed
School of Business & Economics, University of Management & Technology, Lahore, Pakistan

A. Hassan, A. Sharma (eds.), *Overtourism, Technology Solutions and Decimated Destinations*, https://doi.org/10.1007/978-981-16-2474-2_14

Introduction-Overtourism

Tourism is a source of leisure, recreation, and social gathering. This industry integrates societies, cultures and promotes peace and tolerance among countries. Motivational factors that boost this industry are the growing global economy, the emergence of a middle class in developing economies, technological advancement, new business models, low traveling costs, and relaxation in visa policies, foreigner tourist arrivals increase up to 5% in 2018 and achieve 1.4 billion milestones. This figure is ahead of World Tourism Organization (UNWTO) predictions (UNWTO, 2019). In the same vein, tourism significantly contributes to the economies of many countries of the world. This sector has internal and external impacts on the economy to alleviate the poverty of the world. Internally, tourism generates fiscal activities such as food, travel, arts, and crafts; externally, it is a source of transfer of funds from rich to poor countries (Ashley & Mitchell, 2009; Jucan & Jucan, 2013).

The tourism industry in Pakistan has great potential due to diversity in culture, geography, and historical perspectives. Pakistan's tourism can be divided into four types: religious, historical, eco, and natural tourism (Arshad et al., 2018). In religious tourism, people travel to their sacred and worship places to attain spiritual power and relief (Yeoman, 2009). In Pakistan, the followers of different religions are found, such as Muslim, Hindu, Sikh, Buddhist (Arshad et al., 2018). Islamic religious places in Pakistan are Hazrat Faiduddin Ganj-e-shakar, Shah Hussain, Data Ganj Bakhsh, Mian Mir, Bahauddin Zakaria, Lal Shahbaz Qalandar, Shah Abdul Latif Bhattai (Rasul et al., 2016). Religious places of Hindu's are Old Shiv temple of Clifton Karachi, Asthan of Hinglaj, Matajee in Lasbeella, the temple of Baba Bankhandi, the temple of Karsaraj, and many others. Sacred places is Sikh religion are Gurdwara Guru Nanak, Dehra Sahib, Guru Aryan Dev, Panja Sahib, Kartarpur, Chuharkana, and Pati Sahib (Baloch & Rehman, 2015). Religious places of Buddhism are across the Karakoram Highways (Baloch & Rehman, 2015). Every year, Pakistan received thousands of Sikh, Hindu, and Buddhists pilgrims all over the world. These religious places are overcrowded on specific occasions such as Meala (fair), Holi, Diwali, Baisakhi. This situation creates huge problems for residents, the Government of Pakistan, and especially for local and international tourists. Pakistan holds ancient archaeological and historical places such as Indus Valley, Gandhara Civilization, and the Mughal Empire (Fakhar, 2010). The major cities of Gandhara Civilization are Peshawar, Bamyan, Taxila. The belongings of Gandhara Civilization are secured in different museums of Pakistan. The Indus Valley Civilization is recognized as the Harrappan Civilization. These historical locations attract the tourist to visit Pakistan. Ancient buildings and collections of Mughal Empire are great attractions for local and international tourists (Arshad et al., 2018).

Ecotourism is a source to reduce poverty and environmental safety, which served United Nations Sustainable Development Goals (UNSDG) (Arshad et al., 2018). In this type of tourism, the tourists visited the special locations to participate in festivals, museums, events and, purchase local handicrafts, arts, and crafts, and enjoy local food items. Pakistan contains beautiful locations of ecotourism locations of the world such as high mountains, glaciers, cliffs, meadows, forests, deserts, beaches, coastline, and wetlands. The beauty of Pakistan offered different adventurous activities such as mountaineering, trekking, desert jeep race, boar hunting, Indus Dolphin, trout fishing. In Pakistan, there are a lot of festivals such as Shandur, Silk Root, Kalash, and Khanpur festivals. Shandur Polo ground world's highest ground and famous all around the world, which is above 3700-meter sea level. Similarly, the Northern Areas of Pakistan has beautiful tourists spot such as Swat, Malam Jaba, Shangla, Murree, Gilgit Baltistan, Hunza, and Naran. This location of the world contains valleys, rivers, glaciers, and four world highest mountain ranges Himalayas, Hindu Kush, Pamir, and Karakoram. These naturally beautiful sceneries provide amazing attractions to climbers, trekkers, hikers, and general tourists (Khalil et al., 2007). Pakistan has the five highest mountains in the world (K-2, Nanga Parbat, Gasherbrum I, Broad Peak, and Gasherbrum II).

In Pakistan, most of the above-mentioned places are overcrowded. Overtourism is defined as "excessive number of tourists at a specific destination that can result in negative impacts of all types on the community involved" (Dodds & Butler, 2019, p. 519). Overtourism has resulted from a combination of several factors, which together have created a "perfect storm" of visitors to specific sites. This phenomenon is happened due to individual and contextual factors (Butler, 2006; Milano et al., 2019). Personal factors are mobility, leisure, and unique experiences (Milano et al., 2019). The study of Butler (2006) revealed that the natural beauty of the place captures the attention of tourists. The study conducted by Milano et al. (2019) highlighted that mobility, leisure, and unique experience influenced tourists to visit different places. This study also noticed that tourists visit rural places in certain seasons. Excessive gathering on these tourists' spots affects the movement and well-being of residents. In the recent past, the pandemic of COVID-19 broke out. So, there is a need to consider it in tourism to protect the health of local and international tourists. Preceding scholars asserted the need of planning and policy making to clench overtourism (Dredge, 2017).

COVID-19 and Overtourism

Coronavirus as COVID -19 is the fastest emerging disease with a consistent increase in cases and death ratios since its first identification in the city of China, Wuhan, in December 2019. In Wuhan, the rapid increase in the patients of fever, dry cough, throat infection, and decreased in white blood cells consistently were the symptoms of this disease surfaced rapidly. Initially, the patient who possessed all these symptoms diagnosed as "Fever of Unknown origin with Pneumonia". The patient was diagnosed with SAR-COV-2, which transmits human to human and leads to severe pneumonia cause death. The germ of SARS-Cov-2 is aggressively severe and threatening human life. On 30th January 2020, World Health Organization (WHO) has announced severe impacts of this disease and declared public health emergency of international concerns. The statistics about this disease have confirmed 26,359 suspected cases, 31,225 confirmed cases, and the death toll increase up to 639. Different age groups are susceptible to SARS-COV-2 (Lu & Shi, 2020). Coronavirus has close genetic similarities found in bats. At the start, January 2020, scarce information was available about this virus, but it rises from eating some animals in Wuhan markets. COVID -19 is considered to introduce in the human population from the kingdom of the animal. Although the first case was diagnosed in Wuhan in December 2019, the whole virus spread is based on the human to human transmission, not the result of consistent spillover. In weeks, transmission massively increases in Wuhan and people spread this infection through national and international travels to China in the course of the Chinese New Year's holidays. COVID -19 has shown different epidemiological attributes from SARS-COV. As it replicates efficiently in the upper respiratory tract and causes a less abrupt onset of symptoms, similar to conventional human Coronavirus and common colds become the major cause of its rapid spread, especially in the winter season (Heymann & Shindo, 2020).

A sudden outbreak of COVID-19 brought drastic epochal changes worldwide. A few months ago, daily life routines would have been unthinkable to most people. Businessmen, managers, politicians, academicians, and economists were dealing with traditional risks or more thought about the pressing issues of climate change. Within few weeks, COVID-19 grasped attention to the whole world, and the focus has shifted dramatically. All around the world, situations rapidly become critical day by day due to an increase in the ratio of suspected, confirmed, and death cases registered. The uncertain hurricane of COVID-19 has destroyed the economy and the world's whole landscape (Wagner, 2020). Each Government emphasis is stemming the tide with increasingly strong precautionary measures. The extreme wave of uncertainty spell everywhere. The weave of uncertainty fuels confusion, which triggers multiple questions (i.e., how deadly the disease actually is), can it remove from the human body through vaccine and who work in it and when it will be invented and available in the market, how policies effects on it and how people will respond, and so on.

In the situation of the current scenario, the absence of medical treatment and vaccination prevails everywhere. Some behavioral psychologists claim that the unfolding pandemic COVID-19 can only be overcome through massive and rapid behavioral change. According to Betsch (2020), to control this pandemic, we need to systematically monitor an individual risk perceiving tendency (aptitude) and the underlying factors that prompt to act upon it. Behavioral researcher's talk about the influence of pandemic COVID-19 on individual behaviors as it varies from person to person, even country to country worldwide. In some countries, behaving as per standard operating procedures (SOP's) regarding COVID-19, people feel unreasonably embarrassing. Especially in those countries where avoiding physical interaction is voluntary. However, governments took strict measures such as closed schools, shopping plazas, entertainment hubs, and playgrounds (Betsch, 2020).

The current situation requires avoiding handshakes with friends and family and turning down any gathering invitation for lunch & dinner. It is the behavior that we can adopt to save our loved ones. To prevent this pandemic, massive and fast adoptions of behavioral changes are the key to overcoming this crisis. In actuality, people need to admit that understanding behavior is the basis of changing and curb this crisis. People need to learn about each other's behavior in this tragic phase of the crisis (Betsch, 2020). Therefore, some of the countries pool the researchers and try to establish the link between public and Government awareness campaigns through various sources such as media platforms. The Germans initiated to gather responses about COVID-19 crisis from the public, including people's perceptions about risks, protective behaviors, trust, and significant misinformation. The initiative aims to know about common men's psychological understanding, correction of misinformation, and adopting suitable behavioral measures by implementing state policies or restrictions. In short, the fear factor was common worldwide due to the massive increase in COVID-19 suspected and confirmed cases. This fear factor massively has a destructive influence on the whole world's tourism industry (Lu & Shi, 2020). The gradual and consistent increase in COVID-19 circumstances proves an unthinkable disaster for the tourism industry worldwide. Although, the tourism industry was previously facing severe issues of overtourism. People are restricted from staying at home toward Government. Lockdowns and strikes scenarios prevail everywhere. COVID-19 crisis badly impacts social interactions and recreational activities.

The global tourism industry has been faced a wider range of setbacks in the past. Major disruptive events caused many damages for this industry, such as September 11 attacks (2001), the severe acute respiratory syndrome (SARS) outbid (2003), the global economic crisis in 2008/09, and now COVID-19 in 2020 hit hardly. Still, no one shocks are more panicked than COVID-19 pandemic due to non-availability of suitable vaccines and scattering central reason through human to human interaction (Higgins-Desbiolles, 2020). Basically, social interaction is the most central and common point among tourism and COVID-19 protection mechanisms. One of the harsh realities of 21 century is that we are helpless to find tools against fighting with COVID-19 (Baldwin & Weder di Mauro, 2020), badly faced or facing the hardest negative impacts. Although, the virus affected all parts of the value chain of hospitality everywhere in the world. We can use protective measures to save from epidemic due to sudden out bud the wave of COVID-19 tourism industry.

As discussed in the first section, Pakistan is blessed with the unique geographical location and endless beauty of nature with natural diversity, covering the range of seashore to high Himalayan Mountains. People visited Pakistan to amaze with endless scenes of nature and celebrating traditional rituals all over the world. Considering the coronavirus situation, this uncertainty raises multiple questions in the mind of people all around the world. In such a moment of collective confusion, all world economies' economic structures focused a huge radical paradigm shift. Such as ways of doing business changes E-commerce become the priority of each enterprise to ensure its survival. The business model shifted toward E-ways, and the whole market becomes digital media marketing. In short, COVID-19 wild swings hit each economy drastically (Wagner, 2020). So, there is a need to think holistically about the problem of overtourism and COVID-19. We can provide solid solutions to Government, tourist agencies, tourists, local businesses, health care, infrastructure, etc.

Smart Tourism as an Alternative Approach

Smart tourism is a holistic approach to clench overarching issues of overtourism and COVID-19. Chinese experts have established the idea of smart tourism to led the concept of smart planet (Li et al., 2017) to support the urban community in fostering their economic, social and environmental affiliation to provide better infrastructure for tourists. The concrete infrastructure for providing support to tourists is built upon ICT, which provides effective, efficient, and integrated experiences to tourists (Xiang & Fesenmaier, 2017). A progressive change in the societies to prosper new technologies in tourism creates a higher impact on this industry's development (Buhalis & Amaranggana, 2013, 2015). This Smart tourism includes all-new cutting-edge technologies to take part in creating digital effect to fulfilling the gap for providing digital solutions including smart phones, use of radio frequency identification, near field communication technology, social media, websites, augmented and virtual reality (Gretzel et al., 2015). This implementation of the smart move under the umbrella of technology is tilting its context towards creating smart cities and bringing all the stakeholders together to enhance the efficiency of this tourism industry and ecosystems (Buhalis & Amaranggana, 2013). Smart tourism's main goal is to provide strategic solutions to the specific traveling needs (Khan et al., 2017). This includes the usage of applications or technological appliances that guide and identify and select a preferred destination. It also triggers towards I -tour that is another emerging technology stated to use within mobile technologies, Tablets and other resources. Through this idea consumer would have a close look and idea to the preempted virtual applications to the destinations. This will bring into attention that tourists should have installed applications and aware of the use of the technology (Tripathy et al., 2018).

There is a vital role of awareness and knowledge towards technology to become a user of smart tourism. The mobility of an individual and groups has depended on the user. Suppose a user is a knowledgeable, literate and accepts technology. In that case, it will be easier for the user to adapt to smart tourism. In contrast, in traditional tourism, you don't need to use any technology for selecting a destination, hotel, and transport. The role of smart tourism is the transformation from traditional to the smart city (Xiang & Fesenmaier, 2017)in which tourists can approach their destinations with fewer efforts. For example, the study conducted by (Buhalis, 2019) suggested the concept of a smart environment where different industries and stakeholders can facilitate travelers to explore human-computer interaction and feel the exact environment they tend to have at their destination.

Considering the above two sections, smart tourism is a viable solution to handle this industry's multi-faceted problems. This solution is helpful to give directions to government, businessmen, and tourists. There are few reservations of tourists on their agencies while selecting the destination such as trust, services, help in case of emergency and reputation (Tripathy et al., 2018). Smart tourism is a better option to enhance opportunities for all stakeholders such as government, tourists, tourist agencies, and many more. This opportunity alarms the business community to think about the new paradigm in which technology-led business will be more influential than the traditional approach. The government should have to play its role to launch and facilitate smart tourism approaches (Matos et al., 2019), which leads the country towards economic, technological, societal, and environment-friendly benefits.

Conclusion

Tourism is one of the main industries of Pakistan. This sector significantly contributes to the economies of many countries of the world. This sector has internal and external impacts on the economy to alleviate the poverty of the world. Internally, tourism generates fiscal activities such as food, travel, arts, and crafts; externally, it is a source of transfer of funds from rich to poor countries. The tourism industry in Pakistan has great potential due to diversity in culture, geography, and historical perspectives. Tourism of Pakistan can be divided into four types, namely religious, historical, eco, and natural tourism. Every year, Pakistan entertains thousands of international tourists visiting historical places, participating in religious activities, and amusing with nature and Pakistan's ecological locations. Excessive gathering on these tourists' spots affects the movement and well-being of residents. In the recent past, the pandemic of COVID-19 broke out. Considering the situation of the corona virus, this uncertainty raises multiple questions in mind. In such a moment of collective confusion, all world economic structure focused on a huge radical paradigm shift. This study provides some solutions to tackle the problem of overtourism and COVID-19 such that we can provide solid solutions to Government, tourist agencies, tourists, local businesses, health care, infrastructure, etc. The study aims to provide a viable solution to tackle overtourism and COVID-19 holistically. Smart tourism is a solution to handle the multi-faceted problems of this industry. There are few reservations of tourists on their agencies while selecting the destination such as trust, services, help in case of emergency and reputation. Smart tourism is a better option to enhance all stakeholders such as government, tourists, and tourist agencies. It opens the door for the business community to think about a new paradigm in which technology-led businesses are more effective than traditional. This chapter is beneficial for scholars, marketers, tourists, and society as a whole.

References

Arshad, M. I., Iqbal, M. A., & Shahbaz, M. (2018). Pakistan tourism industry and challenges: a review. *Asia Pacific Journal of Tourism Research, 23*(2), 121–132.

Ashley, C., & Mitchell, J. (2009). *Tourism and poverty reduction: Pathways to prosperity*. Taylor & Francis.

Baldwin, R., & Weder di Mauro, B. (2020). *Economics in the time of COVID-19, a VoxEU.org eBook*. CEPR Press.

Baloch, Q. B., & Rehman, A. (2015). Regional integration of Pakistan tourism: Exploring prospects. *Abasyn University Journal of Social Sciences, 8*(2), 405–415.

Betsch, C. (2020). How behavioural science data helps mitigate the COVID-19 crisis. *Nature Human Behaviour, 4*(5), 438–438.

Buhalis, D. (2019). Technology in tourism-from information communication technologies to eTourism and smart tourism towards ambient intelligence tourism: A perspective article. *Tourism Review, 75*(1), 267–272.

Buhalis, D., & Amaranggana, A. (2013). Smart tourism destinations. In Z. Xiang & I. Tussyadiah (Eds.), *Information and communication technologies in tourism 2014* (pp. 553–564). Springer.

Buhalis, D., & Amaranggana, A. (2015). Smart tourism destinations enhancing tourism experience through personalisation of services. In I. Tussyadiah & A. Inversini (Eds.), *Information and communication technologies in tourism 2015* (pp. 377–389). Springer.

Butler, R. (2006). *The tourism area life cycle* (Vol. 1). Channel View publications.

Dodds, R., & Butler, R. (2019). The phenomena of overtourism: A review. *International Journal of Tourism Cities, 5*(4), 519–528.

Dredge, D. (2017). *Overtourism. Old wine in new bottles?*. Retrieved from: https://www.linkedin.com/pulse/overtourism-old-wine-new-bottles-dianne-dredge/. Accessed: 8 Jan 2021.

Fakhar, A. H. (2010). *Factors affecting tourism, tourism potential and strategies for development as an Industry in Pakistan*. Retrieved from: http://hig.diva-portal.org/smash/get/diva2:303282/FULLTEXT01.pdf. Accessed: 8 Jan 2021.

Gretzel, U., Sigala, M., Xiang, Z., & Koo, C. (2015). Smart tourism: foundations and developments. *Electronic Markets, 25*(3), 179–188.

Heymann, D. L., & Shindo, N. (2020). COVID-19: what is next for public health? *The Lancet, 395*(10224), 542–545.

Higgins-Desbiolles, F. (2020). Socialising tourism for social and ecological justice after COVID-19. *Tourism Geographies, 22*(3), 610–623.

Jucan, C. N., & Jucan, M. S. (2013). Travel and tourism as a driver of economic recovery. *Procedia Economics and Finance, 6*, 81–88.

Khalil, S., Kakar, M. K., & Malik, A. (2007). Role of tourism in economic growth: Empirical evidence from Pakistan economy. *The Pakistan Development Review, 46*(6), 985–995.

Khan, M. S., Woo, M., Nam, K., & Chathoth, P. K. (2017). Smart city and smart tourism: A case of Dubai. *Sustainability, 9*(12), 2279.

Li, Y., Hu, C., Huang, C., & Duan, L. (2017). The concept of smart tourism in the context of tourism information services. *Tourism Management, 58*, 293–300.

Lu, Q., & Shi, Y. (2020). Coronavirus disease (COVID-19) and neonate: What neonatologist need to know. *Journal of medical virology, 92*(6), 564–567.

Matos, A., Pinto, B., Barros, F., Martins, S., Martins, J., & Au-Yong-Oliveira, M. (2019). Smart Cities and Smart Tourism: What Future Do They Bring? In Á. Rocha, H. Adeli, L. Reis & S. Costanzo (Eds.), *New Knowledge in Information Systems and Technologies. WorldCIST'19 2019. Advances in Intelligent Systems and Computing* (pp. 358–370). Springer.

Milano, C., Novelli, M., & Cheer, J. M. (2019). *Overtourism and tourismphobia: A journey through four decades of tourism development, planning and local concerns*. Taylor & Francis.

Rasul, F., Fatima, U., & Sohail, S. (2016). Religion Tourism and Entrepreneurial Development (A Case Study Hazrat Data Ganj Bakhsh Shrine). *South Asian Studies: A Research Journal of South Asian Studies, 31*(1), 275–289.

Tripathy, A. K., Tripathy, P. K., Ray, N. K., & Mohanty, S. P. (2018). iTour: The future of smart tourism: An IoT framework for the independent mobility of tourists in smart cities. *IEEE consumer electronics magazine, 7*(3), 32–37.

UNWTO. (2019). *International Tourism Highlights*. UNWTO.

Wagner, A. F. (2020). What the stock market tells us about the post-COVID-19 world. *Nature Human Behaviour, 4*(5), 440–440.

Xiang, Z., & Fesenmaier, D. R. (2017). Big data analytics, tourism design and smart tourism. In Z. Xiang & D. Fesenmaier (Eds.), *Analytics in Smart Tourism Design: Concepts and Methods* (pp. 299–307). Springer.

Yeoman, I. (2009). *Tomorrow's tourist: scenarios & trends*. Routledge.

Chapter 15
Prospects and Challenges in Promoting All-Inclusive Holiday Tourism in Sri Lanka with Special Reference to the Resorts in Kalutara

N. H. D. Kumara, Manori L. Guruge, and D. A. C. Suranga Silva

Abstract This study's main objective is to understand the prospects and challenges in promoting All-Inclusive Holiday Tourism in Sri Lanka in considering the resort/hotel occupancy, revenue, and profitability. On the basis of the review of literature, the conceptual framework has been developed. There were four independent variables and one dependent variable in the Conceptual Framework. To test the conceptual model and comprehend the link between the independent and dependent variables, four hypotheses were established. The study's null hypotheses assumed that four independent variables had no effect on the dependent variable's variance, but the alternate hypotheses suggested that the independent variables do have an effect on the dependent variable's variance. The population for the study consisted of all resort hotels in Kalutara District in Sri Lanka, whilst the sample population for the study consisted of 8 resort hotels located in the Kalutara district of Sri Lanka. A self-administered questionnaire consisting of 69 questions was the primary data collection instrument used in this study. The relationship between the independent variables and the dependent variable was examined using SPSS version 17.0. The correlation between the various holiday packages offered by resort hotels, occupancy, revenue, and profitability was analyzed using Pearson's correlation, whilst linear regression was used to examine the influence of the independent variables on the dependent variable. Data analysis findings indicate that AI holiday package has a strong positive statistically significant correlation with resort hotel occupancy, revenue, and profitability and has a significant impact on the variance in the dependent variable. In terms of the other holiday packages, data analysis indicates that B&B and HB holiday packages have a statistically significant weak negative correlation with the dependent variables, whilst the FB holiday package has a weak negative relationship with the dependent variable, which is not statistically significant. In terms of the unique contribution of these three holiday packages to the vari-

N. H. D. Kumara · M. L. Guruge · D. A. C. Suranga Silva (✉)
Department of Economics, Faculty of Art, University of Colombo, Colombo, Sri Lanka

A. Hassan, A. Sharma (eds.), *Overtourism, Technology Solutions and Decimated Destinations*, https://doi.org/10.1007/978-981-16-2474-2_15

ance in the dependent variable, the contribution of these three holiday packages to the variance in the dependent variable is extremely small (less than 14% in total). Based on the data analysis, it can be concluded that All-Inclusive Holiday Tourism contributes significantly to resort/hotel occupancy, revenue, profitability and are vital for the long performance and viability of these hotels.

Keywords All-inclusive holiday · Resort/hotel occupancy · Revenue · Profitability

Introduction

The tourism industry is a highly competitive industry with various tourism destinations, hotels, and other stakeholders in the industry constantly introducing new tourism and marketing concepts to win customer favor, increase market share and profitability. In this dynamic environment, one concept which withstood the test of time is the All Inclusive (AI), which continues to maintain its position as the most popular tourism holiday package in a majority of beach holiday resorts (Falzon, 2003). The main reason for AI position as the number tourism package in the world can be attributed to the strong demand for this popular in the world's largest tourism source markets such as the United States, German, the United Kingdom, and other highly developed economies (Pangaea Network, 2014). The AI holiday concept which is defined by Travel Intelligence International (2000) as a holiday package where all expenses are included in one pre-paid price, which removes the necessity for the customer to make multiple payments for different tourism products and services such as travel, food, transport with the domestic country, beverages, etc., customers want the entire travel experience organized for them by surrogate parents (i.e., tour operators) (Enoch, 1996). Most tour operators and "surrogate parents" are located in tourist source countries and has resulted in a large percentage of the tourism revenue derived from AI holiday packages being retained by these tour operators and other "surrogate parents" with only a limited percentage of revenue flowing to the tourist destination country (Alegre & Pou, 2006).

Modern technology now offers a way clearer and very perspective market information than what happened in the past. Instead of relying on the workouts of historical data to forecast the future, the more sophisticated and powerful predictive software is used for revenue management of the hotel industry. Such software can now integrate and capture the data from internal, external, and real-time sources. As a result, hotel managers can make better-informed decisions in many guest segments regarding their future demand trends, travel behaviors, trip characteristics, and functionalities. Furthermore, All-Inclusive guests may tend to pay for additional services. They might not confine the promised packages. Sometimes, they may be upsold to buy something additional or more expensive without having them true packages. Such trends and patterns can be accounted for and measure through modern, sophisticated technology.

These types of revenue management technologies can differentiate the customers at the granular level, revealing who pays more for deluxe accommodations or additional bookings like access to special places and cultural trips abroad. Additionally, as increasing profitable revenue is critical to an all-inclusive model, technology makes it easier to use revenue management policies across all revenue streams such as food and drink, spa, sightseeing, and retail buying. By getting an entire view of the number of activities and expected total amount of spending of existing guests, hotel owners can easily identify their most profitable guests and determine the best ways to position for the profit-enhancing category.

In Sri Lanka, the AI concept was introduced in1980s, initially at Robinson Clubs Hotel in Bentota has been modified to a great extent since then by resort hotels particularly in the beach resorts of the country. The AI concept has contributed significantly to the sustainability of the industry despite the frequent ups and down due to the various disruptions on account of the civil war. The AI concept had contributed significantly to the promotion of SL tourism and has become an extremely popular holiday package, particularly amongst European tourists vacationing in the country. In this research, the main objective is to analyze the economic benefits derived from AL holiday packages in comparison to the other holiday packages marketed by resort hotels in SL.

Problem Statement of the Study

At present, different types of tourism promotion have been carried out in many countries in global tourism. Similarly, Sri Lanka has also been promoting more specific interest-based tourism in different destination sites under various themes targeting the potential tourists in different niche tourism market segments undermining the All-Inclusive (AL) Tourism promotion of the country.

Therefore, this study's major research problem is to examine whether AI Tourism has future prospects in Sri Lanka and if so, what are the key challenges to be addressed in promoting the All Inclusive Tourism for making more occupancy, revenue, and profitability to the hotel industry in Sri Lanka.

The Overall Objective of the Study

This chapter analyzes the prospects and key challenges in promoting All-Inclusive (AI) Tourism in Sri Lanka. Some specific objectives of the study are:

(a) Determine the impact of AI holiday packages on the average occupancy of resort hotels in SL.
(b) Determine the impact of AI holiday packages to resort hotel income in SL.

(c) Compare the financial benefits that resort hotels receive from AI holiday packages to the financial benefits that resort hotels receive from other holiday packages.
(d) Make comments and proposals that can help resorts, hotels, the local community, and other hotel stakeholders gain more economic advantages from AI packages.

Scope of the Study

The population for this study consists of all hotels in SL offering AI holiday packages. Given the financial and practical challenges in surveying the entire population, the sample population was limited to resort hotels located in Kalutara, SL. Offering AI holiday packages.In addition, this study only focuses on profitability, turnover, and occupancy in the selected hotels. Given the lack of past empirical research on the issue, this study will fill the gap in knowledge about the economic benefits received by resort hotels in SL from AI packages packages. Researchers will also be able to take inferences from the findings of this study and use them as a guide in future research on AI packages packages, particularly research on the economic benefits obtained by SL resort hotels from AI packages packages. The tourist sector in Sri Lanka, notably resort hotels and tourism agencies, will be able to use the findings of this study to make informed judgments on AI packages packages.

An Overview of Empirical Research

The AI holiday concept first appeared in the 1930s in the Butlin holiday camps in the United Kingdom (Issa & Jayawardena, 2003). These holiday camps' main objective was to provide working-class families in the UK with an affordable holiday option (ibid). This concept was further developed by Club Mediterranee in the 1950s with the establishment of a holiday camp on the Island of Mallorca (ibid). The primary objective of Club Mediterranee in establishing this holiday camp, as per Clark (2000) was to eradicate the additional cost incurred by tourists on their vacations, which was having a negative impact on their tourism experiences. The AI concept has numerous stakeholders ranging from customers to various tourism product and service providers to package producers. AI product packages (i.e., tour operators produce the package by purchasing various tourism products and services from tourism service providers). Customers then purchase these holiday packages, resulting in the consolidated holiday package reverting to its original form of individually separate tourist products and services, which often begins and concludes with a transportation component of the trip (Anderson, 2006). The AI holiday package client, according to Basala and Klenosky (2011), is a customer who enjoys the tourist destination via the protection of an artificial environmental bubble, which restricts his or her engagement and function to a comfortable and secure level.

As per Wong and Kwong (2004), the most popular AI destinations of tourists are the destinations that are in close proximity to them. The findings from this research is in line with findings from the research conducted by Sheldon and Mark (1987) which found that Hawaii is the most popular AI destination for tourist from the USA and Canada. This correlation between geographic proximity and popularity of AI holiday destination is also supported by research conducted by Alegre and Pou (2006), which reported that the Mediterranean is the most popular AI holiday destination for British and German tourists, research conducted by Jorgensen and Solvoll (1996) which reported that Cyprus, Spain, and Turkey were the most popular AI holiday destinations for Norwegian tourist and research conducted by Mbaiwa (2005) which found that Botswana was the most popular AI destination for South African tourist.

Tourism Expenditure is defined by the World Tourism Organization (WTO) (2004) as the total consumption expenditure made by a tourist or on behalf of the tourist at the destination location. Further, according to WTO (ibid) "destination" is defined as any significant place or places visited by the tourist during his or her vacation, whilst "consumption" is defined as the value of products and services used by or on account of the tourist. Tourism expenditure has a direct impact on the economy of the destination (Frechtling, 2006) as the direct beneficiaries of this expenditure utilize these funds to purchase goods and services to support their businesses, remunerate employees, make statutory payments such as taxes.

The main weakness of the AI holiday package, according to most researchers, is the fact that the customer purchases a majority, if not all, of his or her tourism needs from one supplier (Titanont & Chantradoan, 2004). Customers who would normally purchase only the accommodation component from hotels and the rest of their holiday requirements from other sources now purchase the majority of their vacation requirements from the hotel alone, according to these two researchers. The bundling of vacation components into one consolidated package is beneficial to the customer or tourist as the price of the AI holiday package is significantly lower than the price he or she would have to pay if they were to purchase each of the tourism components individually (Guiltinan, 1987). As per Wickens (2002), problems and issues between the tourist and tourism service provides are reduced significantly on account of AI holiday packages as the transaction between tourism service producers and the tourist is extremely low in AI holiday packages.

The Methodology of the Study

The methodology that is utilized in this investigation is described in this chapter. The first section explains the research design that is used in this study, while the second section explains the relationship between the independent variables and the dependent variable, displays the study's conceptual framework, and provides a

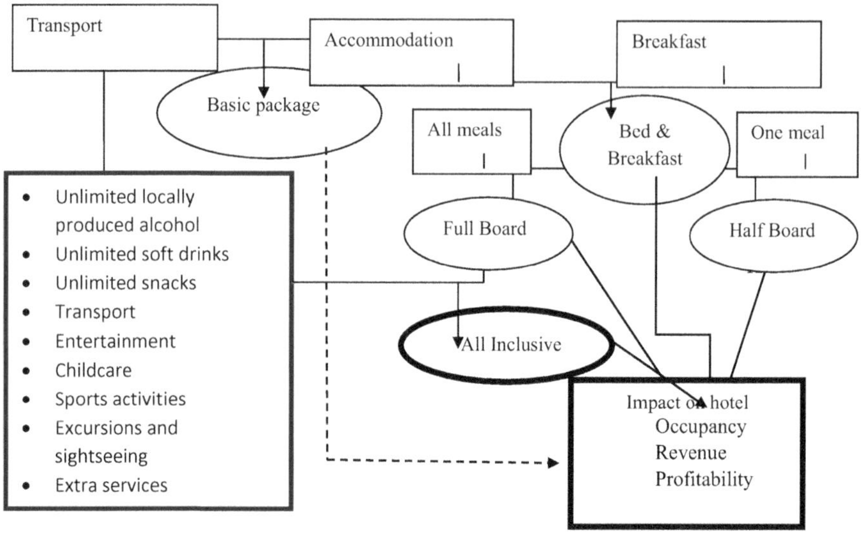

Fig. 15.1 Conceptual framework of the study. (Source: developed by the researchers, 2020)

detailed description of the methods that will be used to collect primary data for both the dependent and independent variables.

The independent factors in this study are full-board, half-board, bed and breakfast, and all-inclusive vacation packages, whereas the dependent variables are hotel profitability, revenue, and occupancy. The independent variables are considered to have a correlation with the dependent variables in this study, and that any change in the state of the independent variable will have a direct influence on the dependent variable's status.

The hypotheses for this study were developed using key results from the literature review (Fig. 15.1).

Hypotheses of the Study

$H1_0$ = Bed and Breakfast holiday packages have no impact on the financial performance of resort hotels in Sri Lanka.

$H2_0$ = Half Board holiday packages have no impact on the financial performance of resort hotels in Sri Lanka.

$H3_0$ = Full Board holiday packages have no impact on the financial performance of resort hotels in Sri Lanka.

$H4_{0=}$ AI holiday packages have no impact on the financial performance of resort hotels in Sri Lanka.

Note: All hypotheses will be tested against a level of significance of 0.05.

Population and Sample

According to Saunders et al. (2007), the sample population for the study should be chosen based on the sample group's capacity to achieve the study's research aims using the information obtained from them. The sample population will be limited to 10 resort hotels in SL due to logistical and financial restrictions in surveying. Resort hotels in the Kalutara District have been selected by the researcher as the sample population on account of the popularity and success of the AI concept in this region of the country.

Data Collection Methods

A structured customer questionnaire was utilized as the major data gathering strategy in this study. The researchers designed the questionnaire and consisted of 65 questions (i.e., one open-ended question and 64 close-ended questions). The 64 close-ended questions have been designed to gather data to test the hypotheses and understand the relationship between the various holiday packages offered by resort hotels in SL, occupancy, revenue, and profitability.

Validity in research can be defined as how accurately a research study answers the research question or, in other words, the strength of the study conclusions. The extent validity and content validity of the research instrument has been tested in this study via a pilot study with 10 survey respondents prior to developing the final self-administered questionnaire.

Reliability can be defined as the degree to which a test instrument gives the same result each time it is used in the same setting with similar types of subjects. This study's internal consistency of measurements will be measured using Cronbach Alpha with an acceptance level of 0.07.

Simple descriptive approaches such as mean, standard deviation, and so on will be used to examine the data. Also, SPSS version 20.0 was applied for analyzing the relationship between the various holiday packages offered by resort hotels, occupancy, revenue, and profitability. The correlation between the various holiday packages offered by resort hotels, occupancy, revenue, and profitability will be analyzed using Pearson's correlation. The impact of various holiday packages on resort hotel occupancy, revenue, and profitability will be measured using linear regression.

Data Analysis and Findings

Profile of Sample Population

Of the sample population of 10 hotels in the Kalutara district, only 80% of the hotels returned the completed questionnaire forwarded to them. All eight were completed satisfactorily, which resulted in responses from 80% of the sample population being used in the data analysis. All Inclusive Holiday Package Contribution to Resort Hotel Occupancy, Revenue, and Profitability.

AI Holiday Package Contribution to Resort Hotel Occupancy

The number of Asian guests on AI holiday packages is extremely low, whilst the number of European guests on AI holiday packages in these hotels is high. As shown in the finding, the mean number of Asian guests on AI holiday packages in SL resort hotels is 1.75. This implies that less than 25% of Asian guests in resort hotels are on AI holiday packages. In terms of European tourists, the mean number of European guests on AI holiday packages is 3.75, which implies that between 40%–55% of tourists in SL resort hotels are on AI holiday packages.

AI Holiday Package Contribution to Resort Hotel Revenue

Mean AI holiday package contribution to room, restaurant, and beverage revenue in SL resort hotels is 3.38, 3.2, and 3.13, respectively. These figures imply that between 25%–40% of the overall room, restaurant, and beverage revenue of SL resort hotels are generated from AI holiday packages. In terms of AI holiday package contribution to other revenue of resort hotels, the mean contribution of AI holiday packages to other revenue of hotels is 2.00. This implies that between 10%–25% of other revenue of resort hotels is generated from AI holiday packages. In terms of AI holiday package contribution to the overall revenue of resort hotels.

AI Holiday Packages Contribution to Resort Hotel Profitability

The mean contribution of AI holiday packages to gross and net profitability of resort hotels is 3.13. This indicates that between 25%–40% of resort hotel gross and net profitability can be directly attributed to AI holiday packages.AI holiday packages contribute significantly to hotel occupancy. In 25% of the hotels which offer AI Holiday packages average annual occupancy exceeds 70%, in 50% of the hotels which offer AI holiday packages average annual occupancy level is between 60%–70%, whilst in 25% of the hotel which offers AI holiday packages average annual occupancy levels are less than 60% (Table 15.1).

In terms of resort hotel profitability, which can be directly attributed to AI holiday packages, as shown in Table 15.2, profitability directly attributed to the AI holiday package is extremely high. In 25% of the hotels surveyed, AI holiday packages

Table 15.1 Revenue from AI holiday package customers

		Hotel revenue directly attributed to AI				
		Less than 10%	25% to 40%	40% to 50%	More than 50%	Total
Type of guest AI package offered to						
	Foreign tourist only	12.5%	25.0%	37.5%	25.0%	100.0%
Total		12.5%	25.0%	37.5%	25.0%	100.0%

Source: survey data, 2020

Table 15.2 Profitability derived from AI holiday package customers

		Net profit from AI				
		Less than 10%	10% to 25%	40% to 50%	More than 50%	Total
Type of guest AI package offered to	Local and foreign tourist					
	Foreign tourist only	12.5%	25.0%	37.5%	25.0%	100.0%
Total		12.5%	25.0%	37.5%	25.0%	100.0%

Source: survey data, 2020

accounted for more than 50% of hotel profitability. In 37.5% of hotels surveyed, AI holiday packages accounted for between 40%–50% of hotel profitability, in 25% of the hotels surveyed, AI holiday packages accounted for between 25%–40% of hotel profitability, whilst in the balance, 12.5% of hotels surveyed AI holiday packages accounted for less than 10% of hotel profitability.

Influence of B&B Holiday Package on Hotel Financial Performance

Table 15.3 describes the relationship between B&B Holiday Package and Hotel Financial Performance. As shown Table 15.3, Pearson Correlation is −0.418, which implies that B&B holiday package has a weak negative relationship with resort hotel financial performance. Furthemore, P-value is 0.027. This is less than the 0.05 level of significance. Given that the p-value is less than the level of significance, the null hypothesis $H1_0$ (i.e., Bed and Breakfast holiday packages have no impact on the financial performance of resort hotels in Sri Lanka) is rejected, and the alternate hypothesis $H1_1$ (i.e., Bed and Breakfast holiday packages have an impact on the financial performance of resort hotels in Sri Lanka) is accepted.

Table 15.3 Influence of B&B holiday package on resort hotel financial performance

		Hotel financial performance	BBNET
Hotel financial performance	Pearson correlation	1	−.418*
	Sig. (2-tailed)		.027
	N	7	7
BBNET	Pearson correlation	−.418*	1
	Sig. (2-tailed)	.027	
	N	7	7

Source: survey data
*Correlation is significant at the 0.05 level (2-tailed)

Table 15.4 Influence of HB holiday package on resort hotel financial performance

		Hotel financial performance	HBNET
Hotel financial performance	Pearson correlation	1	−.447*
	Sig. (2-tailed)		.010
	N	8	8
HBNET	Pearson correlation	−.447*	1
	Sig. (2-tailed)	.010	
	N	8	8

Source: survey data
*Correlation is significant at the 0.05 level (2-tailed)

Influence of HB Holiday Package on Hotel Financial Performance

Table 15.4 describes the relationship between HB Holiday Package and Hotel Financial Performance. As presented in Table 15.4. Pearson Correlation is −0.447 that implies that HB holiday package has a weak negative relationship with resort hotel financial performance. In addition, P-value is 0.010. This is less than the 0.05 level of significance. Given that the p-value is less than the level of significance, the null hypothesis $H2_0$ (i.e., Half Board holiday packages have no impact on the financial performance of resort hotels in Sri Lanka) is rejected, and the alternate hypothesis $H2_1$ (i.e., Half Board holiday packages have an impact on the financial performance of resort hotels) is accepted.

Influence of FB Holiday Package on Hotel Financial Performance

Table 15.5 describes the relationship between FB Holiday Package and Hotel Financial Performance. As shown Table 15.5, Pearson Correlation is −0.433 that implies that FB holiday package has a weak negative relationship with resort hotel financial performance. In addition, P-value is 0.058, which is greater than the 0.05

Table 15.5 Influence of FB holiday package on resort hotel financial performance

		Hotel financial performance	FBNET
Hotel financial performance	Pearson correlation	1	−.339
	Sig. (2-tailed)		.058
	N	8	8
FBNET	Pearson correlation	−.339	1
	Sig. (2-tailed)	.058	
	N	8	8

Source: survey data

level of significance. Given that the p-value is greater than the level of significance, the null hypothesis $H3_0$ (i.e., Full Board holiday packages have no impact on the financial performance of resort hotels in Sri Lanka) is not rejected.

Influence of AI Holiday Package on Hotel Financial Performance

Table 15.6 describes the relationship between AI Holiday Package and Hotel Financial Performance. As shown Table 15.6, Pearson Correlation is 0.868, which implies that the AI holiday package has a strong positive relationship with resort hotel financial performance. In addition, P-value is 0.000. This is less than the 0.05 level of significance. Given that the p-value is less than the level of significance, the null hypothesis $H4_0$ (i.e., AI holiday packages have no impact on the financial performance of resort hotels in SL is reject) and the alternate hypothesis $H4_1$ (i.e., AI holiday packages have an impact on the financial performance of resort hotels in SL is accepted).

Effect of the Dependent Variables on the Independent Variables

Linear regression, as mentioned in the previous section of this chapter is applied to measure the impact of the independent variables on the variance in the dependent variables.

The variation in the dependent variable is described by R square in the aforementioned model (Table 15.7) and may be directly ascribed to the conceptual model in this study. As shown in Table 15.4, the R square is .817, indicating that the conceptual model in this study is responsible for 81.7% of the variation in the dependent variable (i.e., resort hotel financial performance). Furthermore, the adjusted R square in this model is 0.785, indicating that if a different sample is used with this conceptual model, the variance of the dependent variable will reduce slightly.

When all other independent variables have been statistically controlled or partialed out, partial correlation in Table 15.8 reveals the degree of variation in the

Table 15.6 Influence of AI holiday package on resort hotel financial performance

		Hotel financial performance	AINET
Hotel financial performance	Pearson correlation	1	.868**
	Sig. (2-tailed)		.000
	N	8	8
AINET	Pearson correlation	.868**	1
	Sig. (2-tailed)	.000	
	N	8	8

Source: survey data, 2020
**Correlation is significant at the 0.01 level (2-tailed)

Table 15.7 Model summary

					Change statistics				
Model	R	R square	Adjusted R square	Std. error of the estimate	R square change	F change	df1	df2	Sig. F change
1	.904[a]	.817	.785	.213	.817	25.614	4	23	.000

Source: survey data, 2020
[a]Predictors: (Constant), FBNET, AINET, BBNET, HBNET

Table 15.8 Impact of the ındependent variables on the variance in the ındependent variable

		Unstandardized coefficients		Standardized coefficients			Correlations		
Model		B	Std. error	Beta	t	Sig.	Zero-order	Partial	Part
1	(constant)	2.167	.231		9.399	.000			
	AINET	.452	.059	.757	7.620	.000	.837	.846	.680
	BBNET	−.190	.104	−.319	−1.829	.080	−.418	−.356	−.163
	HBNET	−.286	.137	−.443	−2.089	.048	−.194	−.399	−.187
	FBNET	.357	.121	.576	2.952	.007	−.062	.524	.264

Source: survey data, 2020
[a]Dependent variable: hotel financial performance

dependent variable impacted by each independent variable. When all other independent variables have been statistically controlled, 71.57% of the variation in the dependent variable can be ascribed to the independent variable AI vacation package, according to the partial correlation in Table 15.5. When all other independent variables have been statistically controlled, the independent variable B&B holiday package may be ascribed to 12.67% of the variation in the dependent variable.

When all other independent variables have been statistically controlled, the independent variable HB holiday package may be ascribed to 15.92% of the variation in the dependent variable. When all other independent variables are taken into account, the independent variable FB holiday package accounts for 27.45% of the variation in the dependent variable.

Table 15.8 shows how part correlation defines the variance in the dependent variable that can be ascribed to each independent variable separately. On the basis of

Table 15.9 Unique variance in the dependent variable on account of each dependent variable

Independent variable	Unique contribution to the variance in the dependent variable
AI holiday package	46.24%
B&B holiday package	2.65%
FB holiday package	3.49%
FB holiday package	6.96%

Source: survey data, 2020

each independent variable, Table 15.9 displays the unique variance independent variable.

Based on results (Table 15.8), the raw regression equation of the conceptual model is as follows:

Resort Hotel Financial Performance = 2.167 + 0.452(AI Holiday Package) – 0.190 (B&B Holiday Package) – 0.286(HB Holiday Package) + 0.357(FB Holiday Package).

The standardized regression equation of the conceptual model is as follows:

Resort Hotel Financial Performance = 2.167 + 0.757(AI Holiday Package) – 0.319 (B&B Holiday Package) – 0.443(HB Holiday Package) + 0.576(FB Holiday Package).

Conclusion, Practical İmplications, and Recommendations

The main goal of this study was to measure the contribution of AI concept to resort hotels' profitability in SL. Findings from the data analysis indicate that the AI holiday concept is the most significant contributor to resort hotel profitability. According to the study AI holiday concept contribution to the gross and net profitability of resort hotels is significantly higher than the contribution of other holiday packages to these hotels' profitability. The high contribution of the AI concept to hotel profitability can be attributed to a number of factors, namely; Volume, Consumption, Cost.

Another key objective in this study was to measure AI concept contribution to SL resort hotels' revenue. According to the study AI concept contributes significantly to the revenue of SL resort hotels in comparison to the contribution o0f other holiday packages marketed by resort hotels.AI concept is the main contributor to all key revenue categories of resort hotels. AI concept's unique contribution to room, restaurant, beverage, and total revenue of resort hotels is between 25%–40% and is higher than the 10%–25% unique contribution made to these revenue streams of resort hotels by B&B, HB, and FB packages. In terms of other revenue, AI concept unique contribution to the other revenue of resort hotels is between 10%–25%, which is also higher than the unique contribution made by the other holiday packages (less than 10%) marketed by hotels.

Whilst the AI concept is extremely popular amongst European guest and accounts for a high percentage of occupancy in resort hotels, Asian guest on AI holiday packages in SL resort hotels is extremely low. The low percentage of Asian guests on AI holiday packages can be attributed to a number of reasons, namely: Demand, Package Structure, Marketing, and Culture.

AI holiday package contributes the most to occupancy, revenue, and profitability of these hotels. Of the other three holiday packages (i.e., B&B, HB and FB, the next holiday package which contributes greatly to resort hotel occupancy, revenue, and profitability is the HB package, followed by the FB package and the B&B package).

Beach resort hotels in SL benefit greatly from AI holiday packages in terms of occupancy, revenue, and profitability. Although these packages have a little positive impact on local communities because most if not all guest vacation requirements are purchased directly from the hotel, resort hotels will have to continue marketing these packages to ensure their profitability and long-term viability. Any decision on the part of the hotel owners or government authorities to reduce or restrict the marketing of these packages in SL on account of the limited benefits derived by local communities and the local economy from these packages will not only have a drastic impact on these resort hotels, it could also result in a decline in tourist arrivals, particularly tourists from Germany, the United Kingdom, and France that in turn will have a negative impact not only on the tourism industry but also on the economy of the country.

In order to improve occupancy, revenue, and profitability in the resort hotels, the following should be implemented;

Focused Marketing Strategy

Given the high demand for the AI holiday package amongst European tourists and the preference of Asian tourists for other holiday packages, particularly the HB holiday package, resort hotels should develop focused marketing strategies for each customer segment.

Offer AI Holiday Package to Local Tourist

While only 28.57% of the resort hotels offering AI holiday packages market these packages to local tourist. Findings from the data analysis indicate that offering this package to local tourist significantly boost occupancy, revenue, and profitability of these hotels.

Elimination of B&B Holiday Package

Given the B&B holiday package's poor contribution to hotel occupancy, revenue and profitability, the resort should eliminate this holiday package.

Development of AI Package for Asian Tourist

Given the high percentage of Asian tourists visiting the country, resort hotels should focus on developing an AI package which meets the needs of Asian tourist.

Development of AI Package for Local Tourist

The rapid economic development, growing per-capita income of the local population will result in strong growth in local tourism over the next 5–10 years.

Seasonal AI Holiday Packages

Given the significant difference in occupancy during peak season and off-season, resorts and hotels should look towards offering AI packages based on seasonality.

Collaboration with Local Businesses and Communities

One of the biggest criticisms of the AI holiday package it the limited benefit to local businesses and communities from this package.

This study measured the contribution of AI concept to occupancy, revenue, and profitability of SL resort hotels. Further research can be conduct on the impact the AI holiday package has on other key stakeholders such as guests, employees, suppliers, local businesses and communities, etc. Also, while this study focuses on revenue and profitability, it does not consider economic leakage which is a key issue in the tourism industry, which has a significant impact on the profitability of hotels. Further studies should be conducted to ascertain the impact of economic leakage on resort hotels' profitability in the country. Also, further studies should be conducted on what are the key factors that influence tourists to select AI holiday packages. The average PPPN spends of guests on different holiday packages in SL.

References

Alegre, J. & Pou, L. I. (2006, June). *All-inclusive packages: An analysis of their economic implications for the case of the Balearic Islands' (in Spanish)* (DEA Working Paper No 18).

Anderson, L. (2006). Analytic autoethnography. *Journal of Contemporary Ethnography. Sage Journals*. https://doi.org/10.1177/0891241605280449

Basala, S., & Klenosky, D. (2011). Travel style preferences for visiting a novel destination: A conjoint investigation across the novelty-familiarity continuum. *Journal of Travel Research, 40*(2), 172–182.

Clark, J. (2000). *The art of all-inclusive vacation deals: A Caribbean comparison of resorts vs.cruises*. USA Today.

Enoch, Y. (1996). Contents of tour packages: A cross-cultural comparison. *Annals of Tourism Research, 23*, 599–616.

Falzon, J. (2003). The competitive position of Mediterranean counties in tourism: Evidence from the Thomson brochure. *Tourism Management, 33*(5), 1080–1092.

Frechtling, D. (2006). An assessment of visitor expenditure methods and modals. *Journal of Travel Research, 45*, 26–35.

Guiltinan, J. (1987). The price bundling of service. *Journal of Marketing, 51*, 74–85.

Issa, J., & Jayawardena, C. (2003). The "all-inclusive" concept in the Caribbean. *International Journal of Contemporary Hospitality Management, 15*(3), 167–171.

Jørgensen, F., & Solvoll, G. (1996). case. *Tourism Management, 17*(1), 17–24.

Mbaiwa, J. (2005). Enclave tourism and its socio-economic impacts in the Okavango Delta. *Botswana Tourism Management, 26*, 157–172.

Pangaea Network. (2014). *Consultats to the travel industry*. Pangaea Network Ltd.

Saunders et al. (2007). Research methods for business students. *Financial Times/Business & Economics*. Prentice Hall.

Sheldon, P., & Mak, J. (1987). The demand for package tours: A mode choice modal. *Journal of Travel Research, 25*(3), 13–17.

Titanont, A., & Chantradoan, N. (2004). *Packaging phuket: Focusing on price bundling, Travelling trends of the Swedish target market*. Master thesi. Göteborg University.

Tourism Intelligence International. (2000). *How the British will travel 2005*. Tourism Intelligence International.

Wickens, E. (2002). The sacred and the profane. *Annals of Tourism Research, 29*(3), 34–51.

Wong, C., & Kwong, W. (2004). Outbound tourists' selection criteria for choosing all-inclusive package tours. *Tourism Management, 25*(5), 581–592.

WTO. (2004). *Clarifying the treatment of travel agency, tour operator, travel agency services and package Tours in SNA, balance of payments and TSA and their mutual relationship*. WTO.

Chapter 16
Impact of Tourism Accommodation Sector Awareness on Responsible Tourism in Destination Development: A Case Study of Hambantota District

D. A. C. Suranga Silva, Mubarak Kaldeen, and Chinthana Duminduhewa

Abstract The main objective of this study was to analyze the tourism accommodation sector in the Hambantota district awareness of responsible tourism and its benefits, to understand the difficulties and challenges faced by businesses in this sector to adopt responsible tourism practices, and to identify the non-financial and financial benefits derived by these businesses on account of responsible tourism. For the purpose of analyzing the research problem, a total of 32 tourism accommodation institutions from the Hambantota were selected for the study. The main data collection instrument employed in the study was a semi-structured questionnaire. Simple statistical methods such as mean, median, and percentages have been used to analyze the quantitative data obtained from the survey, whilst thematic analysis has been used to analyze the qualitative data obtained from the questionnaire. Findings from the study indicate that all businesses in the tourism accommodation sector of the Hambantota district are aware of the concept of responsible tourism. In terms of businesses in the sector awareness of the benefits associated with responsible tourism, 93.7% of the businesses were aware of the financial benefits, 75% were aware of the social benefits, whilst all businesses in this sector were aware of the environmental benefits associated with this concept. In terms of challenges faced in implementing responsible tourism practices, findings from the study revealed that 43.8% of the businesses in this sector faced challenges and difficulties in implementing responsible tourism practices. Key challenges faced in this regard were lack of capital to implement these practices, limited support from guests for these

D. A. C. Suranga Silva
Department of Economics, Faculty of Art, University of Colombo, Colombo, Sri Lanka

M. Kaldeen (✉)
Department of Marketing Management, Faculty of Management and Commerce, South Eastern University of Sri Lanka, Oluvil, Sri Lanka
e-mail: kmmubarak@seu.ac.lk

C. Duminduhewa
University of Colombo, Colombo, Sri Lanka

A. Hassan, A. Sharma (eds.), *Overtourism, Technology Solutions and Decimated Destinations*, https://doi.org/10.1007/978-981-16-2474-2_16

practices, and lack of support from policymakers and local authorities in this regard. In terms of financial and non-financial profits derived by businesses in the tourism accommodation industry on account of responsible tourism practices, findings from the data analysis revealed that only 37.5% of the businesses had derived financial benefits, whilst only 43.8% of the businesses have derived non-financial benefits.

Keywords Responsible tourism · Responsible tourism awareness · Sri Lanka tourism · Destination developments

Introduction

The Hambantota district, which borders the sacred city of Kataragama and Udawalawa National park, is situated 200 kilometers from Colombo. The district has a number of tourist attractions ranging from the Yala and Bundala national parks to the 'Hummanaya' blowhole, Situlpauwa Rock, Mulkirigala Rock Temple, to Tangolle, Kirinda, and other pristine beaches. Together with a mixed climate, these tourism attractions have made Hambantota a popular destination for both local and international tourists. The popularity of Hambantota as a tourism destination has resulted in the district currently having approximately 41 tourism institutions operating in the district (International Finance Corporation, 2013). In this study, institutions in the tourism accommodation sector are categorized as hotels, boutique hotels, guesthouses, and other institutions that provide accommodation facilities for tourists' (i.e., homestay units, bungalows, rented homes, and apartments). But according to the booking.com, there are 641 properties in Hambantota District as of March 2020, and reasons to visit are nature, safari, and beaches. (www.booking.com). The number of hotels, guest houses, and other tourism institutions in the district is projected to increase exponentially over the next decade on account of recent infrastructural developments, which have made Hambantota easily accessible from any part of the country like the Southern expressway, Maththala airport, and Hambantota port, etc. Also, initiatives by the government, district's business, and tourism chambers to promote tourism in the district. According to these developments, Hambantota District will be the second capital in Sri Lanka in the near future.

The main objectives of this study are: first, to analyze tourism accommodation sector awareness of responsible tourism and its benefits in the Hambantota district. Second, to identify the challenges and difficulties faced by the tourism accommodation sector in following responsible tourism concept in the Hambantota district. Finally, to assess the financial and non-financial benefits derived by the tourism accommodation through the implementation of responsible tourism practices in Hambantota District.

Literature Review

Concept of Responsible Tourism

Responsible tourism can be defined as establishing better locations for individuals to visit and decide to live the place (Camilleri, 2016). Responsible tourism, which occurs in a number of different forms, is based on two key codes that outline the main principles of responsible tourism. These two codes are namely the Globe Code of Ethics for Responsible Tourism developed by the United Nations World Tourism Organization (UNWTO, 2015) and the Principles of Responsible Tourism outlined in the Declaration of Cape Town in 2002 (Cape Town Declaration, 2002). Of these two codes, the latter is considered extremely important as it has been ratified by all stakeholders in the global tourism industry. The key principles of responsible tourism outlined in the Cape Town Declaration include the following; (http://responsibletourismpartnership.org)

- Reduce the negative social, economic, and environmental impacts of tourism,
- Generate greater economic benefits for local communities, enhances working conditions in the industry, and improves access to the industry,
- Contribute to the conservation of the local environment and cultures,
- Provide an enhanced experience for tourists via meaningful connections with local communities,
- Facilitate international tourists better understanding of environmental, social, and economic issues faced by local communities,
- Promote respect between tourists and the local people, and contribute to enhancing the confidence and pride of local communities.

Goodwin (2011) stated that responsible tourism is really about accepting accountability for sustainable development through tourism activities, recognizing and addressing economic, social, and environmental problems that matter locally. Further responsible tourism is not a separate isolated form of tourism, as it connects with all forms of human activity and evolves according to the evolving needs and attitudes of all stakeholders in the industry.

Impact of Responsible tourism on Consumers

Research studies indicate that consumer demand for responsible tourism is increasing. A research study on responsible tourism conducted by the UNWTO in 2012 reported that "tourists' choices are increasingly influenced by sustainability considerations" (UNWTO, 2012), whilst Blackburne (2012) discloses that sustainable tourism is becoming more generally recognized concept in the world similarly

UNESCO, the United Nations Education, Scientific and Cultural Organization are trusting on the sustainable tourism will proceed from 'alternate' to 'mainstream' tourism within a decade. Tip (nd) the founder and CEO of G Adventures attributes his company's excellent performance during the global economic crisis (in contrast to most other travel companies which reported heavy losses, G adventures increased its revenue by 42% (approximately US$ 150 million) during the global economic crisis) to the fact that its business model was aligned with sustainability. In addition, a wire survey conducted by Bricker in 2012 reported that 66% of consumers from around the world expressed a strong preference to buy products and services from organizations that implemented sustainable practices that focus on giving back to society (Bricker, 2013).

This consumer shift towards responsible tourism is also reflected in the changing behavior of consumers. A survey of U.S. holidaymakers by Statistica in 2012 found that a high percentage of hotel guest turn off the air conditioning (72%) and lights (96%) when not in their hotel rooms, participate in the hotel's towel and linen reuse programs, and use the hotel's recycling facilities. Whilst this survey found that most U.S. travelers are focused on sustainability and protecting the environment when on holiday, it also found that most travelers (more than 60%) were not willing to pay extra to vacation in an environmentally friendly hotel (Statistica, 2014).

Travelers' shift towards environmental and social sustainability is also reflected in the holiday destinations selected. Research conducted by the U.S. Department of Commerce (2012) reported that, "a significant number of international travelers seek out nature and culture-based experiences such as visiting historic sites (40% of overseas travelers), cultural sites (23%), and national parks (20%)". This consumer focus on the environment when traveling is also supported by data from TripAdvisor.com, which found that 66% of travelers always take into consideration the environment when selecting hotels, transportation and meals".

Impact of Responsible tourism on Businesses

Research findings indicate that organizations in the tourism industry are becoming increasingly focused on implementing responsible tourism practices. In this regard, a growing number of firms are adopting good environmental and social practices, assessing ways in which they could make their supply chains greener, and seeking certification for their current sustainable practices. One of the key reasons for this increased business focused on adopting responsible tourism practices can be attributed to customer demand. Customers are increasingly shifting towards hotels and other tourism institutions that have adopted responsible tourism practices. The ABTA News (2013) noticed an increasing number of customers, especially young people, who are willing to pay a premium for a more sustainable holiday. Hence it makes economic sense for businesses to have a well-thought-out sustainability strategy that sells customers the positive benefits of sustainability and taps into this growing market.

In addition to customer demand, other key reasons for the increasing adoption of responsible tourism is on account of the economic and strategic benefits that can be derived through these practices. For example, Marriott Hotel was able to reduce water and energy consumption by approximately 25% by adopting the 'LEED Volume Program' of the U.S. Green Building Council (Marriott International, 2011). In addition, Hilton Hotels Worldwide, through the adoption of responsible tourism practices, reduced carbon output by 20.2%, energy consumption by 13.6%, water use by 13.1%, and waste output by 23.1% over the period 2009–2013, whilst TUI Travel by fitting fuel blended winglets on their aircraft were able to reduce fuel consumption by approximately 5% (TUI Travels PLC, 2013).

Tourism destinations Influence on Responsible Tourism

Government, local communities, and business leaders in a growing number of tourism destinations are increasingly realizing the importance of protecting their natural habitats, cultural heritage, scenic appeal, historical appeal, etc., to ensure sustainability and long term viability of their tourism industries. As per Lorimer (2006), tourism can be used as a conserving tool of wildness and heritage areas. It raises the awareness of the moral and environmental value of conserving biodiversity for both travelers and locals. Whilst, Bricker (2013) stated that Without a stable world where economies operate, ecosystem services are preserved, human rights are upheld, life quality is enhanced, long-term services will not really exist.

The massive economic benefits derived by countries through their tourism industry is another key factor driving government, local communities, and business leaders towards adopting responsible tourism practices. The global tourism industry is one of the largest industries in the world, contributing approximately US$ 7.6 Trillion in revenue (Statistica, 2014), with a large number of countries depending on tourism revenue as a key source of foreign exchange income. For example, tourism revenue is the third-largest contribute to the foreign exchange income of Sri Lanka (SLTDA, 2015). Failure to protect the tourism destination via the adoption of responsible tourism practices will have a massive impact on socio-economic development and prosperity in the tourism destination market.

Global warming, extreme weather, unprecedented bio-diversity losses, water scarcity, etc., has also forced tourism destinations to adopt responsible tourism practices in order to protect and ensure long-term viability of the destination. Travel Foundation and Forum for the Future (2012) highlighted the impacts of climate change, extreme weather, water shortages, increasing rates of loss of biodiversity, disease, and rising poverty and inequality are challenging factors for a destination. Further, Sustainable tourism tackles the challenges of protecting the destination while improving brand equity, increasing revenues, cutting costs, and enhancing strategic positioning to attract and retain customers and employ the best talent.

Methodology

Population and Sample

The study was conducted in Hambantota District of Sri Lanka. Population for this study consists of all tourism institutions. Approximately 41 Institutions (International Finance Corporation, 2013) operating in this district were considered as a population to this study the breakup of the population is indicated in Table 16.1. Responses of 32 tourism organizations were collected using semi-structured questionnaires. The questionnaire was developed into three sections as Awareness of responsible tourism and its benefits, Awareness of responsible tourism for community benefits, and Awareness of responsible tourism for environmental benefits. Each section consists of four questions.

The analysis was carried out by stratified randomly selected 32 tourism accommodation institutions located in Hambantota (Table 16.2).

Analysis and Findings

As mentioned in the introduction sample for this study consisted of 32 tourism accommodations from the Hambantota district. All semi-structured questionnaires distributed to the population was properly filled and returned to the researcher with 32 questionnaires with answers. This resulted in data from the sample being used in the study.

Awareness of Responsible tourism and its Benefits

Tourism accommodation sector awareness of the concept of responsible tourism is shown in Table 16.3. As shown in this Table, hotels and other tourism accommodation types are aware of the concept of responsible tourism, whilst 93.7% of the accommodations are aware of this concept.

As shown in this Table, 75% of all respondents are aware of the financial benefits associated with responsible tourism. In terms of the different accommodation types awareness of the financial benefits of responsible tourism, 33.3% of the guesthouses and 66.7% of the hotels are aware of the financial benefits associated with

Table 16.1 Tourism Accommodation Sector in the Hambantota District (source: International Finance Corporation, 2013)

Type of Institution	Hotels	Guesthouses	Other	Total
Number	23	16	2	41
% of population	56%	39%	5%	100%

Table 16.2 Sample (source: International Finance Corporation, 2013)

Type of Institution	Number	% of Population	Sample
Hotels	23	46.3%	19
Guesthouses	16	29.2%	12
Other	02	2.4%	01
Total	41		32

Table 16.3 Survey respondent awareness of responsible tourism and its benefits (source: survey data)

Questions	Answers		Accommodation type			Total	Total %
			Guesthouse	Hotels	Other		
Awareness of responsible tourism	Yes	Count	10	18	2	30	93.7%
		Percentages	33.3%	60.0%	6.7%	100.0%	
	No	Count	2	0	0	2	6.3%
		Percentages	100.0%	0.0%	0.0%	100.0%	
Awareness of financial benefits of responsible tourism	Yes	Count	8	16	0	24	75.0%
		Percentages	33.3%	66.7%	0.0%	100.0%	
	No	Count	4	2	2	8	25%
		Percentages	50.0%	25.0%	25.0%	100.0%	
Awareness of the social benefits of responsible tourism	Yes	Count	10	18	2	30	93.7%
		Percentages	33.3%	60.0%	6.7%	100.0%	
	No	Count	2	0	0	2	6.3%
		Percentages	100.0%	0.0%	0.0%	100.0%	
Awareness of environmental benefits of responsible tourism	Yes	Count	12	18	2	32	100%
		Percentages	37.5%	56.3%	6.3%	100.0%	
	No	Count					
		Percentages					

responsible tourism. What is interesting from this Table is that 25% of the other accommodation category, 50% of the guesthouse category, and 25% of hotels are not aware of the financial benefits associated with responsible tourism.

Table 16.3 shows the tourism accommodation sector's awareness of the social benefits of responsible tourism. As shown in this Table, 93.7% of the tourism accommodation sector is aware of the social benefits of responsible tourism. In terms of the individual accommodation types, all hotels and other accommodation type categories are aware of the social benefits associated with responsible tourism, whilst 60% of the hotels are aware of the social benefits of responsible tourism.

Tourism accommodation sector awareness of the environmental benefits of responsible tourism is shown in the above Table. As shown in this Table, all respondents are aware of the environmental benefits associated with responsible tourism.

Awareness of responsible tourism for community benefits.

Tourism accommodation sector investment in local communities in which they operate is shown in Table 16.4. As shown in this Table, 93.7% of the institutions in the sector invest in the local communities in which they operate. In terms of the

Table 16.4 survey respondent awareness of responsible tourism for community benefits (source: Survey data)

Questions	Answers		Accommodation type			Total	Total %
			Guesthouse	Hotels	Other		
Investment in the local communities	Yes	Count	10	18	2	30	93.7%
		Percentages	33.3%	60.0%	6.7%	100.0%	
	No	Count	2	0	0	2	6.3%
		Percentages	100.0%	0.0%	0.0%	100.0%	
Have programs to minimize resource utilization	Yes	Count	12	18	2	32	100.0%
		Percentages	37.5%	56.3%	6.3%	100.0%	
	No	Count					
		Percentages					
Extent to which individuals from local communities are hired	Yes	Count	12	12	2	26	81.3%
		Percentages	46.2%	46.2%	7.6%	100.0%	
	No	Count	0	6	0	6	18.7%
		Percentages	0	100%	0	100%	
Extent to which resources are sourced from local communities	Yes	Count	12	12	2	26	81.3%
		Percentages	46.1%	46.1%	7.8%	100.0%	
	No	Count	0	6	0	6	18.7%
		Percentages	0	100.0%	0	100.0%	

individual accommodation types, guesthouses and the other accommodation category invest in local communities, whilst 60% of hotels invest in the local communities in which they operate.

Table 16.4 shows tourism accommodation sector strategies to minimize resource utilization. As shown in this Table, all accommodation categories have programs and strategies to conserve the resources which they utilize frequently.

The extent to which the tourism accommodation sector hires individuals from local communities is shown in the above Table as shown all accommodation categories hire individuals from the local communities in which they operate.

Awareness of Responsible Tourism for Environments Benefits

The tourism accommodation sector's willingness to take responsibility for the environmental damage caused by their activities is shown in Table. As shown in this Table, 93.7% of the tourism accommodation sector is willing to take responsibility for the environmental damage caused by its activities. In terms of the individual accommodation categories, whilst all hotels and businesses in the other accommodation category are willing to take responsibility for the environmental damage caused by activities, only 60% of the hotels are willing to take responsibility for the environmental damage caused by their activities.

Table 16.5 survey respondent awareness of responsible tourism for environments benefits (source: Survey data)

Questions	Answers		Accommodation type			Total	Total %
			Guesthouse	Hotels	Other	Total	
Willingness to take responsibility for the environmental damage caused by their activities	Yes	Count	10	18	2	30	93.7%
		Percentages	33.3%	60.0%	6.7%	100.0%	
	No	Count	2	0	0	2	6.3%
		Percentages	100.0%	0.0%	0.0%	100.0%	
Conservation/reduction methods employed in the usage of water, energy and waste	Yes	Count	8	14	2	24	75%
		Percentages	33.3%	58.3%	8.3%	100.0%	
	No	Count	4	4	0	8	25%
		Percentages	50.0%	50.0%	0.0%	100.0%	
Information and training provided to staff on how to be environmentally friendly	Yes	Count	12	18	2	32	100.0%
		Percentages	37.5%	56.2%	6.3%	100.0%	
	No	Count					
		Percentages					
Implementation of an environment conservation policy	Yes	Count	10	14	0	24	75%
		Percentages	41.7%	58.3%	0%	100.0%	
	No	Count	2	4	2	8	25%
		Percentages	25%	50%	25%	100.0%	

Table 16.5 shows the conservation/reduction methods employed by the tourism accommodation sector in the use of ware, energy, and waste disposal. As shown in this Table, 75% of the accommodation sector has implemented conservation/reduction methods in the usage of water, energy, and waste disposal. In terms of the individual accommodation categories, 33.3% of the guesthouses, 58.4% of the hotels, and 8.3% of the businesses in other accommodation categories have implemented strategies to conserve/reduce water, energy, and waste disposal.

Table 16.5 shows the extent to which the tourism accommodation sector has provided employees with training and information on how to be environmentally friendly. All accommodation categories provide employees with training and information on how to be environmentally friendly.

According to the Table, the extent to which businesses in the tourism accommodation sector have implemented an environment conservation policy. As shown in this Table, 75% of the businesses in this sector have implemented an environment conservation policy. In terms of the individual accommodation categories, as shown in this able 41.7% of the guesthouses and 58.3% of the hotels have implemented an environment conservation policy.

Challenges and difficulties faced by the tourism accommodation sector in implementing responsible tourism practices.

Table 16.6 shows the challenges faced by businesses in the tourism accommodation sector in implementing responsible tourism practices. As shown in this Table, 43.8% of the businesses in this sector reported that they faced challenges in implementing responsible tourism practices, whilst the rest of the businesses did not face

Table 16.6 Challenges faced by the tourism accommodation sector in implementing responsible tourism practices (source: Survey data)

			Accommodation type			Total
			Guesthouse	Hotel	Other	
Challenges in implementing responsible tourism practices	Yes	Count	4	10	0	14
		% within Q17	28.6%	71.4%	0.0%	100.0%
		% within Q20	33.3%	55.6%	0.0%	43.8%
	No	Count	8	8	2	18
		% within Q17	44.4%	44.4%	11.1%	100.0%
		% within Q20	66.7%	44.4%	100.0%	56.3%
Total		Count	12	18	2	32
		% within Q17	37.5%	56.3%	6.3%	100.0%
		% within Q20	100.0%	100.0%	100.0%	100.0%

any challenges in implementing these practices. In terms of the individual accommodation categories, 100% of the businesses in the other accommodation category, 66.7% of the guesthouses, and 44.4% of the hotels reported they did not face any challenges in this regard, whilst 33.3% of the guesthouses and 55.6% of the hotels reported that they faced challenges in implementing responsible tourism practices.

Some of the challenges faced in this regard include difficulties in raising capital to implement these practices. Tourists do not understand the concept, lack of support and interest from policymakers and authorities in this regard, and difficulties in convincing local clients about the benefits of responsible tourism.

Financial Benefits Derived from the tourism Accommodation Sector from Responsible Tourism

Financial benefits derived by businesses in the tourism accommodation sector from responsible tourism are shown in Table 16.7. As shown in this Table, only 37.5% of the businesses in the accommodation sector reported deriving financial benefits on account of implementing responsible tourism practices, whilst 62.5% of businesses reported that they do not derive financial benefits from the implementation of responsible tourism practices. In terms of the individual accommodation types, 33.3% of the guesthouses and 44.4% of the hotels reported deriving financial benefits on account of implementing responsible tourism practices, whilst 66.7% of the guesthouses, 55.6% of the hotels, and 100% of the businesses in the other accommodation category said they do not derive benefits on account of implementing responsible tourism practices. Key financial benefits reported by businesses included increased revenue and profits, reduction in utility costs, and waste disposal cost.

Table 16.7 Financial benefits derived by the tourism accommodation sector from responsible tourism (source: Survey data)

			Accommodation type			Total
			Guesthouse	Hotel	Other	
Financial benefits derived by the tourism accommodation sector	Yes	Count	4	8	0	12
		% within Q18	33.3%	66.7%	0.0%	100.0%
		% within Q20	33.3%	44.4%	0.0%	37.5%
	No	Count	8	10	2	20
		% within Q18	40.0%	50.0%	10.0%	100.0%
		% within Q20	66.7%	55.6%	100.0%	62.5%
Total		Count	12	18	2	32
		% within Q18	37.5%	56.3%	6.3%	100.0%
		% within Q20	100.0%	100.0%	100.0%	100.0%

Table 16.8 Non-financial benefits derived by the tourism accommodation sector from responsible tourism (source: Survey data)

			Accommodation type			Total
			Guesthouse	Hotel	Other	
Non-financial benefits derived by the tourism accommodation sector	Yes	Count	6	8	0	14
		% within Q19	42.9%	57.1%	0.0%	100.0%
		% within Q20	50.0%	44.4%	0.0%	43.8%
	No	Count	6	10	2	18
		% within Q19	33.3%	55.6%	11.1%	100.0%
		% within Q20	50.0%	55.6%	100.0%	56.3%
Total		Count	12	18	2	32
		% within Q19	37.5%	56.3%	6.3%	100.0%
		% within Q20	100.0%	100.0%	100.0%	100.0%

Non-financial Benefits Derived by the tourism Accommodation Sector from Responsible Tourism

Non-financial benefits derived by businesses in the tourism accommodation sector on account of responsible tourism are shown in Table 16.8. As shown in this Table, 43.8% of the businesses in the tourism accommodation sector reported deriving non-financial benefits on account of implementing responsible tourism practices, whilst the rest of the businesses in the sector have not derived any benefits in this regard. In terms of the individual accommodation categories, 50% of the guesthouses and 44.45 of the hotels reported deriving non-financial benefits on account of implementing responsible tourism practices, whist 50% the guesthouses, 55.6% of the hotels, and 100% of businesses in the other accommodation category reported zero non-financial benefits on account of implementing responsible tourism practices. Key non-financial benefits gained by businesses in the tourism accommodation sector on account of implementing responsible tourism practices included goodwill, positive impact on brand image, and positive publicity relating to the establishment.

Conclusion and Recommendations

Conclusion

The main objective of this study was to analyze the tourism accommodation sector in the Hambantota district awareness of responsible tourism and its benefits. Findings from the study indicate that all businesses in the tourism accommodation sector of the Hambantota district are aware of the concept of responsible tourism. In terms of businesses in the sector awareness of the benefits associated with responsible tourism, 93.7% of the businesses were aware of the financial benefits associated with this concept, 75% were aware of the social benefits associated with this concept, whilst all businesses in this sector were aware of the environmental benefits associated with this concept.

The second objective of this study was to identify the difficulties and challenges faced by businesses in the tourism accommodation sector of the Hambantota district in implementing responsible tourism practices. Findings from the study revealed that 43.8% of the businesses in this sector faced challenges and difficulties in implementing responsible tourism practices. Key challenges faced in this regard were lack of capital to implement these practices, limited support from guests for these practices, and lack of support from policymakers and local authorities in this regard.

This study's third and final objective was to identify the key financial and non-financial benefits derived by businesses in the tourism accommodation sector of the Hambantota district. Findings from the data analysis revealed that only 37.5% of the businesses had derived financial benefits, whilst only 43.8% of the businesses have derived non-financial benefits. The main financial benefits derived by businesses in this sector on account of implementing responsible tourism practices are increased revenue and profitability and a reduction in utility and waste disposal costs, whilst the main non-financial benefits derived were goodwill and positive publicity pertaining to the establishment.

Recommendations

To increase awareness and implementation of responsible tourism and increase the benefits derived from this concept, tourism authorities, and policymakers in the Hambantota district.

Policymakers and tourism Authorities Should Show More Support and Interest in this Concept

One of the key challenges faced by businesses in the tourism accommodation sector in implementing responsible tourism practices is the limited interest shown by local tourism and government authorities in this regard. Tourism and government authorities should immediately address this shortcoming. Awareness campaigns should be conducted by local authorities in this regard, a forum should be established in which all stakeholders can share and learn best responsible tourism practices, and incentives in the form of tax deductions should be provided for companies that proactively adopted responsible tourism practices.

Financial Constraints in Implementing Responsible tourism Practices

Another key difficulty faced by businesses in the tourism accommodation sector in implementing responsible tourism practices is the lack of capital. In order to address this, issue local government authorities should negotiate with banks in the Hambantota district to provide capital to businesses that face difficulties in this regard.

Implementation of an Environmental Conservation Policy

Findings from the data analysis indicate that 25% of the businesses in the tourism accommodation sector of the Hambantota district do not have an environmental conservation policy. This shortcoming should be immediately addressed, and all companies should be encouraged to establish an environment conservation policy that includes all stakeholders.

Investigate the Reasons for the Low Financial and Non-financial Benefits Derived by Businesses Implementing Responsible tourism Practices

Findings from the study indicate that only an extremely low percentage of businesses in the tourism accommodation sector have derived financial and non-financial benefits on account of implementing responsible tourism practices. This low figure should be immediately investigated as it may indicate a great problem such as the incorrect implementation of responsible tourism practices, implementing practices that are not responsible tourism practices, implementing tourism practices that are not compatible with the organization, etc.

Implementation of Strategies to Reduce Water, Energy Usage, and Waste Disposal

Findings from the study indicate that 33.3% of the guesthouses and 22.2% of the hotels in the Hambantota district do not have strategies to reduce water and energy usage and waste disposal. In order to maximize the benefits from responsible tourism, these businesses should immediately implement strategies in this regard.

Enhance tourism Stakeholders' Awareness about Responsible tourism Concept

According to the research finding tourism, stakeholders do not have the proper knowledge and understand about the responsible tourism concept, its benefits, and future demand. So it is better to conduct training and awareness about the concept and gains the benefits for Sri Lankan tourism by developing responsible tourism.

References

ABTA News. (2013). *ABTA reveals new research on responsible travel*. Retrieved from: http://abta.com/news-and-views/press-zone/abta-reveals-new-research-on-sustainable-holidays. Accessed 2 Jan 2017.

Blackburne, A. (2012). *The Guide to Sustainable Tourism-2013: Annual Report*. Retrieved from: http://blueandgreentomorrow.com/wp-content/uploads/2013/02/BGT-Guide-to-Sustainable-Tourism-10MB1.pdf. Accessed 12 Nov 2020.

Bricker, K. S. (2013). *Trends and issues for ecotourism & sustainable tourism*. Retrieved from: https://sustainabledevelopment.un.org/content/documents/4099Presentation%200.1%20Kelly%20Bricker%20-%20full%20presentation.pdf. Accessed 22 Nov 2020.

Camilleri, M. A. (2016). Responsible tourism that creates shared value among stakeholders. *Tourism Planning & Development, 13*(2), 219–235.

Cape Town Declaration. (2002). *Cape Town Conference on Responsible Tourism in Destinations*. Retrieved from: https://www.capetown.gov.za/en/tourism/Documents/Responsible%20Tourism/Toruism_RT_2002_Cape_Town_Declaration.pdf. Accessed 22 Nov 2020.

Goodwin, H. (2011). *Taking responsibility for tourism*. Goodfellow Publishers Limited.

International Finance Corporation. (2013). *Annual report-2013*. Retrieved from: https://www.ifc.org/wps/wcm/connect/61766bbb-8986-451b-9ca4-57a9466f4e81/AR2013_Full_Report.pdf?MOD=AJPERES&CVID=k5ruszm. Accessed 28 Nov 2020.

Lorimer, K. (2006). *Code green: Experiences of a lifetime*. Lonely Planet.

Marriott International. (2011). *Selected Historical Financial Data: Annual Report-2011*. Retrieved from: https://marriott.gcs-web.com/static-files/46aef136-7335-4f34-89d6-0f07e3e63197. Accessed 26 Nov 2020.

Sri Lanka Tourism Development Authority. (SLTDA). (2015). *Annual Statistical Report, 2015*. Retrieved from: http://www.sltda.lk/sites/default/files/Tourist_Board_Annual_Report_2015.pdf. Accessed 19 Nov 2020.

Statistica. (2014). *Statistics and Facts on Green Tourism*. Retrieved from: http://www.statista.com/topics/1916/green-tourism/. Accessed 2 Nov 2020.

Travel Foundation and Forum for the Future. (2012). *Survival of the Fittest: Sustainable Tourism Means Business*. Retrieved from: http://www.thetravelfoundation.org.uk/images/media/Whitepaper_Survival_of_the_Fittest_2012.pdf. Accessed 2 Nov 2020.

TUI Travels PLC. (2013). *Strategic report-2013: Annual report*. Retrieved from: https://www.tantur.com.tr/pdf/TUI_ARA2013_Interactive.pdf. Accessed 2 Nov 2020.

U.S. Department of Commerce. (2012). *National Travel & Tourism Strategy – 2012*. Retrieved from: http://travel.trade.gov/pdf/national-travel-and-tourism-strategy.pdf. Accessed 2 Nov 2020.

UNWTO. (2012). *Tourism in the Green Economy*. Retrieved from: http://www.unep.org/greeneconomy/Portals/88/documents/gergerfinal_dec_2011/Tourism/in/the/green_economy/unwto_unep.pdf. Accessed 2 Nov 2020.

UNWTO. (2015). *UNWTO Annual Report 2015*. Retrieved from: http://www2.unwto.org/publication/unwto-annual-report-2015. Accessed 2 Jan 2021.

Chapter 17
Empowering the Homestay Operators to Use Social Media Marketing to Solve Overtourism Issues

Cynthia Robert Dawayan and Sharifah Nurafizah Syed Annuar

Abstract The aim of this chapter is to outline the Malaysian tourism industry and to introduce the Malaysian homestay programs. This is to shed light on the current issue of overtourism, particularly in Sabah, Malaysia. This chapter also describes social media usage among homestay operators in Sabah and the impacts of social media competency on homestay performance. The main highlight of this chapter is the discussion on shifting tourists to unexplored destinations and how homestay operators can be empowered to use social media marketing as a tool to redirect tourists' attention from mainstream destinations to less explored destinations, thus diverting tourists to these destinations in order to curb issues of overtourism currently experienced by most tourist destinations. Although homestay operators in Sabah are largely from the rural areas, it is believed that if they have some basic knowledge in using social media and are competent in using social media to market their homestays, they are able to disseminate information about their homestays and create awareness among potential tourists that apart from the mass tourism destination sites, there is much more to be explored when traveling to Sabah, Malaysia. The effort among some homestay operators to promote themselves via social media will help potential tourists to plan their holidays better by having more options to choose from apart from the typical destinations that are often the choice of mass tourists. Simultaneously, this also could be a stepping stone in the effort to divert tourists to the less explored destinations and hence, attempt to tackle overtourism issues currently faced by many mainstream holiday destinations.

Keywords Social media marketing · Homestay operators · Overtourism · Tourism · Malaysia

---

C. R. Dawayan · S. N. S. Annuar (✉)
Faculty of Business and Management, Universiti Teknologi MARA (UiTM), Sabah, Malaysia
e-mail: shari339@uitm.edu.my

A. Hassan, A. Sharma (eds.), *Overtourism, Technology Solutions and Decimated Destinations*, https://doi.org/10.1007/978-981-16-2474-2_17

Introduction

Over the years, the tourism industry has gone through major changes that have brought it to the level it is at today. The development of new technologies has enabled travel makers from all corners of the world to travel to any destination of their choice safely, timely, comfortably, and most of all at a price affordable by many. This development has led to economic growth even in emerging countries. On the other hand, tourism also has brought some problems together with its development. Soaring demands from tourists have created social problems among the communities in major tourist destinations as well as sustainable issues. One of the most prevailing problems is overtourism, where an excessive number of tourists in a destination creates problems for the destination and its communities. Hence, it has been a continuous struggle for tourism players and various parties to search for solutions that may help to mitigate, reduce, and perhaps even to prevent this problem from occurring. As such, this chapter hopes to shed light on the issue by first introducing the concept of the Malaysian homestay and explores using social media marketing among homestay operators in Sabah, Malaysia. Concurrently, the core of this chapter is to discuss ways to shift tourists from mainstream destinations to unexplored destinations by utilizing the powerful tool (i.e., social media). In addition, this chapter is also aimed to highlight the importance of empowering homestay operators to use social media marketing in an effort to divert tourists to the unexplored destination, hence supporting the effort to curb issues of overtourism.

Overview of the Malaysian Tourism Industry

Travel and tourism, as one of the world's major economic sectors, provided 10.3% of global GDP and 330 million jobs, or 10% of total employment, in 2019 (World Travel and Tourism Council, 2020). In 2018, the tourist industry's value-added contribution to Malaysia's GDP was 15.2%, while tourism's direct contribution was 6.5% (Department of Statistics Malaysia, 2019). In 2018, the industry added a total of 23.5% (3.5 million) new employment opportunities, up 0.1 million from 2017. The increase is necessary to accommodate the growing number of tourists, who totaled 26.10 million in 2019, up 0.27 million (25.83 million) from 2018 (Tourism Malaysia, 2020). Before the pandemic, the total number of tourist arrival and receipts to Malaysia was booming and steadily increasing yearly, wherein 2019, a total of 26.10 million tourists arriving in Malaysia, generating RM86.1 billion of tourism receipts (Tourism Malaysia, 2020).

In the state of Sabah, Malaysia, a total of 4.9 million tourists were recorded to arrive in the state, generating approximately RM9.01 million in tourism receipts in 2019 (The Borneo Post, 2020). The majority of tourists worldwide have their eyes on the majestic Mount Kinabalu, the abundance of flora and fauna, the mesmerizing underwater world, and the beautiful white sandy islands that Sabah has to offer.

Nonetheless, Sabah is also rich with a colorful array of cultural attractions, as the state boasts a total of 42 ethnic groups with over 200 sub-ethnic groups, each with its own unique language, culture, beliefs, and lifestyle. Generally, the majority of tourists patronize Sabah to conquer Mount Kinabalu or to explore the underwater ecosystem by diving at the Sipadan Island, which sits in the heart of the Indo Pacific Basin. It is undeniable that these places are among the top must-visit places when in Sabah. Nevertheless, there are still more attractions that Sabah has to offer to tourists, such as the Maliau Basin Conservation Area, the Northern Tip of Borneo, where the South China Sea and the Sulu Sea merge, the various cultural heritage as well as the many rivers, waterfalls, and mountains that are yet to be fully explored.

The Malaysian Homestay Program

The homestay programme is one of the numerous goods supplied by the Malaysian tourism sector, and it provides travellers with an appealing housing option to select from throughout their vacation. Homestays, as opposed to hotels, lodges, resorts, and Airbnb places, provide travellers with not only lodging and food for the duration of their vacation, but also the chance to learn about the indigenous lifestyle and culture of Malaysians living in various states. When in a homestay, tourists can experience the daily livelihood of the local people and participate in their day to day activities such as preparing meals to be eaten together with the hosts, cleaning, washing, and even trying out some of their economic activities such as farming, fishing or even going to the orchards as well as paddy fields. Nonetheless, guests will also be given a chance to experience traditional activities of the ethnic group residing in the area they are in.

Homestays in Malaysia began inadvertently in the 1970s, when local villagers on the coast of Pahang began to offer their houses to visitors who were unable to find lodging in neighbouring beach resorts by charging a minimal price for a night's accommodation and meals in the villager's home (Ibrahim & Razzaq, 2010). This practise of providing lodging and meals in one's own house persisted until the 1980s, when a group of Japanese teenagers were sent to local homestays as part of an exchange programme to experience and learn local Malaysian culture and lifestyle (Ibrahim & Razzaq, 2010). This arrangement was comparable to the homestay concept used in the United States, Australia, Canada, and New Zealand, where international students stayed in the homes of locals to learn English (Kontogeorgopoulos et al., 2014). In contrast, in Australia, this concept is known as farm stay, when tourists stay with host families and receive a taste of ranch life (Tucker & Lynch, 2005). Furthermore, homestays in Indonesia and Thailand are particularly focused on providing cultural heritage learning opportunities to their visitors (Hamzah, 2008). Regardless of the various homestay notions that exist across the world, the basic principle stays the same. The aim is to accommodate visitors in a village or community, where they will stay with a local host family to get a glimpse of and experience the everyday life of the locals in ways that are distinct from traditional tourism interactions and settings (Dolezal, 2011).

In 1993, the Malaysian Ministry of Tourism, Arts, and Culture (MOTAC) established a specific unit to monitor homestay businesses throughout the country. The unit then presented the official definition of Malaysian homestay as a programme in which tourists stay with the host family and experience the family's daily life in both a direct and indirect manner (Ministry of Tourism, Arts and Culture Malaysia, 2020). Because of the potential that the tourist sector has to offer, the Malaysian government devised a number of programmes to assist it and infuse economic stimulus into its development and expansion. The Malaysian homestay programme has been defined as a government-sponsored community-based rural tourism project designed to empower rural communities via their participation in rural tourist activities (Ahmad et al., 2014). The initiative provides local communities with opportunities for social and economic development while protecting the local people's cultural and natural resources and allowing tourists to taste local culture and way of life (Kayat & Zainuddin, 2016).

When the Malaysian government discovered the economic potential of the tourist sector, it committed a significant amount of money and effort in planning and developing tourism in Malaysia. The government therefore determined that the industry needed to be given greater attention by infusing economic stimulus to help it expand and grow (Othman et al., 2013). The government's consistent backing has paid off, as the Malaysian tourism industry has grown to become one of the largest contributors to the Malaysian economy. However, it is clear that the Malaysian homestay business has not grown in tandem with the rest of the tourism industry (Ramli et al., 2018). This can be observed in Table 17.1 (number of tourists visiting Malaysia and number of visitors patronising homestay businesses from 2010 to 2019), where a comparison of total tourist arrivals in Malaysia and statistics on homestay tourists reveals that the number of homestay tourists is still extremely modest. This scenario might be the result of poor advertising, particularly in terms

Table 17.1 Malaysia Tourist Arrival and Malaysia Homestay Tourist Arrival for 2019–2010

Year	Malaysia tourist arrival	Malaysia homestay tourist Arrival
2019	26.10 mil	458,899
2018	25.83 mil	372,475
2017	25.95 mil	382,961
2016	26.76 mil	410,522
2015	25.72 mil	391,225
2014	27.44 mil	367,473
2013	25.72 mil	350,954
2012	25.03 mil	325,258
2011	24.71 mil	254,981
2010	24.58 mil	196,472

Source: Ministry of Tourism Arts and Culture Malaysia (2020) and Tourism Malaysia 2020)

of raising awareness of homestays and disseminating knowledge to help travellers make decisions (Kayat, 2009; Xiang & Gretzel, 2010).

Homestays in Sabah, Malaysia

As of the 31st December 2019, there are a total of 30 homestay communities in Sabah registered under MOTAC. The communities are made of 406 homestay operators from 51 villagers across Sabah, offering a total of 716 rooms to tourists all year round and are situated across the many different districts across Sabah. Table 17.2 shows the total of domestic and international tourists to Sabah patronizing the homestays from 2013 to 2018. It indicates a steady increase in tourist from year to year from both domestic and international tourist markets. This suggests that more and more tourists are becoming aware of the existence of homestays and are willing to give the homestay experience a try. Nonetheless, the number of homestay tourists is left far behind if compared to the number of tourists entering Sabah each year.

Table 17.3 shows the total homestay receipts from 2014 to 2018, which suggests that homestay tourists spend approximately RM50-RM80 per visit, and that amount is considered a quite small amount compared to the amount tourists spend on other tourism activities in the state. This could be an indicator that the potential of the homestay program in Sabah is still not fully explored by both tourism players and

Table 17.2 Sabah homestay tourist arrival for 2013–2018

	Sabah tourist arrival			Sabah homestay tourist		
Year	Domestic	International	Total	Domestic	International	Total
2018	2,517,846	1,361,567	3,879,413	57,852	47,231	105,083
2017	2,449,556	1,235,178	3,684,734	24,182	23,129	47,311
2016	2,299,132	1,128,776	3,427,908	24,623	5,219	29,842
2015	2,197,800	978,426	3,176,226	23,449	4,143	27,592
2014	2,234,123	996,522	3,230,645	17,131	2,999	20,130
2013	2,293,923	1,089,320	3,383,243	13,030	5,002	18,032

Source: Sabah Tourism Board (2020) and Ministry of Tourism Arts and Culture Malaysia (2020)

Table 17.3 Sabah homestay tourist arrival and homestay receipts for 2014–2018

Year	Sabah homestay tourist	Total homestay receipts (RM)
2018	105,083	6,235,688.20
2017	47,311	3,795,065.36
2016	29,842	2,678,772.60
2015	27,592	2,329,718.60
2014	20,130	2,052,034.30

Source: Ministry of Tourism Arts and Culture Malaysia (2020)

the tourists themselves. Consequently, this may suggest that there is a high potential to develop and divert incoming tourists to the various homestay destinations in Sabah rather than having them swarming in mass tourism destinations in Sabah such as the Mount Kinabalu and Sipadan Island areas.

Currently, the homestay operators in Sabah are in the midst of trying to increase awareness of tourists all around the world on the existence of homestays in Sabah. Most of the homestay operators are still learning how to promote their homestays themselves, without being over-dependent to travel agents or the Sabah Tourism Board to promote their homestays for them. According to a recent research conducted among homestay operators in Sabah and Sarawak, marketing capabilities have a major effect on homestay success in terms of improved visitor happiness, reputation among guests, and service quality given to guests (Dawayan et al., 2019). Basically, it indicates that the homestay operators in Sabah do have the capabilities to market their homestays.

Social Media Marketing Competency and Sabah Homestay Performance

In 2017, a research was conducted among 94 homestay operators in Sabah who were enrolled with the Malaysian Homestay Program. A self-administered questionnaire was distributed to homestay operators in the West Coast division (i.e., Papar, Kota Belud, Penampang, and Ranau), the Interior Division (i.e., Beaufort and Kuala Penyu), and Kudat Division of Sabah. The majority of the respondents are those who have been in operations for 1 to 4 years (34 operators; 36.2%). Receiving an average of more than 40 tourists per month (38 operators; 40.4%), their main market is mostly Malaysians (20 operators; 21.3%), closely followed by a mixture of tourists from Malaysia, Asia, Europe, and Australia (19 operators; 20.2%). Results of the study suggest that social media marketing Competence explained 16.7% of the variance in homestay performance where the R^2 was 0.167 ($Q^2 = 0.104$). It is positively related to homestay performance, and thus, supports that social media marketing competence affects homestay performance. This finding appears to be consistent with prior research and supports the notion that social media marketing competency has the potential to improve organisational performance (Parveen et al., 2015; Trainor et al., 2014). Having the ability to employ social media marketing improves an organization's interaction with its consumers by allowing them to communicate in a more convenient and continuous manner, therefore improving both non-financial and financial performance (Ainin et al., 2015; Trainor et al., 2014).

Furthermore, the capacity to use social media has aided companies in lowering marketing expenses, improving customer interactions, and increasing information accessibility, all of which have resulted in enhanced organisational performance (Ainin et al., 2015). According to the previous study, the capacity to use social media has resulted in improved customer relationship management among micro-enterprises in Thailand, which has influenced the organization's overall success (Charoensukmongkol & Sasatanun, 2017), where this expertise has allowed businesses to improve their marketing skills and reach out to clients more effectively.

As for the homestay operators in Sabah, they are currently learning to adapt the use of social media marketing in their promotion efforts. Being from the rural areas, and not having any formal training on the use of social media or marketing, cannot be the reason for operators to ignore this powerful marketing tool as the majority of tourists nowadays refers to the Internet, particularly social media, for information and to help them decide on their travel destination (Azeez, 2020). This is also backed by prior research, which suggests that businesses should learn for using social media for marketing purposes (Luo et al., 2012; Lee et al., 2013). This will enable them to integrate their traditional marketing strategy with social media marketing techniques, allowing them for reaching out to a larger number of customers.

Social Media Usage Among Homestay Operators in Sabah, Malaysia

The Malaysian government, hence, came out with plans and initiatives to boost rural tourism by organizing courses for homestay operators to learn and use technology so that they can reach a wider target market. Nevertheless, approximately 72% of homestay operators in Malaysia are reported hesitant for adopting social media in their businesses (Osman & Zakaria, 2020). In 2017, a study was performed for determining the usage of social media by homestay operators in Sabah, Malaysia. The results in Table 17.4 demonstrate the most common form of social media used by the operators are Facebook and WhatsApp (50.0%). In addition, 93.6% and 97.9% of the homestay operators use social media to search and share information, respectively, while 96.8% of them use social media for communicating their customers. Despite the fact that the majority of them (98.9%) utilise social media to advertise their homestays, they rely largely on personal social media to promote their businesses (79.8%).

A scholar reported that small businesses particularly rural SMEs (Small Medium Enterprises) had not updated their Facebook page regularly, which indicates that they do not put effort in building their social media presence (Pesonen, 2011). The top barriers for SMEs (Small Medium Enterprises) to adopt social media marketing are lack of skills and knowledge in social media marketing for business (Pesonen et al., 2013; Taiminen & Karjaluoto, 2015). In addition, lack of time for social media due to lack of human resources is another hindrance for small businesses to adopt social media (Granger & Reiter, 2015). Small businesses also called out for training and short courses to equip them with the knowledge and skills. Apart from that, access to infrastructure and knowledge and information can pose challenges to homestay operators to adopt social media marketing. Most of the homestay businesses are situated in rural areas where the Internet access is limited. In addition, not

Table 17.4 Sabah homestay operators social media usage profile

Variables	Items	Frequency	Percent
Types of used social media	Facebook Only	4	4.3
	WhatsApp Only	10	10.6
	Facebook and WhatsApp	47	50.0
	Facebook and Instagram	7	7.4
	Facebook, Instagram, and WhatsApp	11	11.7
	Facebook, WhatsApp, and Blog	15	16.0
Use of personal social media for business promotion	Yes	75	79.8
	No	19	20.2
Use of social media for information seeking	Yes	88	93.6
	No	6	6.4
Use of Social Media to Disseminate Information	Yes	92	97.9
	No	2	2.1
Use of social media for communicating customers	Yes	91	96.8
	No	3	3.2
Use of Social Media to Promote Business	Yes	93	98.9
	No	1	1.1

Source: Dawayan (2021)

all of them own smartphones to enable them to use social media. Even if they have smartphones, their usage is limited to read and write texts. Other functions such as downloading and uploading videos would be problematic due to the low-speed internet access. Another issue is the access to knowledge and information. As mentioned above, skills and knowledge have become their top reasons for not using social media marketing. Other than the ability to utilize technology, language can be their utmost concern in social media marketing. The capacity to speak and write in English to reach the international market is decisive. However, most of them operate their own social media pages. Hence, they write posts targetingthe domestic market only.

Shifting Tourists to Unexplored Destinations

The development of digital trends (i.e., social media sites, mobile applications, travel booking platforms) has in a way aids in increasing the number of tourists to popular destinations. The rapid advancement of social media generates a pool of potential travelers not only locally but also globally. Among the benefits of using social media in promoting tourism spots are the ability of the social media to provide real-time information and enable two-way communications not only between service providers and their clients but also between consumers themselves, which helps in reinforce travelers' decision making. In addition, influencers share their experiences in tourists' destinations, and related services are often seen as outstanding compared to any official information from the service providers.

Sabah, or popularly known as North Borneo has received a tremendous increase of tourists' arrival, particularly from China, South Korea, and Japan, and is remarkably known for its Mount Kinabalu and beautiful islands. The influx of tourists at the popular tourist spots in Sabah has much affected the environment and local community. The cleanliness and facilities maintenances due to overcrowded create challenges to the local authorities. In addition, congestion which leads to commuter and traffic woes is also reported. During peak seasons, the locals have to endure difficulty in finding car parking spaces and overcrowded shopping malls and tourism spots owing to tourism. Looking at these phenomena, the local authorities, therefore, try to shift the tourists to unexplored destinations in Sabah by promoting rural tourism. Sabah's Ministry of Tourism, Culture, and Environment has eased the development of community-based tourism (CBT) by assisting the locals in partnering with tour agents in packaging their "products". By doing this, not only the quality of life for the community can be improved, but this would allow tourists to visit destinations other than regular attractions (Ahmad, 2020).

Empowering Homestay Operators to Use Social Media Marketing

Apart from training on technology and language, homestay operators also need education on the correct way to market their homestays using social media. The first step that they need to do is to identify their strengths, resources and capabilities, and competencies. This internal analysis is pivotal so that they can ensure that all their social media marketing plan is aligned with what they can offer in which can be turned as their competitive advantage. Community-based activities are the focal product for homestays in Sabah. The activities can be local dance and singing fiesta, cooking traditional dishes or ethnic food, and handicraft making. Additionally, homestay operators can also identify what are the resources that they have for (e.g., rivers, oceans, farms, lakes etc.) that they can use as highlights in promotional strategies. Nature is a tourist product and a great "tool of persuasion" to attract nature-oriented international tourists (Horáková, 2010). Resources, in essence are the assets to the businesses. Meanwhile, capabilities refer to the ability of these businesses to use the resources that they have effectively. Furthermore, competencies are the combination of specific resources that are directed to achieve specific objectives. The Resource-Based View Theory explains the usage of these resources and capabilities to leverage competitive advantage. In the event where organizations fail to exploit their resources and capabilities, it would be difficult for the organizations to stay relevant in the market. Tourists have many destinations to choose from. Hence, homestay operators must be able to identify their strengths and resources that they have in order to stand out in the market. Moreover, creating values through available resources and strengths to achieve customer satisfaction should be the ultimate goal of business owners (Lee et al., 2020).

Tourists travel because they seek different experiences. They want to discover nature and experience others' culture, food, and traditions. According to Goeldner and Ritchie (2003), there are four motivations for tourists to travel and visit destinations. Firstly, because of physical motivations, for instance, relaxation. Secondly, cultural (i.e., exploring new areas and thirdly is interpersonal which is to socialize and meet new people). Finally, the motivation is prestige like self-esteem and self-actualization. Prior to this, Cha et al. (1995) studied the motivations for Japanese travelers and found there are six motivational factors pushing them to visit destinations. These motivations are relaxation, adventure, knowledge, family, travel bragging, and sports. This is also supported by Pearce and Lee (2005). Meanwhile, Jang and Wu (2006) advocated knowledge-seeking, relaxation and family are the top factors pushing tourists to travel while environment, cost, accommodations, safety, and accessibility are the pull factors to travel. Tourists make their visits to rural areas because they want to adventure or challenge, social interactions, and seek enjoyment (Park & Yoon, 2009). Homestay operators, therefore, need to understand and identify their target market. Market segmentation has become an important tool in a marketing plan. The ability to categorize the market into subgroups based on homogenous characteristics, needs, and motivations. In summary, tourists interested in rural tourism are commonly associated with knowledge-seeking, adventure, self-esteem, and self-actualization traits.

The second step is to build unique selling propositions by highlighting their strengths, resources, capabilities, and competencies. This is imperative as this will serve as a tool to differentiate themselves from others. Homestay operators need to create "motivations" and "reasons" for tourists to visit their places. Social media is a powerful means to persuade travelers to visit destinations. The millennials, particularly, are more adventurous sensation seekers and like to travel bragging. They like to experience new things and share these with their networks online (Buru, 2015). Millennials place Southeast Asia as one of their top three dream destinations. They like physical attributes (i.e., the nature and cultural attractions like food and history). In addition to this, Millennials are also reported that they rely on the Internet to find information about certain destinations and use other tourists' experiences as inputs to their travel decision making. Creating "instaworthy" or instagrammable spots at the homestays would encourage tourists to visit. Huertas (2018) alleged that live videos and stories in social media could influence attitudes and opinions about certain places. By recognizing their uniqueness and strengths, homestay operators should be able to develop their positioning statements. These positioning statements can later be converted into hashtags, which are important for tourists to create creative and attractive content. The use of hashtags, for instance, "unspoiled" or "unexplored" can be a strong positioning for the homestays and thus encourage tourists to visit. This is impressing others and to brag about their experiences can create jealousness to others in their social networks. Furthermore, the "fear-of-missing-out" (FOMO) and "you only live once" (YOLO) attitude among young travelers nowadays can create needs to travel and this should be exploited by homestay operators and local authorities in order to combat overcrowded at popular destinations and shift the tourists to unexplored places.

The third step is for creating a social media marketing plan. Thus, social media is a useful medium to "advise" tourists to visit unexplored destinations. Its ability to provide experiential information can support travelers' decision making. However, postings, copywriting, and contents need to be carefully planned and implemented. According to Kaplan and Haenlein (2010), social media and content are equally important and interdependent. Social media needs good content, and good content without social media will be meaningless. To get engagement, likes and shares, businesses need to ensure that their postings inspire others, shareworthy, fun, and visually attractive (Berthon et al., 2012). In addition, they must see the social media users play the roles of content consumers and content creators. As content consumers, their needs are depending on internal motivations such as openness to experience and risk-taking rather than exposure to external stimuli. With this in mind, social media users can also act as content creators that have impacts on others and influence them to change their attitude towards an object. The number of likes, comments, and shares of certain social media posts determines its "viral-ability". In tourism industry, the viral effect is important to persuade tourists to visit destinations. Furthermore, hashtags (#) and geotags (location-based) would add values to the postings and develop a sense of curiosity in the audience. Attraction aspects, as mentioned in the second step, are the focal point of tourism. Thus, the operators need to leverage their strengths and "Unique Selling Propositions" and position them in consumers' minds in order to create an impression on social media (Fig. 17.1).

Prior to creating persuasive messages, travelers' period is essential to be acknowledged. There are three phases in traveling, and they are pre-trip, during the trip, and post-trip. In the first phase, potential tourists would normally search for information and gather as many materials as possible to their dream destinations. Therefore,

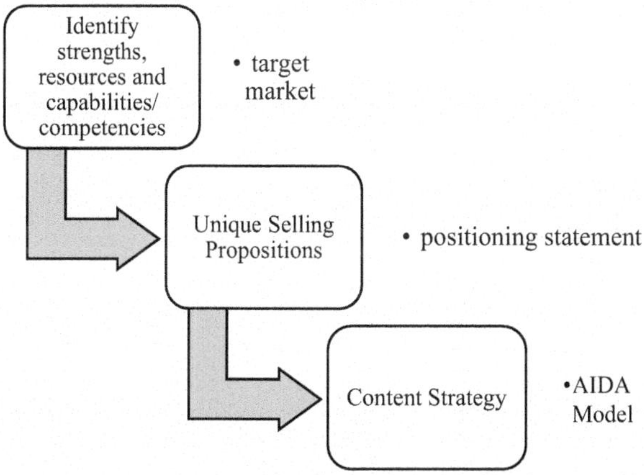

Fig. 17.1 Proposed solutions on empowering the homestay operators to adopt social media marketing. (Source: developed by the authors, 2020)

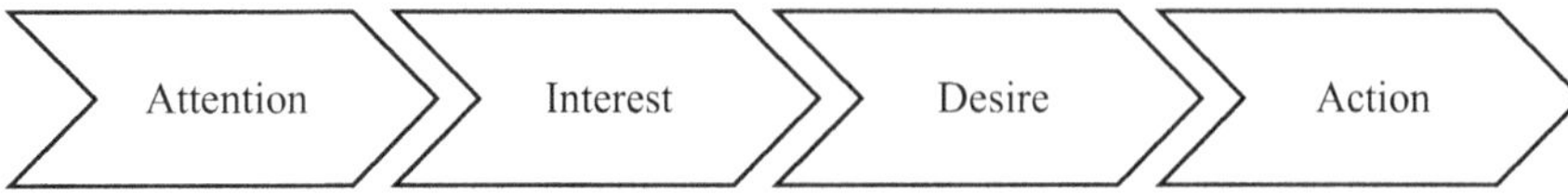

Fig. 17.2 AIDA model. (Source: adapted from Barry, 1987)

homestay operators must provide all details related to their homestays in their social media as normally, these channels will be their main reference. In addition, previous tourists' testimonials and reviews are also important to be made available on their social media pages. This will aid in reinforcing their decision to visit the destinations. Homestay operators must ensure that their social media are updated and respond to queries promptly. Pages that are not updated and take time to address concerns by potential tourists are deemed as unattractive. During the travel phase, tourists must be able to be furnished with information such as local attractions, activities, souvenirs stores and etc. This would enrich their experience during their visits. Tourists often turn to social media, especially when they are visiting unfamiliar destinations (Liu et al., 2020). Accordingly, accessibility to such information through social media would reduce the time of the tourists to search for information and hence, would help in avoid resentment. The final phase, which is post-travel is to encourage the tourists to share their experiences by contributing testimonials, photos, and videos that would benefit future tourists. These would be useful for future tourists and, therefore, must be uploaded and shared on the official social media pages of the homestays. Besides, studies have also evidenced that sharing such information has a positive effect on repeat travel behavior (Ruhanen, 2019; Katsoni & Dologlou, 2016).

AIDA (Attention-Interest-Desire-Action) model has extensively linked with persuasion. Marketing scholars recognize the profound impact of the model in "copywriting". Tourists specify that the selection of their destinations was mostly contributed by factors like scenery, nature, and culture, and by looking at photos and videos on social media, their awareness about the destinations are enhanced and thus creates interest for them to visit the destinations (Liu et al., 2020). Although there is a debate that social media in tourism can only form attention and generate interest towards a destination (Kane et al., 2012), however, for those who are planning for a trip and looking for "approval", these social media sites can certainly provide an immediate effect (i.e., taking action) (Liu et al., 2020) (Fig. 17.2).

Conclusion

This chapter gives some insight into Sabah's homestay operators' social media competence as well as their current level of social media activities. Additionally, suggestions were also presented on how homestay operators can fully utilize social media marketing in their business operations. This information is imperative as it leads to the possibility of using social media marketing in resolving issues of overcrowding faced by many mainstream tourists' destinations.

Social media has changed the way businesses interact with the marketplace and society, opening up a new universe of opportunities and problems across the board (Aral et al., 2013). As posited by many past researchers, social media marketing has changed the ways businesses are operated, where it has simplified communication between the business and its customers, despite their geographic distance by allowing online engagement between businesses and consumers for creating and sharing opinions, information, insights, and experiences with other users through numerous social media platforms (Kietzmann et al., 2011; Gupta & Brooks, 2013). Essentially, social media has allowed businesses to expand their network and connect with a larger audience, as well as improve information distribution and marketing efficiency (Gabrielsson & Gabrielsson, 2011) as well as the organizations' performance (Liang et al., 2010; Dodokh & Al-Maaitah, 2019).

Although homestay operators in Sabah have begun to use social media to communicate with their guests, it has been discovered that owing to a lack of skills and understanding, they are still unable to grasp and fully profit from the benefits of social media marketing. Even if they have had some social media training, it is regarded insufficient because the bulk of the operators are peasants who are illiterate and lack the necessary skills to utilise technology efficiently and effectively. Furthermore, owing to language issues, the operators may have difficulty interacting via social media. The majority of homestay operators are unable to speak effectively in English. They prefer to communicate in their native tongue, which limits the number of people who may hear them on social media.

In view of this situation, it is suggested that homestay operators try to adopt some strategies as discussed in this chapter, which is believed to be useful for them in their effort to better reach out and promote their homestays to potential tourists. Simultaneously, when they are able to grasp the techniques as suggested in this chapter, the operators can also use their social media marketing competence to help the local tourism industry by diverting tourists from overcrowded mainstream tourist destinations to destinations that are more pristine, unique and most of all unexplored by the mass tourist.

References

Ahmad, A. (2020). Moving travel industry forward. *The Star.* Retrieved from: https://www.thestar.com.my/metro/metro-news/2020/06/04/moving-travel-industry-forward. Accessed: 5 Jan 2021.

Ahmad, S. Z., Jabeen, F., & Khan, M. (2014). Entrepreneurs choice in business venture: Motivations for choosing home-stay accommodation businesses in Peninsular Malaysia. *International Journal of Hospitality Management, 36*, 31–40.

Ainin, S., Parveen, F., Moghavvemi, S., Jaafar, N. I., & Mohd Shuib, N. L. (2015). Factors influencing the use of social media by SMEs and its performance outcomes. *Industrial Management and Data Systems, 115*(3), 570–588.

Aral, S., Dellarocas, C., & Godes, D. (2013). Introduction to the special issue—Social media and business transformation: a framework for research. *Information Systems Research, 24*(1), 3–13.

Azeez, Z. A. (2020). The Influence of Social Media on Tourism Marketing: Field Study on Travel Agency in the City of Baghdad. *Journal of Talent Development and Excellence, 12*(2s), 2684–2708.

Barry, T. E. (1987). The development of the hierarchy of effects: An historical perspective. *Current issues and Research in Advertising, 10*(1-2), 251–295.

Berthon, P. R., Pitt, L. F., Plangger, K., & Shapiro, D. (2012). Marketing meets Web 2.0, social media, and creative consumers: Implications for international marketing strategy. *Business Horizons, 55*(3), 261–271.

Buru, L. (2015). *Millennials – Cine sunt şi cum îi abordăm în campaniile noastre?.* Retrieved from: http://romaniancopywriter.ro/millennials-cum-ii-abordam/. Accessed 23 Dec 2020.

Cha, S., McCleary, M., & Uysal, M. (1995). Travel motivations of Japanese overseas travelers: A factor-cluster segmentation approach. *Journal of Travel Research, 33*(2), 33–39.

Charoensukmongkol, P., & Sasatanun, P. (2017). Social media use for CRM and business performance satisfaction: The moderating roles of social skills and social media sales intensity. *Asia Pacific Management Review, 22*(1), 25–34.

Dawayan, R. C. (2021). *Enhancing Malaysian homestay performances through resources and capabilities*. Unpublished Doctoral dissertation. Universiti Malaysia Sabah.

Dawayan, C., Sondoh, S. L., Jr., Tanakinjal, G. H., Nasip, S., & Boniface, B. (2019). Homestay Resources, Marketing Capability and Homestay Performance: A Study Among Homestay Operators in Sabah and Sarawak. *Malaysian Journal of Consumer and Family Economics, 21*(S1), 86–109.

Department of Statistics Malaysia. (2019). *Tourism Satellite Account 2018*. Retrieved from: https://www.dosm.gov.my/v1/index.php?r=column/cthemeByCat&cat=111&bul_id=Wk1KWlpxZTRDWnVhVWNMV21ZVVY3Zz09&menu_id=TE5CRUZCblh4ZTZMODZIbmk2aWRRQT09/. Accessed: 28 Dec 2020.

Dodokh, A., & Al-Maaitah, M. A. (2019). Impact of social media usage on organizational performance in the Jordanian dead sea cosmetic sector. *European Journal of Business and Management, 11*(2), 75–91.

Dolezal, C. (2011). Community-Based Tourism in Thailand: (Dis-) Illusions of Authenticity and the Necessity for Dynamic Concepts of Culture and Power. *ASEAS - Austrian Journal of South-East Asian Studies, 4*(1), 129–138.

Gabrielsson, M., & Gabrielsson, P. (2011). Internet-based sales channel strategies of born global firms. *International Business Review, 20*(1), 88–99.

Goeldner, C. R., & Ritchie, J. R. (2003). *Tourism: principles, practices, philosophies*. Wiley.

Granger, M. M., & Reiter, L. (2015). An exploratory analysis of entrepreneurs' social media usage in small business practices. *Journal of Marketing Management (JMM), 3*(2), 11–26.

Gupta, R., & Brooks, H. (2013). *Using social media for global security*. Wiley.

Hamzah, A. (2008). Malaysian homestay from the perspective of young Japanese tourist: The quest for furusato. In J. Cochrane (Ed.), *Asian tourism: Growth and change* (pp. 193–207). Routledge.

Horáková, H. (2010). Post-communist transformation of tourism in Czech rural areas: New dilemmas. *Anthropological Notebooks, 16*(1), 59–77.

Huertas, A. (2018). How live videos and stories in social media influence tourist opinions and behaviour. *Information Technology and Tourism, 19*(1), 1–28.

Ibrahim, Y., & Razzaq, A. R. A. (2010). Homestay program and rural community development in Malaysia. *Journal of Social Sciences and Humanities, 2*, 7–24.

Jang, S., & Wu, C.-M. E. (2006). Seniors' travel motivation and the influential factors: An examination of Taiwanese seniors. *Tourism Management, 27*, 306–316.

Kane, K., Chiru, C., & Ciuchete, S. G. (2012). Exploring the eco-attitudes and buying behaviour of Facebook users. *Amfiteatru Economic, 14*(31), 157–171.

Kaplan, A. M., & Haenlein, M. (2010). Users of the world, unite! The challenges and opportunities of social media. *Business Horizons, 53*(1), 59–68.

Katsoni, V., & Dologlou, N. (2016). ICT applications and Web 2.0 components for tourism in protected areas. In V. Katsoni & A. Stratigea (Eds.), *Tourism and culture in the age of innovation* (pp. 563–575). Springer.

Kayat, K. (2009, 4–5 August). *Community based tourism in developing countries*. Paper presented at the *International Seminar on Community Based Tourism*, Shah Alam, Malaysia.

Kayat, K., & Zainuddin, N. F. A. (2016). Community-based tourism initiative in rural Malaysia: Is it a success? *International Review of Management and Marketing, 6*(7S), 242–249.

Kietzmann, J. H., Hermkens, K., McCarthy, I. P., & Silvestre, B. S. (2011). Social media? Get serious! Understanding the functional building blocks of social media. *Business Horizons, 54*(3), 241–251.

Kontogeorgopoulos, N., Churyen, A., & Duangsaeng, V. (2014). Success factors in community-based tourism in Thailand: The role of luck, external support, and local leadership. *Tourism Planning and Development, 11*(1), 106–124.

Lee, K., Oh, W. Y., & Kim, N. (2013). Social media for socially responsible firms: Analysis of Fortune 500's Twitter profiles and their CSR/CSIR ratings. *Journal of Business Ethics, 118*(4), 791–806.

Lee, C. S., Martin, D., Hsieh, P. F., & Yu, W. C. (2020). Principles of value creation in event tourism: Enhancing the competitiveness of regional clusters. *Journal of Global Scholars of Marketing Science, 30*(4), 437–453.

Liang, T. P., You, J. J., & Liu, C. C. (2010). A resource-based perspective on information technology and firm performance: a meta-analysis. *Industrial Management and Data Systems, 110*(8), 1138–1158.

Liu, X., Mehraliyev, F., Liu, C., & Schuckert, M. (2020). The roles of social media in tourists' choices of travel components. *Tourist Studies, 20*(1), 27–48.

Luo, J., Fan, M., & Zhang, H. (2012). Information technology and organizational capabilities: A longitudinal study of the apparel industry. *Decision Support Systems, 53*(1), 186–194.

Ministry of Tourism, Arts and Culture Malaysia (MOTAC). (2020). *Malaysian homestay programme*. Retrieved from: http://www.motac.gov.my/en/faqs/malaysian-homestay-programme. Accessed: 3 Jan 2021.

Osman, Z., & Zakaria, Z. (2020). Factors influencing intention to adopt social media platform among rural homestay operators in Malaysia. *Journal of Technology Management and Business, 7*(1), 25–35.

Othman, F., Sazali, F., & Mohamed, B. (2013). Rural and community based tourism development in Malaysia: Prospects for Homestays as a Social Economy Enterprise. *TEAM Journal of Hospitality and Tourism, 10*(1), 65–76.

Park, D. B., & Yoon, Y. S. (2009). Segmentation by motivation in rural tourism: A Korean case study. *Tourism Management, 30*(1), 99–108.

Parveen, F., Jaafar, N. I., & Ainin, S. (2015). Social media usage and organizational performance: Reflections of Malaysian social media managers. *Telematics and informatics, 32*(1), 67–78.

Pearce, P., & Lee, U. L. (2005). Developing the travel career approach to tourist motivation. *Journal of Travel Research, 43*, 226–237.

Pesonen, J. (2011). Tourism marketing in Facebook: Comparing rural tourism SME's and larger tourism companies in Finland. In R. Law, M. Fuchs, & F. Ricci (Eds.), *Information and communication technologies in tourism 2011* (pp. 537–546). Springer.

Pesonen, J., Mikkonen, J., Makkonen, M., & Honkanen, A. (2013). Barriers affecting social media adoption in Finnish tourism businesses. In Z. Xiang & I. Tussyadiah (Eds.), *Information and communication technologies in tourism 2014* (pp. 723–731). Springer.

Ramli, R., Md Daud, S., Mat Kasim, M., & Kayat, K. (2018). Success index for promoting Malaysian homestay programmes based in multi criteria analysis. *Journal of Telecommunication, Electronic and Computer Engineering, 10*(1-10), 25–31.

Ruhanen, L. (2019). The prominence of eco in ecotourism experiences: An analysis of post-purchase online reviews. *Journal of Hospitality and Tourism Management, 39*(39), 110–116.
Taiminen, H. M., & Karjaluoto, H. (2015). The usage of digital marketing channels in SMEs. *Journal of Small Business and Enterprise Development, 22*(4), 633–651.
The Borneo Post. (2020). *Sabah's tourism industry in dire straits*. Retrieved from: https://www.theborneopost.com/2020/07/28/sabahs-tourism-industry-in-dire-straits/. Accessed: 18 Jan 2021.
Tourism Malaysia. (2020). *My tourism data*. Retrieved from: http://mytourismdata.tourism.gov.my/. Accessed: 23 Jan 2021.
Trainor, K. J., Andzulis, J., Rapp, A., & Agnihotri, R. (2014). Social media technology usage and customer relationship performance: A capabilities-based examination of social CRM. *Journal of Business Research, 67*(6), 1201–1208.
Tucker, H., & Lynch, P. (2005). Host-guest dating. *Journal of Quality Assurance in Hospitality and Tourism, 5*(2e4), 11–32.
World Travel and Tourism Council (WTTC). (2020). *Economic impact reports 2019*. Retrieved from: https://wttc.org/Research/Economic-Impact. Accessed: 23 Jan 2021.
Xiang, Z., & Gretzel, U. (2010). Role of social media in online travel information search. *Tourism Management, 31*(2), 179–188.

Chapter 18
Technology Application to Manage Overtourism in Turkey

Aysen Ercan İştin and Eray Turpcu

Abstract The presence of special attractions motivates potential tourists especially to visit these areas. This situation increases the density of tourists' excessive numbers in popular destinations. Overtourism, which causes many negative economic, social, cultural and physical effects, puts pressure on local people living in the region and affects their daily lives. Therefore, it is important for sustainability to manage the carrying capacity and to find out and solve the reasons in order to minimize or eliminate these negative effects that cause negative perceptions on tourists. In this context, technological innovations should be used as one of the most helpful areas to combat overtourism as well as movements that increase overtourism. This chapter of the book discusses the uses of technological applications on overtourism in Turkey. Finally, in the context of technological applications, it is presented facilitating solutions and suggestions for both the tourists coming to the region and the local people living in the region.

Keywords Technology · Application · Overtourism · Destination · Turkey

Introduction

The attractions of certain destinations increase the flow of tourists to these destinations. Especially positive posts made through social media channels or platforms for tourism destinations increase the popularity of these regions and make these destinations more attractive. Thus, overtourism events occur in these destinations, and the negative effects of overtourism emerge. The negative effects of overtourism

A. E. İştin (✉)
Department of Gastronomy and Culinary Arts, School of Tourism and Hotel Management, University of Şırnak, Şırnak, Turkey

E. Turpcu
Department of Tourism Management, Bulancak Kadir Karabaş School of Applied Sciences, Giresun University, Giresun, Turkey

A. Hassan, A. Sharma (eds.), *Overtourism, Technology Solutions and Decimated Destinations*, https://doi.org/10.1007/978-981-16-2474-2_18

such as overcrowded facilities, noise pollution, environmental pollution, visual pollution, deterioration of the infrastructure, excessive water and energy consumption, waste, loss of cultural identity damage the environment and cultural heritage and negatively affect the life quality of the local people.

New technologies that offer low cost and easy travel fail to manage the increasing flow of visitors to the popular destinations. On the other hand, new technologies now offer smart solutions to deal with overtourism in the popular destinations and to prevent congestion in city centers and thus to get rid of the negative effects of overtourism. Technological applications used to direct tourists to less-visited destinations are now important to ensure the sustainability of the destinations. This chapter examines the overtourism concept, the effects of overtourism, and the Technologies used to reduce overtourism and overtourism in Turkey. Technological applications used in overtourism are mentioned, especially in the context of Istanbul and Antalya provinces, and technology-related suggestions are offered to reduce the negative effects of overtourism.

Overtourism

The development of new technologies in many areas, the widespread use of low-cost airline transportation and the increase in comfort and quality in the transportation sector, shortening of working hours, increase in leisure time, increase in income have contributed to the increase in travel mobility and thus to the increase of tourism revenues due to the development of the tourism industry when the developments in the global tourism industry are evaluated. This situation has led to growth in other developing economies. Although the tourism industry provides significant advantages to the economies of the country, the concentration of tourism movements in certain destinations and the uncontrolled demand for destinations have negatively affected both the regions and local communities (Capocchi et al., 2019). Especially local people living in smaller destinations have felt the effect of overtourism more due to the pressure caused by the tourist influx (Anuar et al., 2019). Because overcrowding negatively affects the daily life of local residents and communities, the natural and socio-cultural assets of the destinations, the infrastructure and superstructure of tourism, and the holiday experience of tourists visiting the region. However, in some cases, due to overtourism, the residents of the region have turned to civil protests to stop tourism and realized an anti-tourism phenomenon (Kuščer & Mihalič, 2019).

Considering the negative effects of overtourism in this way, it is an undeniable fact that socio-culturally, it affects the local people living in the region, employees working in the tourism sector and tourists who go to the region for holiday purposes. For this reason, this concept, which is increasingly important in the tourism industry, has been the subject of study for many researchers and has been tried to be defined in different ways with different perspectives. Before mentioning the definition of the concept of overtourism, when the meaning of the word "over" is examined, it is defined as "more than the usual or tolerable degree, overflow" and "more than necessary" in the Turkish Language Association's Contemporary Turkish dictionary. When the concept of overtourism is evaluated in general, concepts such as exceeding the carrying capacity, overcrowding, damaging, negative impact, congestion, and density are used as key words in the definitions. UNWTO (2018) defines the concept of overtourism as tourism mobility that negatively affects the life quality of local people living in all or part of a touristic destination or the tourists' holiday experiences, who visiting that destination. CREST (2018) defines overtourism as the deterioration of the environment and infrastructure in the current destination, decrease in travel experience, wear and tear on the built cultural heritage and result in negative effects on the local population with tourism movements moving beyond the limits of acceptable change in a destination due to the visitors' increasing number. However, Goodwin (2017) defined the concept of overtourism, which he thinks is the opposite of responsible tourism, as the feeling that the life quality and experience in the region has deteriorated unacceptably when there are too many local people or visitors. Perkumienė and Pranskūnienė (2019) pointed out that overtourism is characterized by the visitors' excessive number that affect human's quality of living space; it was also considered a tourism movement where the human's life quality assurance (both travelers and locals) was violated and environmental changes occurred.

It is possible to talk about seven elements about the scope and meaning of the concept of overtourism. These elements are specified by Koens et al., as follows (2018):

- Although the interest in overtourism has increased recently, the main problems on which it is based are not new.
- Overtourism and mass tourism are different concepts.
- Although the effects of overtourism are predominantly in popular and well-known areas of cities and in certain time periods, all residents living in the area can perceive these negative effects.
- Overtourism is not only a problem of the tourism industry but also affects conjunctural events around the world.
- Technological or smart solutions alone cannot solve overtourism problems.
- Since the way overtourism manifests itself and the possibilities to cope with the problems depends largely on the urban environment, solutions that suit this environment should be offered. So there is no one-size-fits-all solution to solving overtourism problems.
- Since overtourism is not a problem that only affects cities, rural areas and other regions may also be affected.

When the concept of overtourism is evaluated in general, it is seen that the negative effects of overtourism are a cause of discomfort not only by the local people living in the region but also by the tourists visiting the region. While the local people turned their perspective towards tourism negatively due to the deterioration of the daily life order, the inability to meet their expectations for tourism, and the injustice in the distribution of benefits (Yüncü, 2019), tourists could not benefit from their holidays or travels efficiently due to the crowded environments they encountered in the holiday regions where they went for rest, entertainment and recreational purposes during their annual leaves, which were limited to get rid of the noise, traffic and the intensity of the city they live in all year round (Aylan & Aylan, 2020). For this reason, it has become necessary to develop and implement new strategies in order to protect the daily life order of the local people and to meet the guests' expectations, who come to the region for holiday purposes for the destinations that are getting crowded due to the overtourism movement (Çetinoğlu & Meydan Uygur, 2020).

The fact that there is a tourist flow that is much more than the destination can handle in a certain period, the natural, cultural, and historical attractions of the region and touristic facilities provide an environment for businesses to be negatively affected. For this reason, determining the factors that cause overtourism and preventing the negative consequences of overtourism by determining these reasons will be beneficial in ensuring the sustainability of touristic destinations (Doğan, 2013).

Achieving sustainability for touristic destinations will be possible with the growth of tourism qualitatively rather than quantitatively. Paying attention to the socio-cultural, economic, and environmental sustainability of a growing tourism industry should be one of the issues that should be carefully handled by the private and public sectors that have a say in the destination. For this, new generation tourism policies should be evaluated, and complex and multidimensional problems addressed within the scope of overtourism should be tried to be solved with institutional and behavioral approaches. These policies can provide solutions not only for destinations that are affected today and need them urgently but also for other less-visited destinations trying to avoid the harmful effects of overtourism in the first place. Therefore, it is of great importance to determine the reasons that cause overtourism in touristic destinations and prevent them from the very beginning. The factors causing overtourism in touristic destinations are listed as follows by different researchers (Duyar & Bayram, 2019; Nilsson, 2020; Goodwin, 2019; Koens et al., 2018; Weber, 2017):

- Diversification and cheapening of accommodation and transportation opportunities encourage people to participate in tourism activities.
- The internet makes all parts of the tourism industry (i.e., hotels, cruise lines, airlines, and digital reservation systems) more accessible.
- Public spaces to serve tourists need to be free of charge, seasonal tourism, and extension of the tourism season.
- Tourism enterprises and trade unions emphasize the economic importance of tourism and employ a large number of people from outside the city in the industry.
- Excessive use of destination resources, infrastructure, and facilities or some of them
- The lack of strategic approach of the private and public sector in the context of bad management, the lack of penalty systems, the presence of unskilled personnel, stakeholders' insufficient coordination, lack of transparency, and pressure exists.

- Due to social differences, cultural conflicts between people, low-income levels, high unemployment, and low tourism awareness.
- Large tourist groups come together and visit the region.
- It is the uncontrolled increase of touristic accommodation opportunities in tourism destinations.

Depending on the concept of overtourism, it can be observed that the reasons stated above are generally composed of a supply and demand framework. When analyzed from the demand side, it is seen that these reasons are due to a higher spending tendency, better economic conditions, travel technology, globalization, complexity, changing lifestyle, and human liberation. When evaluated in terms of supply, it can be expressed as lack of resources, poor management, the vulnerability of destinations that accept tourists, lack of planning, lack of technical expertise, and lack of sustainability sensitivity (Avond et al., 2019).

As a result, tourism destinations consist of many stakeholders, especially the public and private sectors, that foster harmonious and conflicting interests. When decisions are made about overtourism that affects the destination, and some stakeholders may be willing to reduce tourism-related investments, others may not necessarily support the initiative given the loss of revenue. Therefore, overtourism can still continue if all stakeholders do not align their decisions on overtourism. In this case, it is of great importance to understand the expectations of both the tourists coming to the region and the local people living in the region and to prevent the reasons causing overtourism in this direction by the stakeholders in order to minimize the negative effects (Gowreesunkar & Seraphin, 2019).

Negative Effects of Overtourism

Tourism is one of the largest and fastest-growing industries in the global economy, bringing new opportunities for many countries, cities, and regions, as well as many new challenges (OECD, 2020). The most important of these difficulties is the characteristic of the tourism industry that is always in fierce competition at all levels. While the tourism industry provides the emergence and marketing of a new destination with a significant number of different tourism types and supply methods, it also provides the opportunity for established and more mature destinations to create a more competitive market environment to combine and develop existing markets with new ones. Tourism enterprises also compete with the emergence of new competing enterprises from other industries. However, in this developing competitive model, the higher the demand, the more likely it will be out of competition, and the variety of supply generation capacity will increase. In today's globalizing world, when it comes to the choice and demand of tourism destinations, it is seen that travel agencies and tour operators offer tourists an international selection range (Moreno-Gil & Coca-Stefaniak, 2020).

The wide range of tourists'destination selection and the convenience of their preferences increase the density of popular destinations. When the overtourism impact is felt economically, socio-culturally, and environmentally (positively or negatively), it puts pressure on the infrastructure, superstructure, environment, local communities, economic sectors, and local people living in the region. When this situation is left unchecked, it can have significant effects on local people's the daily lives as well as cultural heritage and environmental areas of critical importance, often leading to negative perceptions and even more resentment towards tourists and tourism (OECD, 2020). When overtourism is evaluated from this point of view, it is seen that there are negative effects in touristic destinations due to exceeding the carrying capacity and the resulting excess. When the effects of overtourism in touristic destinations are evaluated in general, it is an undeniable fact that it provides economic benefits, causes socio-cultural and environmental degradation, and that local community are affected both positively and negatively. It is predicted that alternative tourism plans, which will be carried out consciously by paying attention to the carrying capacity, will benefit both the destinations and the development of the country's tourism (Paslı, 2020).

Negative Economic Effects of Overtourism

Today, income from both domestic and foreign tourism not only provides employment and income in many ways but also contributes significantly to the socio-economic and cultural development of many cities and their surroundings, often including transport infrastructure and new business ventures. However, the income generated from tourism benefits the growth of urban tourism, the use of natural resources that enable sustainability and tourism development, and the increase of socio-cultural impact. On the other hand, this situation creates significant difficulties in providing applications that minimize the negative effects on infrastructure in terms of pressure, mobility, and congestion in touristic destinations. Such contradictions have been part of tourism since early times, and as a general rule, as with all forms of economic development, economic returns have prevailed over social and environmental concerns (Dodds & Butler, 2019; UNWTO, 2018). However, due to the effect of overtourism, it will be possible to encounter some negative effects besides the positive effects in terms of economy in the destinations where touristic activities are carried out. One of the negative effects caused by overtourism is that it causes housing price increases and, therefore, problems in acquiring housing. Another negative effect is the increase in the prices of many services sold in the region due to the increase in inflation due to overtourism and the decrease in the local people's purchasing power. Also, overtourism can create part-time, seasonal, and low-budget jobs. It can be seen that the regional economy is overly dependent on the tourism sector with the tourists' increasing number . Investment expenditures may increase by making new or extra investments to meet the tourists' needs, who coming to the region. Finally, the opportunity cost may also create problems for the region and local people, instead of investments that may be socially beneficial for the regional economy (hospitals, schools, etc.), by using most of the income for the supply of scarce resources due to tourism resources or excessive density (Mason, 2003; Harcombe, 1999; Weaver, 2001; Page & Connel, 2006; Milano, 2017; Kumar et al., 2015).

Negative Socio-Cultural Effects of Overtourism

While it becomes easier to evaluate the economic leg of the development of tourism with income, foreign exchange, and employment, and its environmental impact with concrete data such as sustainability, wear, aging, and deterioration, there is no such thing in socio-cultural ethics. According to Pizam and Milman (1986), the social and cultural effects of tourism can be measured by the ways tourism contributes to value systems, individual behaviors, family relationships, collective lifestyles, moral behavior, creative expressions, traditions, and customs and changes in community organization. In other words, it is considered as the effects of direct and indirect links with tourists on local people. Therefore, among the economic, environmental, and socio-cultural effects of overtourism, socio-cultural effects are identified as the most difficult and uncertain of the effects of overtourism (Żemła, 2020).

It is one of the socio-cultural effects of overtourism that make daily life activities (traffic congestion, crowding, increased local people's need for parking areas, density in public transportation areas, excessive demand for public spaces such as beaches, cafes, restaurants, etc.) . In addition, the local people may experience unhappiness due to exposure to overtourism. Conflicts may arise between local people and tourists due to language differences. The crime rate may increase due to overtourism in the touristic destination. With the people's arrival from different countries to touristic destinations, the spread of diseases and exchange of diseases between societies may occur. The increase in the consumption of products such as junk food, alcohol, and tobacco by the local people in line with the tourists' needs due to overtourism may cause the changes in the norms and cultures pre-determined in the touristic destination. With the increasing dependence on tourism, most of the society may lose their traditional activities and self-confidence, and this may cause cultural degeneration. There may be fewer housing opportunities for local residents due to online platforms such as Airbnb. There may be religious conflicts between tourists coming to the region and local people. In interpersonal relations, economic interest groups can replace moral values. Due to the differences in the level of life between the people of the region and the tourists coming from outside, the local people may start to perceive themselves as people who are only responsible for serving tourists, with the tourists' envy (Kuščer & Mihalič, 2019; Gössling, 2002; Mckercher & Cros, 2002; Koens et al., 2018; Ratz, 2000; Kozak et al., 1997; Jordan et al., 2018).

Negative Environmental Effects of Overtourism

The problem of overtourism (i.e., tourist congestion in the places visited and the negative effects on the environment caused by tourists and residents) is becoming more pronounced with the increasing popularity of tourism (Szromek et al., 2019). The environment quality, in which both natural resources and human influence play an important role, and the sustainability of this quality are very important for tourism. However, the relationship of tourism with the environment is complicated because it involves many activities that may have negative environmental effects. Many of these effects are linked to the construction of tourism facilities, including roads, airports, resorts, hotels, restaurants, shops, golf courses, and marinas that need a solid infrastructure. The negative effects of tourism development can gradually destroy the environmental resources it depends on and even destroys them (Sunlu, 2003). Depending on the population increase due to overtourism, the need for excessive use of energy resources, food resources, and scarce resources in the region may increase. Increasing sewage, garbage, and waste pollution in accommodation facilities in regions where touristic activities are intensely used and where interesting natural attractions are located can damage natural resources. The occurrence of air pollution, water pollution, soil pollution, and noise pollution in places with high tourist density may cause damage to livable resources for the region. Due to the construction of accommodation facilities on natural resources to meet the demand in the tourism season, many ecological problems such as cutting down trees, increasing concreting, damaging wildlife and vegetation, and visual pollution may occur in these regions. In addition to these, the fact that overtourism is not prevented and its realization in destinations for a long time may cause degradation or even disappearance of natural and built resources that depend on infrastructure and superstructure over time. Endangered fauna species and endemic flora species may also be damaged due to overtourism. In addition, the problem of destruction of archaeological sites and historical sites may arise due to overtourism (Kolawole et al., 2016; Rath & Gupta, 2017; Rabbany et al., 2013).

Prevention of Overtourism

The congestion and overcrowding in touristic areas is not a new event, but since 1960s, when mass tourism began to emerge in certain destinations, it has been addressed with a theoretical and practical perspective. With the congestion and crowding experienced in touristic areas, many different techniques and tools have been tried in order to manage the carrying capacity of the regions in question and to solve them by revealing the reasons. Overtourism creates undeniable effects and consequences that could potentially worsen in tourism resources that overcrowded processes and activities of touristic destinations (Ivars-Baidal et al., 2019).

Since the concept of overtourism and its effects are multidimensional, there is no clear responsibility for its solution. It is an undeniable fact that each destination has its own solution paths and that these solutions vary from destination to destination. Considering that the problem in most destinations is due to the lack of planning, it can be easily provided to establish solutions for the destinations by establishing the studies on sound policies to prevent overtourism and to eliminate the planning deficiencies. Although many tourist groups visit some popular areas of the destinations at the same time, they will face an unlimited tourist growth for the local people living there, but at the end of this situation, negative attitudes towards tourism and tourists will occur among the local people due to the failure to manage sustainable growth. With the proper implementation of planning and policies, measures can be taken for destinations with potential exposure to overtourism, along with the management of destinations that will be affected by overtourism (Duyar & Bayram, 2019).

In order to eliminate the negative effects of overtourism on touristic destinations, the local peopleand tourists coming to the region, comprehensive cooperation can be achieved in which the local people are at the center of every stage of tourism development. Investments suitable for the local texture can be encouraged in the investments to be made in the region, taking care not to damage the social, cultural, architectural, and natural environment. In addition, while appropriate integrated approaches to tourism development have a critical role in preventing overtourism, an appropriate balance must be struck in meeting the key stakeholders' needs such as local businesses, local residents, and tourists. Economic instruments such as taxes, fees, and variable pricing imposed on tourists are seen as important policies to combat overtourism. However, such taxes and associated practices are only reasonable if used to manage the environmental, social, or cultural negative impacts of overtourism. Additionally, housing management and regulation may be introduced for local residents. Destinations that have not yet been exposed to overtourism can be made more attractive. It may be helpful to provide information that encourages tourists to respect local culture and the environment and to inform them of the negative effects of not following these principles. In order to prevent overtourism, plans and policies involving the local people can be made and implemented in a participatory process in cooperation with both private and public sectors. The number of tourists can be limited by determining the carrying capacity of the touristic destination. The tourist density can be distributed, and the crowding of tourists in certain places can be prevented by diversifying the demand in time and place in a touristic destination, creating innovative products and different experiences. In the investments to be made in the region, the benefit for the local people should be prioritized, and at the same time, the local people's active participation and support , which can be a tourist product, can be provided. Finally, daytime intensity can be avoided by providing 24-h service for museums, which are an important resource in the touristic destination, and by offering the advantage of wages at night (Duran & Özkul, 2012; Nepal & Nepal, 2019; Szromek et al., 2019; Maingi, 2019; Çeti, 2018; Aylan & Aylan, 2020; UNWTO, 2018).

As a result, it seems that there is no single-item magic wand that can solve the problem of overtourism. What is important here is that the problems that may cause overtourism are addressed objectively, the local people's welfare level is kept high, and the long-term strategic planning for the tourism industry is combined with a horizontal and vertical inclusive governance model (Joppe, 2019). It will enable the implementation of plans and policies, taking into account the economic and environmental, and socio-cultural effects of the measures related to overtourism in touristic destinations where touristic activities take place intensively and the preservation and preservation of existing tourist resources in a recyclable and sustainable manner.

Technological Applications in Overtourism

The negative impact of overtourism on destinations has increased especially in the last decade, and has turned into anti-tourism and anti-tourist actions on behalf of the local people living in those destinations. Overtourism has mainly benefited the travel industry, which encourages tourists in the name of economic growth while also causing an increase in the number of tourists in destinations. Although the increase in the number of tourists is perceived as positive, overtourism has caused economic, socio-cultural and environmental problems as a result of mismanagement of destinations and lack of coordination in implementation. Of course, it is unthinkable to link the source of these problems only with local governments. In addition, social media, mobile applications, the emergence of travel booking platforms, advertising, promotion, marketing, and sales strategies, which are included in digital technology in the twenty-first century, have also played an important role in the increase the number of tourists coming to popular destinations (Malani et al., 2020).

Considered as such, the combined effect of technological applications in increasing the number of tourists is very complex and difficult to fully understand. However, technological applications that provide many innovations and comfort areas in the formation of price competition between destinations, facilitating access to information, providing reservation opportunities and payment solutions, price reductions in transaction costs for tourists have affected tourists to be more comfortable in creating their own routes. At the same time, technological applications have affected the tourism culture of the tourist destination by providing tourists to travel irregularly. In short, the applications of technological changes and developments in the tourism industry have increased the supply of tourism services due to the increase in demand in the touristic destination along with social and cultural changes, and this has led to the emergence of overtourism (Nilsson, 2020).

As it can be seen that technological developments are wanted to be used in the tourism industry, it has also affected tourist behavior and tourism culture. However, it is an undeniable fact that technological innovations are used as one of the most helpful areas to combat overtourism, as well as movements that increase overtourism (Malani et al., 2020). One of the technological applications carried out in order to eliminate the negative effects of overtourism and to make positive effects of digital developments on the touristic destination is the use of technological devices such as special sensor cameras or heat map to provide information about the density of the destinations and offer alternative visits. In addition, within the framework of sustainable Urban Mobility Plans, real-time information can be provided to users, and the use of public transport networks can be encouraged depending on the traffic density. Intelligent analytical systems can be used to determine the number of people coming to the destination and to take measures, as well as the use of tourist cards at destinations, wi-fi networks, monitoring the location-notification activities in social networks, and controlling the expenditures made by credit cards. Smart city projects can be widespread in touristic destinations exposed to overtourism. Well-known travel agencies can promote tourists to less popular tourist destinations both on their social networks and on their websites, providing tourists to change their holiday routes. This may reduce the demand load of popular tourist destinations. Mobile applications can be installed on the tourists' mobile phones , who coming to the touristic destination to check their mobility within the destination. In order not to create crowds in popular areas of touristic destinations, a mobile application can be made that shows crowding rates and alternative tourism routes with low density. In addition, with the application of smart parking systems, tourists coming to the region can prevent parking problems. This situation may also contribute to the reduction of traffic congestion (Vargas-Sánchez, 2020; Ivars-Baidal et al., 2019).

The aforementioned technological developments for the prevention of overtourism can be evaluated as short, medium, and long term for touristic destinations. While technological developments are evaluated as solution-oriented to overtourism, they should be able to offer facilitating solutions for both the tourists coming to the region and the local people living in the region. Since the costs of smart systems created by both private and public sectors will vary depending on the technology chosen, facilitating methods such as GPS, search engines (Google etc.), mobile applications on mobile phones, digital panels can be used to reduce costs in touristic destinations. Thanks to the data obtained from the technological solutions in these platforms, visitors and the local people will have an innovative application. In addition, their visits and lives at the destination will also be of high quality. Directing people to alternative areas without being crowded in the touristic destination, determining the traffic density and parking areas, directing local law enforcement officers to the place of the incident and providing immediate intervention to the incidents, developing interesting places of less visited areas and turning them into touristic demand, developing new attraction centers and determining visitor routes according to need can also be considered as the common benefits of technological data. While it is important that technological data are an indisputable issue in preventing overtourism, it is very difficult to collect and evaluate data in tourist destinations. It is also possible to encounter many problems such as expensive installations in technologies applied against overtourism, the need for infrastructure and superstructure, difficulties in obtaining permissions for the protection of personal data, and the cultural structure in the settlement (Zubiaga et al., 2019).

In order to ensure the competitiveness of a touristic destination and to protect it from the negative effects caused by overtourism, decision makers belonging to private and public institutions in the destinations allocate great resources to smart system technologies and initiatives. Moreover, tourist cities offering tourism supplies around the world are increasingly accepting the concept of smart tourism and related strategies as a means of optimizing sustainable environments. Current technological solutions, especially for tourist destinations that face problems such as exposure to overtourism and negative perceptions of the local people towards overtourism, encourage a tourist destination to overcome this challenge by creating urban areas where the local people and visitors can have fun together and there is no density (Lee et al., 2020).

Overtourism and Technological Applications in Turkey

One of the factors that create the smart city concept is the phenomenon of "sustainability". Sustainability was first expressed in the UN Brundtland Report published in 1987 as "the harmonious integration of economic life and environment". Sustainability is expressed as keeping the negative impacts on the ecosystem in balance below the level that the world cannot bear. The essence of this statement is not to be a consumer society but to be in universal solidarity and to think with a focus on environmental management, social responsibilities, and economic solutions (Mirghaemi, 2019: 39). Using renewable energy resources that do not restrict life on the basis of technological developments will contribute positively to people and ecosystems in order to make human welfare sustainable on the basis of development.

As a result of the transformation of the demand for urban life, which is seen as the main source of innovation, especially after the 2000s, rapid increase in the rate of urbanization in Turkey brings with it many problems such as traffic, housing, air pollution, education, health, safety, communication, infrastructure, energy, environment, and unemployment. By solving these problems, increasing the life quality and improving services seem possible only with information and communication technologies supported solutions. There are some studies of central administrations and local governments based on geographical information systems about smart cities in Turkey, and however, comprehensive targets have not been determined yet. Especially in big cities, it is seen that smart city solutions have started in partial areas such as urban services, transportation, energy, and water services (Örselli & Akbay, 2019). In this context, especially in the period when the carrying capacity of tourism in Turkey, which is dense cities that exceeded the cities of Antalya and Istanbul are examined, and the emphasis is on what are the technological applications for overtourism.

Structural change and tourism development in Turkey are based on the 1980 (Istanbul Kalkınma Ajansı, 2012). In this development and change, the Tourism Incentive Law numbered 2634, which was enacted in 1982, which enabled the working style and conditions of the Turkish tourism industry to reach international standards (Gökdeniz et al., 2009) and started the liberal period in the history of Turkish Tourism (Akdağ et al., 2019). An almost irrevocable decision is given on the tourism industry in Turkey as a result of the economic, social, and political crises in 1980. Thus, generous incentives were given to the tourism industry with the Tourism Incentive Law enacted. Consequently, Turkey unexpectedly rapid tourism growth was experienced in the absence of proper planning and development principles (Mercan, 2015).

Turkey's tourism is concentrated mostly on the southern and western coasts. Turkey's rapid growth in the demand and supply capacity has been observed between 1980 and 1990 with the Tourism Promotion Law. However, the tourism development mostly in coastal regions, has caused regional differences, and many problems have emerged (Tosun et al., 2006). Antalya, which was an agricultural city in the early 1980s, became one of the world's leading destinations (Istanbul Kalkınma Ajansı, 2012). Turkey is one of the provinces most exposed to internal migration because the job opportunities provided by the tourism industry attract people to Antalya where is specifically described as Turkey's tourism capital. The increase in the number of accommodation facilities and potential tourists in Antalya causes the unplanned use of the coastal areas and the rapid loss of natural values. Especially with the allocation of golf courses in the Lara-Kundu region, a significant amount of forest areas are destroyed. Thus there is a loss of natural values and biological diversity. In addition, excessive use of water during the tourism season causes the disruption of the water supply process and water scarcity (Arüv et al., 2021). In this context, Antalya, which hosts millions of tourists from all over the world every year, benefits from smart city solutions. Examples of smart cities that stand out in Antalya are smart city management platform, solar power plants, solid waste integrated facilities, smart lighting, and irrigation system (T.C. Çevre & Şehircilik Bakanlığı, 2019).

Another important tourist destination is benefiting from smart city solutions in Istanbul. Istanbul, which was the capital of the Roman, Byzantine, Latin, and Ottoman Empires and whose history goes back 300 thousand years, is Turkey's most important urban megalopolis in terms of economy, finance, population (14,800 thousand people). Istanbul is not only in Turkey but also in many important cities in the world in terms of its historical, touristic and geographical features. Istanbul is a center of attraction in terms of tourism with its palaces, mosques, walls, baths, bridges, the unique beauty of the Bosphorus, in short, in every aspect. However, the emerging urban problems (traffic, housing, etc.), the increase in migration originating from rural areas and Syria in recent years pose threats to Istanbul in the context of overtourism with the ever-increasing population in Istanbul, which is an important center of attraction on a global scale. On the other hand, the increased security, housing, infrastructure requirements, the risk of air pollution (especially particulate matter), the inadequacy of the roads in parallel with the increase in the number of vehicles on the traffic, the insufficient rate of green areas per person due to increasing housing and infrastructure investments pose threats to Istanbul tourism in the context of overtourism (Küçükaltan, 2017).

The city of Istanbul, which has a large population for all time, benefited from advanced technology in every period. Thus, It has been a pioneer in the field of smart cities in Turkey. The city of Istanbul, whose water resources are extremely limited, has had a highly developed water management system throughout history with structures such as aqueducts, cisterns, arches, fountains, and distribution places (water distribution stations). The city of Istanbul, which built the world's second underground subway in 1875 after London, started a natural gas transformation in order to cope with the air pollution that peaked in the 1980s and successfully implemented it in the following years. Istanbul is one of the cities where the traffic density map, which has become indispensable for the local people in the world, was first put into service. Given that the return of citizens with relevant services effectively transmitted to the unit received 153 White Table line has been a pioneer in Turkey. In addition to these, Istanbul Metropolitan Municipality Smart City Directorate, established in 2016, combined all smart city stakeholders and initiatives in line with a single vision with the Istanbul Smart City Road Map prepared in 2017. In addition, a Smart City Manifesto has been prepared to address all Istanbulites. Prominent examples of smart cities in Istanbul are environmental control center, transportation management center, adaptive traffic signal control, i-taxi, floating solar power plant, smart recycling container, domestic waste incineration, and energy generation facility, mobile eds, energy production from landfill gas, smart park management, ground Istanbul (T.C. Çevre & Şehircilik Bakanlığı, 2019).

Conclusion and Recommendations

The experienced seasonal tourism activities in Turkey and more tourism activities in the Mediterranean, Aegean, and Marmara coast of condensation in the summer causes more cases of overtourism and be experienced in these areas. To provide a significant economic contribution to the tourism industry in Turkey is seen as a priority target as a developing country. For this reason, it can be said that the works on measuring the negative effects of the overtourism phenomenon are in the background. Apart from some smart cities such as Istanbul and Antalya, it is seen that technological applications for the negative effects of overtourism are not widespread due to the seasonality of the tourism events in Turkey.

The concept of overtourism, which emerged in the 1960s to determine the limit of the number of visitors that destinations can overcome, became very popular in the 1970s. However, interest in the concept of overtourism has decreased to date. It is said that are limited applications for the prevention of overtourism in Turkey, and most of the current applications are seen in coastal areas and smart cities. Overtourism, which is related to the number of tourists, the flow of visitors, the protection of nature, and the experiences of visitors, is actually an important event that should be addressed in all destinations where tourism activities are intense.

The direction of tourism mobility can be changed by proposing alternative destinations that are similar to the regions where tourism activities are intense in order to reduce the social, cultural, physical, environmental, and similar negative effects of overtourism and to ensure the sustainability of the destinations on social media and digital platforms related to travel agencies. Even with digital technologies such as augmented reality applications that direct low-crowded areas as a selection criterion, tourists can be directed to alternative destinations. Apart from these, technological applications such as intelligent transportation systems can be used in order to control the traffic density created by the crowd and to monitor public transportation vehicles, and also, technological applications such as a tourist management system or a tracking system can be used to monitor the number of tourists in order to prevent the carrying capacity of certain regions. Thus, thanks to smart transportation systems, congestions in the transportation network are reduced, relocation can be done more easily and comfortably, negative environmental effects and energy consumption are reduced, and high efficiency is achieved. As a result of this, overtourism, which adversely affects the life quality of the local people during the peak periods of tourism activities, can be controlled. At the same time, tourists do not experience the adverse effects of overtourism that affect their life quality for tourism activities, thus creating a positive image for the destination. This contributes to the sustainability of the destination in terms of tourist satisfaction and tourist loyalty.

The acceleration of tourism activities due to the increase in the world population, the diversity of tourism activities due to the spread of the modern lifestyle, and the demand for more popular destinations can be seen as the reasons that cause overtourism. It can be said that this situation makes the problems such as environmental pollution, visual and noise pollution, crowding, traffic density, unplanned urbanization, increase in greenhouse gas emissions, which are closely related to overtourism, make them important issues that occupy the world agenda. Therefore, in the context of efforts to tackle environmental problems caused by overtourism. Innovative solutions and applications based on information and communication technologies are a priority in both national and international platforms in terms of ensuring the sustainability of destinations, increasing the welfare level of the local people, and ensuring the availability of existing resources for future generations.

References

Akdağ, G., Ozan, G., & Çakıcı, A. C. (2019). Turizme Genel Bakış. In A. C. Çakıcı (Ed.), *Kavramlar ve Örneklerle Genel Turizm* (pp. 1–26). Detay Yayıncılık.

Anuar, A., Ridzuan, F., Jaini, N., Sulaiman, F., & Hashim, I. (2019). The impact of overtourism towards local community in heritage city. *Journal of Tourism & Hospitality, 8*(3), 1–5.

Arüv, C., Yıldırım, M., & Topkaya, B. (2021). *Antalya Kentindeki Turizm Faaliyetlerinin Kıyı Alanlarına Etkilerinin Belirlenmesi*. Retrieved from https://www.imo.org.tr/resimler/ekutuphane/pdf/11094.pdf. Accessed 12 Feb 2021.

Avond, G., Bacari, C., Limea, I., Seraphin, H., Gowreesunkar, V., & Mhanna, R. (2019). Overtourism: A result of the Janus-faced character of the tourism industry. *Worldwide Hospitality and Tourism Themes, 11*(5), 552–565.

Aylan, F. K., & Aylan, S. (2020). Aşırı Turizm Kavramının Farklı Boyutlarıyla İncelenmesi: Instagram'da Paylaşılan Fotoğraf ve Yorumların İçerik Analizi. *Journal of Tourism and Gastronomy Studies, 8*(1), 584–607.

Capocchi, A., Vallone, C., Pierotti, M., & Amaduzzi, A. (2019). Overtourism: A literature review to assess implications and future perspectives. *Sustainability, 11*(12), 3303.

Center for Responsible Travel (CREST). (2018). *The case for responsible travel: Trends & statistics 2018*. Retrieved from https://www.responsibletravel.org/docs/The_Case_for_Responsible_Travel_2018_FINAL_FOR_WEB.pdf. Accessed 14 Dec 2020.

Çeti, B. (2018). Turizm Politikası ve Planlaması Kapsamında Taşıma Kapasitesi Kavramının Değerlendirilmesi. *Sosyal, Beşeri ve İdari Bilimler Dergisi, 1*(2), 1–17.

Çetinoğlu, D., & Uygur, S. M. (2020). Aşırı Turizm Sorununa Çözüm Olarak Pazarlamama Stratejileri Kullanılabilir Mi? Pamukkale Örneği. *İşletme Araştırmaları Dergisi, 12*(3), 3008–3026.

Dodds, R., & Butler, R. (2019). The phenomena of overtourism: A review. *International Journal of Tourism Cities, 5*(4), 519–528.

Doğan, M. (2013). Bir Kriz Faktörü Olarak Taşıma Kapasitesi Aşımı ve Bozcaada Örneği. *Trakya Üniversitesi Sosyal Bilimler Dergisi, 15*(1), 85–108.

Duran, E., & Özkul, E. (2012). Yerel Halkın Turizm Gelişimine Yönelik Tutumları: Akçakoca Örneği Üzerinden Bir Yapısal Model. *International Journal of Human Sciences, 9*(2), 500–520.

Duyar, M., & Bayram, M. (2019). Aşırı Turizm ve Turizm Korkusu: Yerel Halk ve Turizm İlişkisinin Evrimi. *International Journal of Geography and Geography Education, 40*, 347–362.

Goodwin, H. (2017). *The challenge of overtourism* (Responsible tourism partnership working paper no. 4). Retrieved from https://haroldgoodwin.info/pubs/RTP'WP4Overtourism01'2017.pdf. Accessed 14 Jan 2021) (pp. 1–19).

Goodwin, H. (2019). *Overtourism: Causes, symptoms and treatment, tourismus wissen – Quarterly April 2019*. Retrieved from https://responsibletourismpartnership.org/wp-content/uploads/2019/06/TWG16-Goodwin.pdf. Accessed 12 Jan 2021.

Gowreesunkar, V., & Seraphin, H. (2019). Introduction: What smart and sustainable strategies could be used to reduce the impact of overtourism? *Worldwide Hospitality and Tourism Themes, 11*(5), 484–491.

Gökdeniz, A., Dinç, Y., Aşık, N. A., Münger, L., & Taşkır, H. (2009). Türkiye'de İç Turizm Kavramı ve İç Turizmde Önemli Bir Destinasyon Olan Ayvalık'ta Müşteri Arz ve Talebine Yönelik Ampirik Bir Araştırma Işığında Geliştirme Stratejileri. *Balıkesir Üniversitesi Sosyal Bilimler Enstitüsü Dergisi, 12*(22), 216–231.

Gössling, S. (2002). Global environmental consequences of tourism. *Global Environmental Change, 12*(4), 283–302.

Harcombe, D. (1999). The economic impacts of tourism. *ABAC Journal, 19*(2), 10–22.

Ivars-Baidal, J., García Hernández, M., & Mendoza de Miguel, S. (2019). Integrating overtourism in the smart tourism cities agenda. *e-Review of Tourism Research (eRTR), 17*(2), 122–139.

İstanbul Kalkınma Ajansı. (2012). *Türkiye ve İstanbul Bölgesinde Turizm*. Retrieved From https://www.istka.org.tr/media/1074/t%C3%BCrkiye-ve-%C4%B0stanbul-b%C3%B6lgesi-nde-turizm.pdf. Accessed 12 Feb 2021.

Joppe, M. (2019). 17. The roles of policy, planning and governance in preventing and managing overtourism. In R. Dodds & R. Butler (Eds.), *Overtourism* (pp. 250–261). De Gruyter.

Jordan, P., Pastras, P., & Psarros, M. (2018). *Managing tourism growth in Europe*. European Cities Marketing.

Koens, K., Postma, A., & Papp, B. (2018). Is overtourism overused? Understanding the impact of tourism in a city context. *Sustainability, 10*(12), 4384.

Kolawole, I. O., Mbaiwa, J. E., & Mmopelwa, G. (2016). The environmental impacts of tourism on community people's quality of life in Maun. Botswana. *Journal of Hospitality, Tourism and Leisure, 5*(4), 1–14.

Kozak, N., Akoğlan, M., & Kozak, M. (1997). *Genel Turizm*. Anatolia Yayınları.

Kumar, J., Hussain, K., & Kannan, S. (2015). Positive vs negative economic impacts of tourism development: A review of economic impact studies. *Developments of the New Tourism Paradigm in the Asia Pacific Region*, 405–413.

Kuščer, K., & Mihalič, T. (2019). Residents' attitudes towards overtourism from the perspective of tourism impacts and cooperation-the case of Ljubljana. *Sustainability, 11*(6), 1823.

Küçükaltan, H. (2017). *Turizm Sektörü Raporu (TÜRSAB & TUADER)*. Retrieved from https://www.tursab.org.tr/dosya/16736/turizm-faaliyet-kitapcigi_16736_492081.PDF. Accessed 12 Feb 2021.

Lee, P., Hunter, W. C., & Chung, N. (2020). Smart tourism city: Developments and transformations. *Sustainability, 12*(10), 1–15.

Maingi, S. W. (2019). Sustainable tourism certification, local governance and management in dealing with overtourism in East Africa. *Worldwide Hospitality and Tourism Themes, 11*(5), 532–551.

Malani, P., Gupta, S., & Chaturvedi, A. (2020). Digital trends: A problem or a solution for Overtourism. *Edrag & Organisatie Review, 33*(3), 309–326.

Mason, P. (2003). *Tourism impacts, planning and management*. Butterworth–Heinemann.

Mckercher, B., & Cros, H. (2002). *Cultural tourism, The Partnership between tourism and cultural heritage management*. The Hawort Pres.

Mercan, Ş. O. (2015). Turizm Politikasının Tanımı, Özellikleri ve Turizm Politikasını Oluşturan Unsurlar. In D. Küçükaltan, H. Çeken, & Ş. O. Mercan (Eds.), *Değişik Perspektifleriyle Turizm Politikası ve Planlaması* (pp. 1–10). Ankara.

Milano, C. (2017). *Overtourism and tourismphobia: Global trends and local contexts*. Ostelea School of Tourism & Hospitality. https://doi.org/10.13140/RG.2.2.13463.88481

Mirghaemi, S. A. (2019). Akıllı Kentler Üzerine Bir İnceleme: Türkiye Örneği. *Beykent Üniversitesi Fen ve Mühendislik Bilimleri Dergisi, 12*(2), 37–46.

Moreno-Gil, S., & Coca-Stefaniak, J. A. (2020). Overtourism and the sharing economy–Tourism cities at a crossroads. *International Journal of Tourism Cities, 6*(1), 1–7.

Nepal, R., & Nepal, S. K. (2019). Managing overtourism through economic taxation: Policy lessons from five countries. *Tourism Geographies*. https://doi.org/10.1080/14616688.2019.1669070

Nilsson, J. H. (2020). Conceptualizing and contextualizing overtourism: The dynamics of accelerating urban tourism. *International Journal of Tourism Cities, 6*(4), 657–671.

OECD. (2020). *Rethinking tourism success for sustainable growth*. Retrieved from https://www.oecd-ilibrary.org/docserver/6b47b985en.pdf?expires=1613830237&id=id&accname=guest&checksum=55AB162C8CFFA16230DEA87F0F9F43EB. https://doi.org/10.1787/6b47b985-en. Accessed 27 Dec 2020.

Örselli, E., & Akbay, C. (2019). Teknoloji ve Kent Yaşamında Dönüşüm: Akıllı Kentler. *Uluslararası Yönetim Akademisi Dergisi, 2*(1), 228–241.

Page, S., & Connell, J. (2006). *Tourism a modern synthesis* (2nd ed.). Thomson.

Paslı, M. M. (2020). Hırçın Karadeniz'in Yavaş Şehri Perşembe. In F. Başar & T. Türkmendağ (Eds.), *Yavaş Şehirler ve Alternatif Turizm* (pp. 79–95). Detay Yayıncılık.

Perkumienė, D., & Pranskūnienė, R. (2019). Overtourism: Between the right to travel and residents' rights. *Sustainability, 11*(7), 1–17.

Pizam, A., & Milman, A. (1986). The social impacts of tourism. *Tourism Recreation Research, 11*(1), 29–33.

Rabbany, M. G., Afrin, S., Rahman, A., Islam, F., & Hoque, F. (2013). Environmental effects of tourism. *American Jurnal of Environment, Energy and Power Research, 1*(7), 117–130.

Rath, N., & Gupta, R. (2017). Environmental impact of tourism. *International Journal of Advance Research and Innovative Ideas in Education, 2*(3), 50–53.

Ratz, T. (2000). The socio cultural impacts of tourism case of Lake Balaton. *Research Support Scheme*, 1–20.

Sunlu, U. (2003). Environmental impacts of tourism. In D. Camarda & L. Grassini (Eds.), *Local Resources and global trades: Environments and agriculture in the mediterranean region conference on the relationships between global trades and local resources in the mediterranean region* (Options Méditerranéennes: Série A. Séminaires Méditerranéens; n. 57) (pp. 263–270). CIHEAM.

Szromek, A. R., Hysa, B., & Karasek, A. (2019). The perception of overtourism from the perspective of different generations. *Sustainability, 11*(24), 7151.

T.C. Çevre ve Şehircilik Bakanlığı. (2019). *Akıllı Şehirler*. Retrieved from https://webdosya.csb.gov.tr/db/cbs/menu/akillisehirler-kitap_20190311022214_20190313032959.pdf. Accessed 10 Feb 2021.

Tosun, C., Fletcher, J., & Fyall, A. (2006). Turkey: EU membership implications for sustainable tourism development. In D. Hall, M. Smith, & B. Marciszewska (Eds.), *Tourism in the Europe: The challenges and opportunities of EU enlargement* (pp. 270–287). CAB International.

UNWTO. (2018). *Overtourism? Understanding and managing urban tourism growth beyond perceptions: Executive summary* (pp. 1–12) [online]. : UNWTO. Retrieved from https://www.e--unwto.org/doi/book/10.18111/9789284420070. Accessed 2 Jan 2021.

Vargas-Sánchez, A. (2020). Overtourism and smart cities: Present and future. In J. A. Oskam (Ed.), *The overtourism debate* (pp. 267–278). Emerald Publishing Limited.

Weaver, D. B. (Ed.). (2001). *The encyclopedia of ecotourism*. CABI.

Weber, F. (2017, June 14–17). Overtourism. An analysis of contextual factors contributing to negative developments in overcrowded tourism destinations. In *BEST EN Think Tank XVII: Innovation and progress in sustainable tourism* (pp. 315–320). .

Yüncü, H. R. (2019, October 16–19). Turizmin Karanlık Yüzü: Şehirlerde Aşırı Turizm. 20. In *Ulusal Turizm Kongresi* (pp. 1026–1030). .

Żemła, M. (2020). Reasons and consequences of overtourism in contemporary cities-knowledge gaps and future research. *Sustainability, 12*(5), 1729.

Zubiaga, M., Izkara, J. L., Gandini, A., Alonso, I., & Saralegui, U. (2019). Towards smarter management of overtourism in historic centres through visitor-flow monitoring. *Sustainability, 11*(24), 7254.

Chapter 19
Tourism Development in Italian Natural Protected Areas: Traditional and Innovative Tools to Face Overtourism and Other Possible Negative Effects

Sonia Ferrari and Nicolaia Iaffaldano

Abstract Today natural protected areas are frequently successful tourist destinations worldwide. However, because of their nature and functions, they should develop sustainable forms of tourism and avoid attracting mass tourism. For this reason and also because the green tourism segment is growing very fast, the management of protected areas has to face several problems linked to different aspects of tourism sustainability. In fact, together with tourism development, what can arise is overtourism and its consequences, such as environmental damages, negative reactions of the local community, or other sustainability problems. As for socio-cultural sustainability, it is related to the impact of tourism development on the socio-cultural fabric of the place and the optimization of the relationship between hosts and guests. Overtourism is probably the most problematic aspect to evaluate, predict and plan in terms of possible negative impact on the natural and social-cultural environment. In the paper, the authors will identify and describe the variables that could significantly influence tourism sustainability in green destinations, above all in protected natural areas. They will present the case of Tremiti Islands Marine Reserve in order to identify and analyze the more effective instruments that can be used to cope with these problems in a natural protected area, considering the various aims and expectations of the stakeholders involved. Qualitative research is conducted interviewing different categories of stakeholders. Findings identify tools to prevent harmful behavior and for monitoring the impact of tourist flows and the traditional and innovative measuring instruments to reduce the impact of the flows of visitors which are on Tremiti Islands.

S. Ferrari (✉)
Department of Business and Law Science, University of Calabria, Arcavacata, Italy
e-mail: sonia.ferrari@unical.it

N. Iaffaldano
Ionian Department of Law, Economics and Environment, University of Bari, Bari, Italy

A. Hassan, A. Sharma (eds.), *Overtourism, Technology Solutions and Decimated Destinations*, https://doi.org/10.1007/978-981-16-2474-2_19

Keywords Tourism development · Sustainability · Natural protected areas · Overtourism · Italian protected areas

Introduction

Terrestrial and marine protected areas currently cover about 15% of the world's surface. They receive 8 billion visits per year, and United Nations World Tourism Organization (UNWTO) estimates that tourism in protected areas will continue to grow by a 3.3% annual rate through 2030 (UNEP-WCMC and IUCN, 2016). Nature-based tourism is one of the tourist segments that is increasing more in the last decades, and nowadays, natural protected areas have become important destinations (Eagles & McCool, 2002; Frost & Hall, 2010; Leiper, 1990; Manning & Anderson, 2012; Wall-Reinius & Fredman, 2007). Tourism is considered important for the development of the host community; in fact, it can help promote the place as a destination and strengthen local brands and productions, and thus increasing job opportunities and incomes. In addition, it enhances local residents' quality of life, especially in inner and peripheral areas frequently affected by unemployment and depopulation and where parks very often arise. The risk could be a form of tourism dependence, that is, excessive dependence of the economy of the communities that host nature-based tourism attractions on the income from the tourism industry (Sisneros-Kidd et al., 2019).

The main aim of the first natural protected areas, born after the year 1870, was the preservation of their high natural value (Hall & Frost, 2010). In fact, in that period, protected areas were seen as sanctuaries of nature. Since the 1970s, a new scope was born, and it was linked to the need to stop the loss of biodiversity, which is connected to the concept of environmental sustainability (Cimini, 2011). Today, natural protected areas have become tourist destinations of increasing importance, and they are seen as effective instruments for growth (Ferrari & Pratesi, 2012).

However, because of their role and institutional goals, natural protected areas should focus on forms of sustainable tourism (Fredman & Tyrväinen, 2014; Haukeland et al., 2014; Newsome & Hughes, 2018; Wearing et al., 2016). Actually, high visitor numbers and overtourism cannot occur without creating ecological impacts, and the over-development of tourism in many national parks is likewise causing serious worries about the natural environment and biodiversity conservation. As a consequence, an effective governance model is needed to monitor, halt or reduce the potential impacts of this phenomenon, thus enjoying the economic, social, and conservation benefits of nature-based tourism as well. Strategic approaches are necessary to maintain a balance between the human presence and environment conservation, and tourism, especially sustainable tourism, is often considered as a potential mechanism in these terms.

Natural Protected Areas and Sustainable Tourism

In the tourism industry, the concept of sustainability has evolved over time. The term "sustainable tourism" is complex and difficult to define (Butler, 1999). It was first used in the late 1980s as a result of the growing awareness of the often negative tourism development impact. In those years, it was mainly perceived as the exact opposite of "mass tourism", which was considered the source of all the negative effects of tourism (Clarke, 1997; Mowforth & Munt, 2003; Swarbrooke, 1999). From the initial concept of a form of tourism that improves the quality of life and preserves ecosystems, nowadays, there is a new idea of tourism development that equally distributes the generated value among the stakeholders. The main aim of sustainable tourism development is to promote the growth of an area, avoid the destruction of its resources for future generations and minimize negative environmental impacts (Buckely, 2012; Stoddard et al., 2012; Tyrrell et al., 2013; Weaver, 2006,2010). Currently, in literature, the three main dimensions of sustainability are those of the so-called "triple bottom line" (TBL): environmental, socio-cultural, and economic (Boley & Uysal, 2013; Elkington, 1997, 1998, 2004; Farrell, 1992).

Obviously, the first aspect must aim at promoting environmental biodiversity, ecosystems, and heritage conservation. As for the socio-cultural dimension, sustainable development is generally viewed as a matter of quality of life, which involves multiple elements of the host community's living conditions, such as the enhancement of civic pride and place identity and image, the increasing of welfare, the revitalization of local culture, the reinforcement of social fabric, the satisfaction of tourists and residents, and the improvement of the relationship between them (Bartelmus, 1986; Clarke, 1997; Pearce et al., 1990; Swarbrooke, 1999). The involvement of key stakeholders, primarily the local residents, in the processes of planning and management of tourism development strategies is necessary to avoid negative attitudes and give the host community a sense of control of local resources. Therefore, place stakeholders' involvement is essential for the sustainability of tourism. For this reason, any analysis of sustainability should not overlook the political aspects (Ferrari and Gilli, 2018; Jamal and Getz, 1995; Ostrom et al., 1999; Peng et al., 2016). Referring to economic sustainability, all forms of development should maximize the benefits for the destination in terms of income and jobs, thus concurrently favoring the growth of local enterprises through the promotion of place brands and area products and services as well (Choi & Sirakaya, 2006; Garay & Font, 2012; Roberts & Tribe, 2008).

The development of the new segment of "slow tourism"(Dickinson & Lumsdon, 2010; Heitmann et al., 2011; Matos, 2004; Meng & Choi, 2016) is favoring patterns of local development based on the described dimensions of sustainability, especially in green destinations. This tourist segment involves visitors that are looking for an immersion in everyday life of local residents and that are eager to discover authentic aspects of the destination's culture and heritage, avoiding standardized tourist packages (Cohen, 1988; MacCannell, 1973). These visitors are curious to discover tastes and scents, prefer to stay in accommodations linked to the host culture (such as

Bread and Breakfast; Diffuse hotels), buy local products, and enjoy “zero-kilometer” food (i.e., the Italian for the English “zero-mile” food), thus showing a high awareness about natural environment conservation. All that favors local productions and the diffusion of economic benefits among the stakeholders. Therefore, we can say that this tourism segment seems more environmentally oriented, respects hosts and promotes attention to local culture, thus strengthening place identity, social cohesion and civic pride.

Impacts of Tourism Development in Natural Protected Areas and Sustainability Problems

Tourism development can bring many different positive environmental, socio-cultural, and economic effects. As to the environmental aspect, it can provide resources for monitoring, research, and educational initiatives, increasing the awareness about the value of natural resources and the important role of natural protected areas. The economic benefits are linked to the increase of jobs and income, the promotion of local productions and destinations, the birth of new enterprises, the improvement of infrastructures and services, etc. Finally, also the potential socio-cultural positive effects could be numerous: for example, the improvement of quality of life and local pride, the enhancement and conservation of place culture, arts and crafts (Leung et al., 2018).

Despite these positive implications, the growing number of visitors in protected areas and the rising interest of park managers in tourism are creating a growing concern regarding the natural environment, biodiversity, and ecosystems conservation besides the welfare of species targeted for tourism. The problem is not new; in fact, the effects of mass tourism on destinations have been a broadly researched theme over the last years (Butler, 1980; Ferrari et al., 2018).

However, the effects of tourism in natural protected areas are originated not only by the large extent of tourist flows, which in some cases cause the phenomenon of overtourism, but also by the incorrect behavior of many visitors (Capocchi et al., 2019; Cheer et al., 2019; Peeters et al., 2018). Actually, tourists are not often respectful of nature, consequently creating temporary or permanent damages to the delicate equilibrium of natural protected areas ecosystems. As described by Heslinga et al. (2019), the main aspects of tourist behavior that can be harmful both to visitors and the environment are three. The first one is the low level of preparation and experience that can create troubles. The second aspect is the every day more and more frequent underestimation of the risks related to an experience in nature that can also give rise to serious dangers. Finally, there is the disrespectful behavior towards natural resources in a host area. All that often led to local residents’ unfavorable attitudes towards tourism, with consequent frictions and tensions among hosts and guests and social sustainability concerns. The instruments of park management to face these problems and prevent their negative consequences are monitoring, environmental education, and teaching, in order to show future visitors how to be respectful of protected areas and their treasures.

Overtourism is the phenomenon that occurs when tourist flows are too high with reference to the local situation (dimension of the area, number of residents, etc.). In the last decades, numerous researchers have studied overtourism and how destinations should manage it. Two important indicators of the capacity of the host area to welcome visitors and of social sustainability are the "tourism intensity" (the visitor to resident ratio) and the "tourism function index" (ratio of tourist beds to residents) (Hernandez-Maskivker et al., 2019; Van der Borg et al., 1996; Vargas-Sánchez et al., 2011). The first indicator is related to the carrying capacity and indicates residents' perceptions of tourism. According to UNWTO, carrying capacity is "the maximum number of people that may visit a tourist destination at the same time without causing destruction of the physical, economic or socio-cultural environment and an unacceptable decrease in tourist satisfaction" (1997: 5). The second indicator is related to the social carrying capacity, that is, the residents' attitude towards tourism development (Graefe & Vaske, 1987; Van der Borg, 1992). In natural protected areas, both indicators are important to understand the possible attitudes of the local community and whether the tourist flows are socially and environmentally sustainable.

The negative effects of overtourism and of unsustainable tourism development could affect natural environment elements such as flora, fauna, habitat, and landscape. These impacts can be associated with the construction of private and public infrastructures (for sports, accommodation, transport, recreation). The social-cultural negative impact can be related to the commodification or disruption of local traditions, the conflict among tourists and residents, the destabilization of communities, and so on. Finally, the negative economic effects could be linked to the request for low wages and low skills, the loss of seasonal jobs, the dependency on tourism with a higher economy vulnerability, an unequal distribution of generated benefits among the stakeholders, etc. (Eagles et al., 2002; Leung et al., 2018).

To prevent and limit all the negative aspects of excessive tourism development and of overtourism, the management of natural protected areas has to program continuous monitoring, with the aim of controlling the flows of visitors (Burns et al., 2020) and being aware of the kinds of effects that are arising and their possible consequences on the physical environment and on the host community. The first step in this process is visitor use monitoring. In fact, "determining the amount of use, use trends, characteristics of use and users, and qualities of the experience (e.g., opportunities for privacy and solitude) can provide important information for managers" (Andrew et al., 2021: 2). As Andrew et al. (2021) explain, there are many different methodologies to make it, and the main are: self-counting (visitors register themselves voluntarily), direct-counting (through many instruments as aerial photography, surveys, direct observation, satellite statistics, etc.), and indirect counting. Regarding direct counting, many innovative tools are provided by Information and communications technology (ICT)s also thanks to the widespread use of smartphones and other mobile devices and the increasing offer of Wi-Fi facilities. For example, by means of photos posted on social media, it is possible to estimate the number of visitors. Other useful instruments are Global Positioning System (GPS) visitor tracking and the data about mobile phones, but also drones, electronic counting devices, and the Vessel Monitoring System (VMS) that it transmits the position of vessels. GPS are particularly important in the case of visitors moving away from the designed trails and official Personnel Office Identifier (POI)s and they can provide data to specific computer models that describe the movements of visitors of protected areas (D'Antonio et al., 2010; Ferrari & Inglese, 2019).

Besides, park managers have to act both on the demand and the supply side. As for the demand side, visitors' number and behavior must be controlled to verify possible problems and damages and to prevent negative impacts, educating and guiding tourists as well. In addition, it is necessary to avoid overtourism and peaks of presences, trying to better distribute the flow of visitors in spatial and temporal terms.

It is possible to reach these aims in different ways (Eagles et al., 2002). Some of the most used instruments are the following: trying to reduce demand seasonality (through the offer of events and discounts in low season, the imposition of fees, the adoption of specific communication policies) also investing to develop social tourism (schools, elderly and disabled groups that visit the park staying overnight during the low season); using restrictions of length of stay, total numbers, or type of use; encouraging visits in different areas through the creation and promotion of new points of interest (i.e., visitors centers, museums, etc.) and of various resources (villages, historic and artistic heritage, sports); charging differential visitor fees; establishing prohibitions to enter some areas in certain periods of the year (for example, during the reproduction of some species) and protecting some more sensible areas (establishing very strict rules, especially in the SICs, the "sites of community importance"); favoring a better distribution of visitors in the area through facility design or information.

What park managers can do to influence tourists' behavior and attitude is: discouraging or banning certain practices and/or equipment; encouraging or requiring some kinds of behavior, skills, and equipment; trying to attract more environmental conscious tourist segments; investing in environmental education programs, especially in the schools, and other activities (i.e., guided tours, laboratories, conferences, training courses) and teaching a wilderness ethic, informing visitors about appropriate natural protected areas fruition and promoting forms of responsible tourism (keeping woods clean, etc.). Another segment, slow tourism, should be favored by investing in infrastructures such as trails for hiking, biking, horse-riding, and cross-country skiing, accessible to disabled people as well.

Finally, it is necessary to create among local residents a high awareness about the importance of the protected area, its role, and the advantages that tourism can bring to the community. It can improve the capacity to welcome visitors and the desire to develop sharing accommodation initiatives, putting tourists in closer contact with the local way of living, thus avoiding new invasive structures and favoring the reuse and conversion of existing buildings at the same time. It is also important to use the resources generated by the natural protected area to make investments at a local level and to promote sustainable projects with the aim of getting young people to stay. All that can be obtained by creating craft schools and sustainable activities, such as B and B, artisan productions with high added value, enhancing local area brands and quality products, and encouraging cooperation among local firms and the development of supply chains. These investments in social capital can help create and strengthen a network of subjects interested in a sustainable and successful future for the natural protected areas, often forgotten and neglected in the past.

The Case of the Tremiti Islands Marine Reserve

In Italy, there are 871 protected areas, covering a total of 3 million hectares, more than 11% of the country; in addition, there are 2,850 thousand hectares of marine areas and 658 km of protected coastline. The peculiarity of the Italian natural protected areas system, which distinguishes it from those of other countries, is the presence of towns and villages in these areas, reaching a total population of around two million inhabitants (Ministero dell'Ambiente e della Tutela del Territorio e del Mare, 2017). As far as the number of tourists is concerned, the data are difficult to calculate, because Italian parks have no entrance gates and there is no ticket to pay. For this reason, and considering the high percentage of excursionists among visitors, the data related to the nights of stays in local accommodation are not sufficient to estimate the total number of visitors. Moreover, many municipalities have only a part of their territory inside the park, while tourist data are collected at the municipal level. It is therefore difficult to calculate the two indicators of environmental and social sustainability previously mentioned: tourism intensity and tourism function index.

Today, about 25 million tourist arrivals are estimated in Italian protected areas, for a total of more than 94 million overnight stays (ISPRA, 2018). In our country there are about 30.5 millions of tourist overnight stays related to naturalistic tourism and the potential of natural protected areas is even greater, representing the true heart of naturalistic tourism offer in Italy: 42.5% of the total tourists present in the country declare that they would like to go on holiday in a natural park, even 44.5% if the destination is in Italy (Ministero dell'Ambiente e della Tutela del Territorio e del Mare, 2017). As a matter of fact, the most popular protected areas are located close to large urban centers; they are affected by significant peaks in tourist presences, especially on weekends and holiday periods. Nowadays, many managers of Italian protected areas are working to increase tourism environmental and social sustainability, adopting many of the initiatives illustrated above.

The object of our study is the Tremiti Islands Marine Reserve that is part of Gargano National Park. The reserve was established in 1989 and covered 1,466 hectares. Tremiti archipelago is made up of 4 islands: San Domino, San Nicola, Caprara, Pianosa, and the small islet of Cretaccio. The inhabited islands are San Domino, where most of the accommodation facilities are located, and San Nicola, with about 500 residents. In 2018, tourist arrivals in the archipelago reached 22,696 and overnight stays 91,556 (Puglia Regional Tourism Agency, 2018).

Methodology

To carry out this study, we conducted 31 face-to-face in-depth interviews with stakeholders, each lasting approximately 30–50 min. These were conducted between July and September 2019, using a purposive sampling method. The respondents were: staff of the Tremiti Islands Marine Reserve and of Gargano National Park (N = 4; Group A), representatives of Public Administrations (N = 6; Group B), cultural, business, and tourist operators (N = 15; Group C), and local associations (N = 6; Group D).

The main aim of the research was to analyze the attitude of the host community towards the following issues: the advantages and disadvantages of tourism development and of overtourism, and the sustainability in the archipelago. Data analysis followed an inductive approach. We managed data by hand through a thematic content analysis, identifying three main themes: the use of tools to prevent harmful behavior, the adoption of instruments for monitoring the impact of tourist flows, the use of traditional and innovative measuring instruments to reduce the impact of the flows of visitors.

Findings and Discussions

Regarding the first aspect, the use of tools to prevent harmful behaviors by visitors, the interviewees explain that there are not environmental education courses organized in the area. However, a representative of the Public Administrations (group B) told us: "The islanders are sensitive to environmental problems because they love their land and have always lived here. Besides, there are initiatives that are carried on by private subjects, such as the diving operators". In fact, a diving operator (C) explained to us: "Every time I go out in the sea for underwater activities with my clients, I simultaneously do environmental education activities, especially in the field of the plastic collection in the sea". However, also the Gargano Park organizes initiatives of this type, such as the Clean Up Campaign in 2019. In general, what is evident is the need and desire to develop a greater number of activities, aimed both at residents and tourists, in order to increase awareness about sustainability issues. Moreover, there is also the necessity to remove this burden from the local operators as it does not belong to them.

Referring to the second theme, the adoption of instruments for monitoring the impact of visitors' flows, a member of the staff of the Tremiti Islands Marine Reserve and Gargano National Park (group A) told us that they conduct real-time monitoring when they allow ships coming from Termoli and Rodi to enter their ports. The other two interviewees (B and C) had a different opinion and explained to us that there are neither systems nor control bodies to monitor the impact of the visitors' flow. A member of a local association (D) added that the percentage of excursionists that get there to go around the islands by boat and go back home at night is very high, around 70%. For this reason, the monitoring activities are more difficult and the data about tourist arrivals are not complete. One tourist operator (C) told us something interesting about the positioning of the local tourist offer: "The economy of the archipelago is centralized on day trips. 90% of the population works by the sea with boats to accompany the excursionists on tours. They finish the working day at 5 in the afternoon, also because many tourists have to leave on the ferry to Termoli. During the evening, which is the most beautiful moment to enjoy the islands, there is nobody". In addition, a diving operator (C) told us: "We monitor the behavior of tourists, but the checks run by local authorities are poor". Besides, a subject of the group D said: "Fishermen are not aware of the importance of the monitoring tools in this area and today on Tremiti islands fishermen work as boatmen, thus allowing looting tourism".

A tourist operator (C) explained to us: "Connections are targeted at commuting tourism". The problem of transport that penalizes the area is also underlined by others (1B and 1C): "Tremiti Islands suffer because of a peripheral position. The closest port is that of Termoli, which is in Molise, another region", one of them said. Today excursionists are the main market target. But the park authority (A) had a different opinion: "Today the impact of the flow of excursionists is not strong. It has decreased compared to the 80s–90s, when many more people came, up to 7,000 people a day, even cruise ships with thousands of people arrived. Unfortunately, people no longer go on vacation for a long time, and in the summer the island is expensive, considering that there are not many structures, so everything that is brought here costs a lot because transport costs are spread over few structures".

All that suggests different and sometimes conflicting attitudes among the various segments of stakeholders and the need to intensify the monitoring activities, while the sensitivity to sustainability issues appears already quite high.

As to the last theme, the use of traditional and innovative measuring instruments to reduce the impact of visitors' flows, the first issue is the signposting in the sea. Three members of different groups (A, B, C) explained that the signposting in the sea is effective, thanks to the use of large yellow buoys, as required by law. Four people (2 A, C, and D) explained to us that there is a need for new and technological tools to monitor the number of visitors but also their behavior, preferences, and movements, such as GPS, drones, and others. But, according to the point of view of a local association representative (group D): "There is the need to reduce the impacting flow that in a marine reserve must be controlled, discouraged, spread and diluted also offering tourists different opportunities out of season". In fact, the main problem seems to be the high seasonality of tourism, which is concentrated during the summer months (June, July, and August). The interviewees are aware of the need to reduce seasonality and to lengthen the tourist season. Four people (one in every different group) told us that they are trying to act in this direction, focusing on the island of San Nicola. On this island, it has been programmed a renovation project with the creation of museums, laboratories, exhibitions, guided walks, a luxury hotel, and other attractions. "The island is an open-air museum that is not currently valued", someone told us (D). "The aim is", a representative of group B said, "to extend the season towards October and November, because people who love cultural heritage will come all year round". Others (2C) explained that: "More attention needs to be paid to other points of interest, such as the forest in San Domino Island. So they have to organize tours and trekking in the forest, that is a way to decongest tourism". A diving operator (C) said that there are other points of interest to enhance, such as: "A millennial abbey with an incredible story, that is an open-air museum and is the only Abbey in the middle of the sea existing in the Mediterranean". He added: "They need to better organize these resources through the creation of services, museums, and a multifunctional center. Since there are no services for activities that do not include the sea, it is impossible to reduce tourism seasonality, because in October and November there is nothing to do". In addition, he said (C) "On the Tremiti Islands there are many walking and cycling routes, and therefore I have traced a route in San Nicola, in San Domino, where there are often small plates imprinted with the QR codes. The QR code can store a lot of information and provides tourists with them, such for example the history of the Abbey. Visitors can read the information on their smartphones and other mobile devices. So I made a QR code tour project that gives specific information about every point of interest in the Tremiti Islands in different languages, German, English, Russian, French, and also the language of the signs for deaf people. We must understand that it is time to change direction adapting ourselves to the new technologies because the tourist must be able to choose the most suitable tool, so it is necessary to offer an app or a guide. In the Abbey, there are no tourist services with the exception of an app that gives minimal information, trying to fill a gap and pushing visitors to know more about local history. We must try to understand that the tourist is now more and more autonomous. He/she does not expect to be included in a group to benefit from the guided tour because the whole holiday has already been planned by him/herself. Since most of the tourists are independent, then let's give that app-service". Another tourist operator (C) explained: "On the Tremiti tourists mainly perform underwater activities, because the sea is extraordinary, but also because there are no other services for them".

A member of group D pointed our attention to an interesting aspect: "The diving activities distribute the flows and reduce the seasonality; in addition, they require 'correct' and 'ecological' behaviors and a patent. But, above all, people interested in diving are more sensitive to ecological problems. They are nature lovers compared to hit and run tourists because they know that climate change is also impacting the coralligenous system. It is a cultural issue. Therefore, the negative pressure is encouraged by mass tourism, but the experiential and less seasonal tourism is a more cultured segment".

Conclusion

In conclusion, in the archipelago, there is now widespread awareness of the problems related to a short and very concentrated tourist season. Many stakeholders interviewed also know the most suitable tools to reduce the spatial and temporal concentration of flows. All that is very important because, in the next future, the pressure of the stakeholders and the public opinion could facilitate the adoption of the right measures to reduce the tourist impact on the reserve and to prevent over-tourism. Furthermore, some initiatives have already been planned.

The measures to be taken are still numerous, both for monitoring and for the seasonal adjustment and reduction of the tourist impact. However, there are not any tangible forms of unsustainability in social terms, with negative attitudes of the residents towards visitors. This is certainly positive because it shows that the social carrying capacity (Graefe & Vaske, 1987) of the area has not been exceeded yet. Furthermore, 30 years after the establishment of the reserve, it is evident that the awareness of people living and operating within a protected natural area is high and many of the stakeholders are able to identify and adopt, whenever possible, even independently, interesting initiatives to increase the sustainability of the tourist flows.

The territory is therefore mature to take the qualitative leap that is necessary to significantly improve the tourism sustainability without generating an attitude of refusal of the protected area, that is no longer seen as a limit but rather as an important development opportunity for the whole area.

References

Andrew, R. G., Burns, R. C., Schwarzmann, D., Allen, M. E., & Moreira, J. C. (2021). Blue water visitor monitoring potential: A literature review and alternative proposal. *Water, 13*(305), 1–22.

Bartelmus, P. (1986). *Environment and development*. Allen and Unwin.

Boley, B. B., & Uysal, M. (2013). Competitive synergy through practicing triple bottom line sustainability: Evidence from three hospitality case studies. *Tourism and Hospitality Research, 13*(4), 226–238.

Buckley, R. (2012). Sustainable tourism: Research and reality. *Annals of Tourism Research, 39*(2), 528–546.

Burns, R. C., Andrew, R. G., Allen, M. E., Schwarzmann, D., & Moreira, J. C. (2020). Conceptualizing the national marine sanctuary visitor counting process for marine protected areas. *Journal of Ecotourism, 19*(4), 362–372.

Butler, R. W. (1980). The concept of a tourist area cycle evolution: Implications for the management of resources. In R. W. Butler (Ed.), *The tourism area life cycle* (pp. 5–12). Chanel View Publications.

Butler, R. W. (1999). Sustainable tourism: A state-of-the-art review. *Tourism Geographies, 1*(1), 7–25.

Capocchi, A., Vallone, C., Pierotti, M., & Amaduzzi, A. (2019). Overtourism: A literature review to assess implications and future perspectives. *Sustainability, 11*(12), 1–18.

Cheer, J. M., Milano, C., & Novelli, M. (2019). Tourism and community resilience in the Anthropocene: Accentuating temporal overtourism. *Journal of Sustainable Tourism, 27*(4), 554–572.

Choi, H. C., & Sirakaya, E. (2006). Sustainability indicators for managing community tourism. *Tourism Management, 27*, 1274–1289.

Cimini, N. (2011). *L'economia della vita. Il valore del parco Nazionale della Majella*. Documenti scientifici del Parco Nazionale della Majella, Rocca S. Giovanni.

Clarke, J. (1997). A framework of approaches to sustainable tourism. *Journal of Sustainable Tourism, 5*(3), 224–233.

Cohen, E. (1988). Authenticity and commoditization in tourism. *Annals of Tourism Research, 15*(3), 371–386.
D'Antonio, A., Monz, C., Lawson, S., Newman, P., Pettebone, D., & Courtemanch, A. (2010). GPS-based measurements of backcountry visitors in parks and protected areas: Examples of methods and applications from three case studies. *Journal of Park and Recreation Administration, 28*(3), 42–60.
Dickinson, J. E., & Lumsdon, L. (2010). *Slow travel and tourism*. Earthscan.
Eagles, P. F. J., & McCool, S. F. (2002). *Tourism in national parks and protected areas. Planning and management*. CABI Publishing.
Eagles, P. F. J., McCool, S. F., & Haynes, C. D. (2002). *Sustainable tourism in protected areas guidelines for planning and management*. IUCN – The World Conservation Union.
Elkington, J. (1997). *Cannibals with forks, the triple bottom line of 21st century business*. Capstone Publishing.
Elkington, J. (1998). Partnerships from cannibals with forks: The triple bottom line of 21st-century business. *Environmental Quality Management, 8*(1), 37–51.
Elkington, J. (2004). Enter the triple bottom line. In A. Henriques & J. Richardson (Eds.), *The triple bottom line: Does it all add up* (pp. 1–16). Earthscan.
Farrell, B. (1992). *Tourism* as an *element* in *sustainable development*: *Hana*, *Maui*. In V. L. Smith & W. R. Eadington (Eds.), *Tourism alternatives* (pp. 115–134). University of Pennsylvania Press.
Ferrari, S., & Gilli, M. (2018). Sustainable tourism in natural protected areas: The points of view of hosts and guests in Sila National Park. In I. Azara, E. Michopoulou, A. Clarke, D. Taff, & F. Niccolini (Eds.), *Tourism, health, Well-being and protected areas* (pp. 39–55). CABI.
Ferrari, S., & Inglese, R. (2019). Innovative web accessibility instruments as tourism marketing tools: The case of Sila National Park. *Proceedings of the 14th IFKAD, international forum on knowledge asset dynamics 2019 "knowledge ecosystems and growth"*. Matera, the 5th–7th June.
Ferrari, S., & Pratesi, C. A. (2012). National Parks in Italy – Sustainable tourism marketing strategies. *Matkailututkimus (Finnish Journal of Tourism Research), 8*(1), 7–23.
Ferrari, S., Bollani, L., & Gilli, M. (2018). Analysis and segmentation of visitors in a natural protected area. Marketing implications. In D. Batabyal (Ed.), *Managing sustainable tourism resources* (pp. 39–62). IGI Global.
Fredman, P., & Tyrväinen, L. (2014). *Frontiers in nature-based tourism: Lessons from Finland, Iceland, Norway and Sweden*. Routledge.
Frost, W., & Hall, C. M. (2010). American invention to international concept: The spread and evolution of national parks. In W. Frost & C. M. Hall (Eds.), *Tourism and National Parks* (pp. 30–44). Routledge.
Garay, L., & Font, X. (2012). Doing good to do well? Corporate social responsibility reasons, practices and impacts in small and medium accommodation enterprises. *International Journal of Hospitality Management, 31*(2), 329–337.
Graefe, A. R., & Vaske, J. J. (1987). A framework for managing quality in the tourist experience. *Annals of Tourism Research, 14*(3), 390–404.
Hall, C. M., & Frost, W. (2010). The making of national park concept. In W. Frost & C. M. Hall (Eds.), *Tourism and National Parks* (pp. 3–15). Routledge.
Haukeland, J. V., Grue, B., & Veisten, K. (2014). Turning national parks into tourist attractions: Nature orientation and quest for facilities. In P. Fredman & L. Tyrväinen (Eds.), *Frontiers in nature-based tourism: Lessons from Finland, Iceland, Norway and Sweden* (pp. 76–99). Routledge.
Heitmann, S., Robinson, P., & Povey, G. (2011). Slow food, slow cities and slow tourism. In P. Heitmann & S. Dieke (Eds.), *Research themes for tourism* (pp. 151–178). CABI Publishing.
Hernandez-Maskivker, G., Ferrari, S., & Cruyt, A. N. J. (2019). Exploring community stakeholders' perceptions of mass tourism: The case of Bruges. *Tourismos: An International Multidisciplinary Journal of Tourism, 14*(1), 77–94.

Heslinga, J. H., Hartman, S., & Wielenga, B. (2019). Irresponsible responsible tourism; observations from nature areas in Norway. *Journal of Tourism Futures*. https://doi.org/10.1108/JTF-10-2019-0118

Iaffaldano, N., & Ferrari, S. (in press). Applying sustainability indicators in the analysis and segmentation of tourist demand – Implications for marine visitor experiences on the Tremiti Islands, Italy. In J. N. Albrecht (Ed.), *Managing visitor experiences in nature-based tourism*. : CABI Books.

ISPRA. (2018). *Annuario dei dati ambientali. Stato dell'Ambiente, n.84/2010*. ISPRA.

Jamal, T. B., & Getz, D. (1995). Collaboration theory and community tourism planning. *Annals of Tourism Research, 22*(1), 186–204.

Leiper, N. (1990). Tourist attraction systems. *Annals of Tourism Research, 17*(3), 367–384.

Leung, Y. F., Spenceley, A., Hvenegaard, G., & Buckley, R. (Eds.). (2018). *Tourism and visitor management in protected areas: Guidelines for sustainability*. IUCN – The World Conservation Union.

MacCannell, D. (1973). Staged authenticity: Arrangements of social space in tourist settings. *American Journal of Sociology, 79*, 589–603.

Manning, R. E., & Anderson, L. E. (2012). *Managing outdoor recreation: Case studies in the National Parks*. CABI Publishing.

Matos, R. (2004). Can slow tourism bring new life to alpine regions? In K. Weiermair & C. Mathies (Eds.), *The tourism and leisure industry: Shaping the future* (pp. 93–103). The Haworth Hospitality Press.

Meng, B., & Choi, K. (2016). Extending the theory of planned behavior: Testing the effects of authentic perception and environmental concerns on the slow-tourist decision-making process. *Current Issues in Tourism, 19*(6), 528–544.

Ministero dell'Ambiente e della Tutela del Territorio e del Mare (Ministry of the Environment and Protection of Land and Sea). (2017). *Natura e cultura. Le aree protette, luoghi di turismo sostenibile*. Ministero dell'Ambiente e della Tutela del Territorio e del Mare.

Mowforth, M., & Munt, I. (2003). *Tourism and sustainability. Development and new tourism in the third world* (2nd ed.). Routledge.

Newsome, D., & Hughes, M. (2018). The contemporary conservation reserve visitor phenomenon! *Biodiversity and Conservation, 27*(2), 521–529.

Ostrom, E., Burguer, J., Field, C. B., Norgaard, R. B., & Policansky, D. (1999). Revisiting the commons: Local lessons, global changes. *Science, 284*, 278–282.

Pearce, D., Barbier, E., & Markandya, A. (1990). *Sustainable development, economics and environmental in the third world*. Edward Elgar.

Peeters, P., Gössling, S., Klijs, J., Milano, C., Novelli, M., Dijkmans, C., Eijgelaar, E., Hartman, S., Heslinga, J., Isaac, R., Mitas, O., Moretti, S., Nawijn, J., Papp, B., & Postma, A. (2018). *Research for TRAN committee – Overtourism: Impact and possible policy responses, European Parliament*. Policy Department for Structural and Cohesion Policies.

Peng, J., Chen, X., & Wang, J. (2016). Applying relative deprivation theory to study the attitudes of host community residents towards tourism: The case study of the Zhangjiang National Park, China. *Current Issues in Tourism, 19*(7), 734–754.

Puglia Regional Tourism Agency. (2018). *Osservatorio Agenzia Puglia Promozione – Destination Management System*. Retrieved from: https://www.agenziapugliapromozione.it/portal/osservatorio-del-turismo. Accessed 28 Dec 2020.

Roberts, S., & Tribe, J. (2008). Sustainability indicators for small tourism enterprises – An exploratory perspective. *Journal of Sustainable Tourism, 16*(5), 575–594.

Sisneros-Kidd, A. M., Monz, C., Hausner, V., Schmidt, J., & Clark, D. (2019). Nature-based tourism, resource dependence, and resilience of Arctic communities: Framing complex issues in a changing environment. *Journal of Sustainable Tourism, 27*(8), 1259–1276.

Stoddard, J. E., Pollard, C. E., & Evans, M. R. (2012). The triple bottom line: A framework for sustainable tourism development. *International Journal of Hospitality and Tourism Administration, 13*(3), 233–258.

Swarbrooke, J. (1999). *Sustainable tourism management*. CABI.
Tyrrell, T., Paris, C. M., & Biaett, V. (2013). A quantified triple bottom line for tourism: Experimental results. *Journal of Travel Research, 52*(3), 279–293.
UNEP-WCMC and IUCN. (2016). *Protected planet report 2016*. UNEP-WCMC and IUCN.
United Nations World Tourism Organization (UNWTO) (UNWTO). (1997). *What tourism managers need to know* (p. 5). UNWTO.
Van Der Borg, J. (1992). Tourism and urban development: The case of Venice, Italy. *Tourism Recreation Research, 17*(2), 46–56.
Van der Borg, J., Costa, P., & Gotti, G. (1996). *Tourism in European heritage cities*. Pergamon Press.
Vargas-Sánchez, A., Porras-Bueno, N., & de los Ángeles Plaza-Mejía, A. M. (2011). Explaining residents' attitudes to tourism: Is a universal model possible? *Annals of Tourism Research, 38*(2), 460–480.
Wall-Reinius, S., & Fredman, P. (2007). Protected areas as tourism attractions. *Annals of Tourism Research, 34*(4), 839–854.
Wearing, S. L., Schweinsberg, S., & Tower, J. (2016). *Marketing national parks for sustainable tourism*. Channel View Publications.
Weaver, D. (2006). *Sustainable tourism: Theory and practice*. Elsevier.
Weaver, D. (2010). Geopolitical dimensions of sustainable tourism. *Tourism Recreation Research, 35*(1), 45–51.

Part VII
The Future with Technology Solutions

Chapter 20
Technology and Overtourism: A Paradoxical Perspective

Pinaz Tiwari, Nimit Chowdhary, and Snigdha Kainthola

Abstract The unprecedented growth of the tourism and hospitality sector globally correlates with the advancement of digital media and technological tools. The dominance of information and communication technology is prevalent at every stage of travelers' decision making process (i.e., from searching for a suitable destination to posting feedback on social media platforms). Not only the travelers' behavior patterns are influenced by technology, but destinations also utilize technology for marketing and enhancing consumers' experience. Nevertheless, the advancement of technology has acted like a double-axed sword for the tourism sector. Frequently, digital media is held accountable for popularising a destination to an extent that it becomes a hub for mass-tourism. Issues like tourismphobia, anti-tourism movements, and touristification etc. are gaining hype through technology and online social platforms. Alternatively, destination managers utilise technological tools to sustain tourism growth and visitor experience for better management. Information and communication technology (ICT) has played a key role in influencing tourists to visit popular destinations that led to the issue of overtourism. Likewise, the incorporation of technology is equally vital in managing the tourists' flow, and subsequently, avoiding crowding and overtourism. The chapter aims to highlight the ambidextrous role of technology in overtourism. The study is conceptual and uses short cases of various popular destinations affected by overtourism and how technology served as an emancipator to combat the unsustainable consumption patterns. The chapter discusses the practical implications of utilising technology to combat issues leading to unsustainability in tourism. It also highlights the emerging role of technology in enhancing visitors' experience in the post-COVID-19 scenario. This study presents a holistic perspective and the relationship between technology and tourism. Several studies have discussed the bright side of technology in the tourism and hospitality sector. However, the darker side is less acknowledged.

Keywords Overtourism · Technology · Destination management · Digital media · Crowding · Ambidextrous

P. Tiwari (✉) · N. Chowdhary · S. Kainthola
Department of Tourism and Hospitality Management, Jamia Millia Islamia, New Delhi, India

A. Hassan, A. Sharma (eds.), *Overtourism, Technology Solutions and Decimated Destinations*, https://doi.org/10.1007/978-981-16-2474-2_20

Introduction

The twenty-first century has witnessed unprecedented growth in the global tourism sector. This growth is often measured in the number of tourist arrivals at a destination. The UNWTO reported that the number of annual international trips has increased from 25 million in the 1950s to 1.4 billion in 2018 (UNWTO, 2019). Consequently, several famous destinations faced the reverberation of negative impacts of tourism such as congestion, noise pollution, damage to heritage sites and natural resources, among others. As a result, discourses on unsustainable tourism practices began to take place. Notions like overtourism, anti-tourism movements, and tourismphobia emerged in the academic literature to emphasize the ill-effects of "unmanageable" growth of tourism (Dodds & Butler, 2019; Milano, 2017; Milano et al., 2019a, b). Stating that overtourism is a "consequence" than a "cause" (Seraphin et al., 2019), various researchers have identified the antecedents that lead to the phenomenon. These antecedents include a change in demographics, the behavioral pattern of consumers, the influence of social media, increase in income and the proportion of the middle-income population, increased use of smart technologies, cost-effective packages, and increasing dominance of gig economy in the tourism sector (Agyeiwaah, 2020; Oklevik et al., 2019; Tiwari et al., 2020; Zmyślony et al., 2020). Information and communication technology has played a crucial role in influencing tourists to visit popular destinations that led to the issue of overtourism. The dominance of information and communication technology is prevalent at every stage of travelers' decision-making process (i.e., from searching for a suitable destination to posting feedback on social media platforms). Not only the travelers' behavior patterns are influenced by technology, but destinations also utilize technology for marketing and enhancing consumers' experience. In the past few years, several discourses related to the relationship between technology and tourist consumption pater have taken place in academic and corporate spheres (Choi et al., 2019; Falcao et al., 2019; Rajasekaran et al., 2019). Of all these, the current study focuses specifically on the rise of information and communication technology as a cause and cure of overtourism.

Overtourism and its related issues are often considered complicated due to the lack of a suitable conceptual framework to study. The available literature has limited their studies to single destinations (Barbhuiya, 2020; Dai et al., 2019; Martín et al., 2018; Seraphin et al., 2018). While those studies adopting a subjective perspective to analyze the issue have overlooked the objective perspective and vice versa (Pasquinelli & Trunfio, 2020). Furthermore, the role of technology in acting as an antecedent to overtourism is less explored in the existing literature. Thus, it is imperative to understand the paradoxical role of technology at destinations affected by overtourism through the case-by-case approach. This approach would provide a holistic view to address the challenges created by unsustainable tourism practices, and subsequently, introduce innovative technological solutions to manage and regulate such practices at a destination.

The relationship between the technology and tourism industry has acted as a double-edged sword. Indeed, it has led to the advancement in the operationalization of the travel sector at a global level. However, several platforms related to the gig economy, like Airbnb are held responsible for spatial concentration and congestion in urban tourist destinations (Nathan et al., 2020). Similar factors like social media and other mobile applications that make tourism affordable and accessible for tourists, and create conflicts among stakeholders are considered as the driving force of overtourism (Cesarani & Nechita, 2017). Besides, digital media is held accountable for popularizing a destination to the extent that it becomes a hub for mass-tourism (Alonso-Almeida et al., 2019). Alternatively, destination managers utilize technological tools to sustain tourism growth and visitor experience for better management. It widely includes the use of robotics, artificial intelligence, and big data analytics to track the visitor flow at a site, offer convenience, and thereby enhance the visitors' experience. Even though recent studies have discussed the smart cities framework to manage overtourism and overcrowding at tourist destinations (Gowreesunkar & Seraphin, 2019; Pasquinelli & Trunfio, 2020; Vargas-Sánchez, 2020), this chapter aims to bridge the gap in the overtourism literature by analyzing the technological paradox at tourist destinations. Thus, the chapter attempts to explore the ambidextrous role of technology in overtourism. Furthermore, the chapter discusses the significance of technological tools in preventing the decimation of destination in the post-COVID-19 era.

This chapter is divided into five sections. The first section is introduction followed by an integrated literature review on the ambidextrous role of technology in overtourism. Short cases of destinations adopting technology to resolve overtourism and related issues are represented to support the argument. The third section focuses on technological solutions to prevent the decimation of destinations. The fourth section discusses the incorporation of technology in the post-COVID-19 era as a solution for robust and sustainable growth of tourism at destinations.

Ambidextrous Role of Technology in Overtourism

Technology and innovation are at the core of increased globalization and emancipation in contemporary times. The breakthrough of technology is equally evident in the highly competitive environment of the tourism and travel sector (Tscheu & Buhalis, 2016). The surge in tourism activities at popular destinations is also a result of increased competition and technological advancements. Overtourism is often defined as a situation of concentrated tourism activities at popular destinations leading to a conflict of interest among diverse stakeholders (Koens et al., 2018). Both overtourism and technological advancements in tourism are correlated. For instance, the use of smartphone applications like Google Maps, Airbnb, Uber, and others, the convenience of online booking, social media influences, widespread phenomena of "millennial instragrammability" (Vargas-Sánchez, 2020), online travel blogs that persuades people to 'must visit' popular destinations, are some of the causes overtourism. Indeed, employing information and communication technology tools is suggested to increase the tourist arrivals at destinations that are witnessing lesser growth in the tourism sector over the period (Seraphin & Ivanov, 2020).

Conversely, technological tools have addressed and resolved the overtourism issues at many destinations. These issues include a ticketing system, utilizing digital footprints of travelers and promoting alternative destinations, regulating traffic and parking systems, and big data analysis to compute tourism impacts, among others. In the current scenario, destination management organizations utilize technological tools to enhance tourists' experience at destinations (Neuhofer & Buhalis, 2018). Accordingly, this section discusses the paradoxes in the role of technology at destinations affected by overtourism.

The Flipside of Technology: An Antecedent of Overtourism

The issue of overtourism is observed when too many tourists visit the same place at a time and are encouraged by similar factors (Gowreesunkar & Seraphin, 2019). These factors include low-cost carriers, low-cost packages, easy access to destinations, cost-effective accommodation, excessive marketing by destination management organizations, social and digital media, and other smartphone applications that facilitate and encourage consumption of tourism activities (Milano et al., 2018). Seraphin et al. (2019) also identified that socio-economical, business, and technological factors (development of technology and low-cost transport) as the roots of overtourism. The evolution of technology is considered as one of the dominant factors. The chapter has identified *2S's* to discuss the negative impact of technology on the consumption of tourism. Figure 20.1 is the graphical representation of the flipside of technology that acts as an antecedent to overtourism. These are sharing economy and social media.

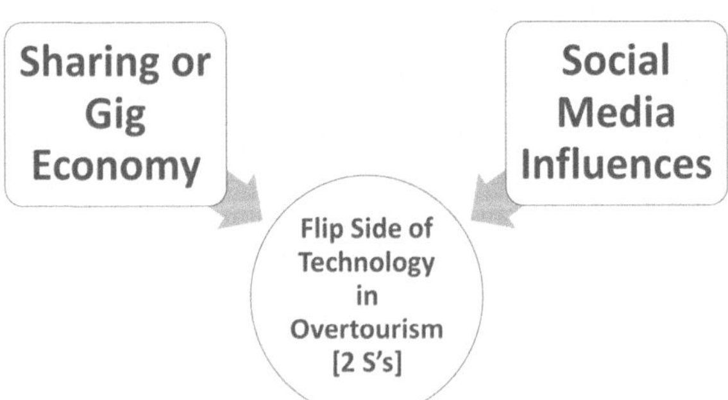

Fig. 20.1 Flip side of technology in Overtourism. (Source: Developed by the authors)

Sharing Economy

Out of numerous factors that leading overtourism, several authors have identified the increasing proportion of the "sharing economy" at tourist destination as one of the primary reasons (Arias Sans & Quaglieri Domínguez, 2016; Nilsson, 2020; Renau, 2018). Sharing economy, especially asset-free companies like Airbnb, is blamed for causing tourist gentrification at destinations like Barcelona, Amsterdam, Reykjavik, and Venice and leads to concentrated tourism in cities (Goodwin, 2017).

With the wide-scale popularity and development of the sharing economy, majorly platforms like Airbnb and Uber cabs, the negative impacts of tourism at destinations are noticeable (Bourliataux-Lajoinie et al., 2019). Local communities have raised their concern regarding irresponsible conduct by tourists in their neighborhoods. They hike the prices of rent in their cities. The situation worsened when cities like Venice and Barcelona reported that locals had to displace themselves to other cities due to the increased cost of living (Gutiérrez et al., 2017). Besides, Airbnb has ignored the situation that property managers, real estate organizations, and other individuals can list as many properties as they want. These companies captured the travelers' desire to stay at "home away from home" and gradually buy properties in the city Centre. Thereby, this phenomenon decreases the supply of rental properties for locals, and increases the rental prices, and ultimately forces residents to move out of the cities (Nomadic Matt, 2020).

In some cases, destination management organizations and government bodies have taken regulatory measures. In contrast, in others, the use of such platforms is restricted. For instance, the local government of Barcelona decided to stop issuing new licenses to short-term rental accommodation in 2017 (Nomadic Matt, 2020). Likewise, in November 2019, London's administrative body rescinded UBER's license to operate as a private transport provider owing to loopholes in its services in the city (Topham, 2019).

Social Media

The unprecedented use of social media and similar platforms have brought a transition in the behavior and consumption pattern of visitors at a destination. The relentless use of social media through blogs, networking websites, and peer opinion platforms has persuaded people to share their travel experiences online (Ayeh et al., 2012). Social media can influence and inspire people to choose a particular destination and make decisions accordingly (Gowreesunkar & Dixit, 2018). In the context of overtourism, the predominance of social media platforms is considered as a vital factor in the creation of "hotspots" and "instagrammable locations". Bourliataux-Lajoinie et al. (2019) emphasized that a sense of social compliance is created on social networks, which results in identical tourist expectations among millions of them connected. While conducting a study on Chinese tourists visiting Barcelona, Alonso-Almeida et al. (2019) reported that social media is a cause of overtourism. Tiwari et al. (2020) and Gowreesunkar & Dixit (2018) emphasized that the overexposure of destinations in social media platforms, marketing of destination through social media influencers, and film-induced traveling aggravated the unsustainable tourism at destinations. For instance, a large influx of people could be seen at Maya beach after the film "The beach" hit theatres or increased visitation to Dubrovnik after the famous series Game of Throne was released (Kraus, 2018). Moreover, the continuous use of social media by visitors induces irresponsible behavior that is frowned upon by locals at a destination. For example, clicking selfies with locals without asking for their permission, making videos at prohibited sites etc. cause annoyance to the locals.

The Bright Side of Technology: Addressing and Resolving Overtourism

The adoption of technology is widely considered useful for resolving the problem of overtourism and crowding at different destinations (Hospers, 2019). For instance, the first few recommendations are given by McKinsey & Company and World Travel and Tourism Council (WTTC) (2017) proposed a set of five strategies for addressing overcrowding. These recommendations included smoothing visitors' flow over time, spreading visitors across different sites, adjusting the price to match demand and supply, regulating accommodation supply, and limiting access and tourism activities. While implementing these tactics to combat overcrowding and overtourism, various destinations incorporated technological tools. For example, online booking of tickets, limiting tourist numbers using technology, introducing travel apps that could suggest alternative destinations to tourists, tracking digital footprints of visitors, and inform them about crowded sites, among others. Likewise, the recommendations suggested by the World Travel & Tourism Council (Dichter & Gloria, 2017) also listed the adoption of technological tools for the management of destinations and enhancing the tourists' experience.

- Using *technology* to enhance tourists' experience, manage destinations, and communicate with the local community.
- Long-term *planning* and coordination among the public sector, locals, and private sector
- *Diversification* of tourism *product* and introduction of alternative products; encouraging visitors to experience different (or less visited) locations to avoid concentrated tourism

Case of Dubrovnik

1. ***Du Visitors***: It is the Dubrovnik visitor counting system that provides information on the number of visitors at a site. Became operational in 2019, the system is composed of cameras and uses machine learning to gather data and predict visitor flow.
2. ***Du Card***: A tourist card partnered with a mobile application. It provides discounts to tourists, unlimited use of public transport, and access to the

city's top nine attractions. Combining the Du visitor and du card systems, the visitors are notified of the most visited areas and suggest alternative attractions to prevent overcrowding.

3. ***Dubrovnik Eye***: It is an interactive online platform that facilitates communication between the administrative sector of the city and the local community. Locals are allowed to report any tourism-related issue and send pictures through this platform so that the problem gets addressed in real-time.
4. ***Rural Dubrovnik-Neretva***: It is an online platform that aims to promote 84 agro-tourism destinations. It includes wine bars, rural stays, local events, thematic trails, among others. This platform will get operational soon and focuses on offering alternative tourism spots to tourists visiting Dubrovnik.

(Source: adapted from Camatti et al., 2020)

Technology Solutions for Preventing Decimation of Destinations

There are different approaches suggested by several authors to conceptualize and combat the issue of overtourism at popular destinations. Such as degrowth to ensure sustainable development of tourism (Cheung & Li, 2019; Milano et al., 2019a), Trexit (Seraphin et al., 2018), dispersal of tourists in space and time (Hardy et al., 2020; Peeters et al., 2018; UNWTO, 2018), demarketing in marketing mix framework for crowded destinations (Tiwari et al., 2020), and balance between the rights' of tourists and residents (Perkumienė & Pranskūnienė, 2019), among others. However, the solutions for every destination are different, and there is no 'one size fits all' solution to combat the overtourism and related issues. Owing to the predicaments in conceptualizing the phenomena, Pasquinelli & Trunfio (2020) has divided these approaches into three categories namely, marketing, regulation, and management. Figure 20.2 demonstrates the category and its features:

To prevent the decimation of destinations from overtourism or undertourism, incorporation of technological tools in these three categorizations becomes vital. Various destinations have utilized technology at different steps, which have been categorized below in Table 20.1.

The above table reveals the predominance of technological tools in management and marketing approaches to prevent overtourism issues at destinations. Even though destinations like Venice, Barcelona, Croatia, Amsterdam, and Kyoto have adopted various "regulatory" strategies such as mandatory licensing and taxation on tourists, the incorporation of technology in this approach is less evident. One probable reason could be the predicaments in bringing the fragmented sub-sectors under a single umbrella or the Janus-faced character of the tourism industry (Sanchez & Adams, 2008). This implies that when a country attempts to achieve economic

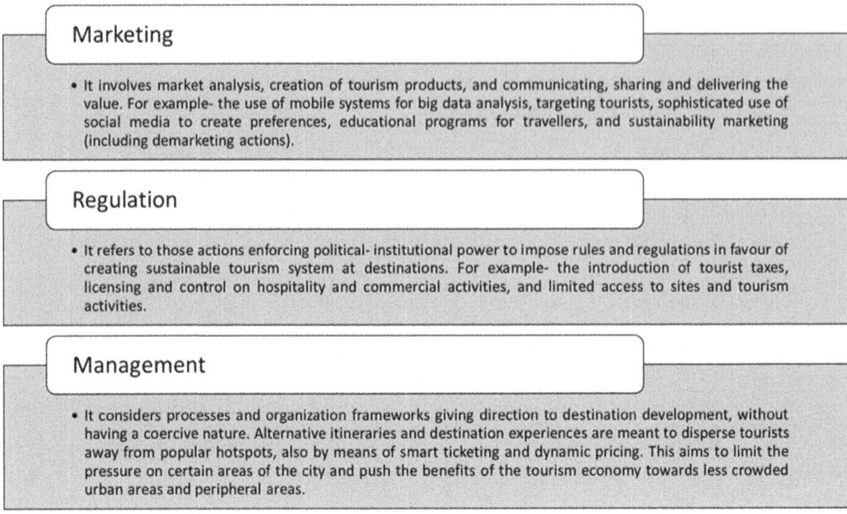

Fig. 20.2 Approach to prevent the decimation of destinations. (Source: Adapted from Pasquinelli & Trunfio, 2020)

Table 20.1 Classification of approaches and technological strategies implemented by decimated destinations

Destination	Technological strategy adopted to prevent overtourism	Approach
Ha'ena Park	Geo-fenced boundary; collects user-enabled location data through smartphone apps; Visitor limit supported by a web-based advanced reservation system and associated shuttle system	Management
Cinque Terre, Italy	An app for visitors that shows real-time congestion on the trails or nearby crowded villages. The mobile app directs them to less crowded local sites	Management
Trento	Created Smart POIs (points of interest) outside the overcrowded areas and encourages tourists to enjoy them using digital technologies Introduced an app "Trento Percorsi" which allows visitors to choose a different category of routes	Management
Valencia	It launched a new project known as 'Alter Eco' which aims to collect data in order to reduce the concentration of tourists at popular sites. The app aims to strengthen sustainable local development. It allows for city tours with augmented reality games, shares selfies on social networks, and promotes local identity	Management
Amsterdam	Advanced analytics to predict people in an area, camera and GPD trackers at popular sites to manage the crowd It introduced robotics innovations that allow autonomous robot tour guides Smartphone apps combined with iBeacons Redesigning of city cards for tourists	Marketing and management

(continued)

Table 20.1 (continued)

Destination	Technological strategy adopted to prevent overtourism	Approach
London	A free mobile gaming app ('Play London with Mr Bean') shows several attractions to visitors. After visiting the place, the travelers earn points that can be redeemed for vouchers and discounts during their stay in London	Marketing and management
Barcelona	Install SmartCitizen app that allows environmental monitoring sensors to track noise and pollution levels;	Marketing
Dubrovnik	Introduction of Du Card, Du visitors, Dubrovnik Eye and Dubrovnik Neretva	Marketing and management
Kumbha festival	Invented a mat that counts tourist footfall at the event and avoids the situation of a stampede by monitoring the flow of visitors	Marketing

Source: Developed by the authors

growth by contradicting the social equity risk at a destination, which might pose myriad challenges in implementing the framed regulatory policies.

The Future Ahead: Incorporating Technology at Destinations in COVID-19 Era

In the year 2017, Harold Goodwin (2017) cautioned that overtourism and related issues like tourist gentrifications, tourism phobia, crowding, and others, would be a common sight if tourism continues to grow unmanaged. A substantial number of researches were instituted to decipher the phenomenon of overtourism, both objectively and subjectively (Veríssimo et al., 2020). Various managerial interventions were being experimented with to control the situation (Joppe, 2019). However, the year 2020 demonstrated the unpredictable nature of the world as COVID-19 took millions of lives and disrupted global businesses. Besides, the phenomenon of overtourism has gone missing, at least temporarily from the face of the Earth (Niewiadomski, 2020), and rather a metamorphosis from overtourism to zero tourism surfaced within a short period. The change in expectations and perception of both the tourists and the locals are the most apparent predictions by the scholars. The fear created by the life-threatening COVID-19 along with hedonic and utilitarian motives (Addo et al., 2020), influenced both the demand and supply sides simultaneously. Therefore, the change in the consumer behavior of tourists are natural outcomes and needs consideration while formulating future strategies for a destination. This change in consumer behavior and consumption patterns were predominantly observed in the adoption of technology. Technological tools, especially smartphone apps, became an integral part of people's daily routine after the COVID-19 breakout.

Travel and tourism, being the worst-hit sector in the world, is seeking revival with the help of technological tools in their operations (Gretzel et al., 2020). The

COVID-19 has pushed the tourism industry towards adopting digitalization and advanced ICT to create as well as cater to the demand. Several webinars and seminars during the initial phases of COVID-19 spread discussed strategies to revive the tourism sector and highlighted the thriving role of technology to bring back the lost fortune. For creating a resilient tourism and hospitality sector, numerous destinations have incorporated advanced technology and related innovations to attract travelers' post-pandemic. The global health crisis has made contactless transactions a pre-requisite for tourism and hospitality to cater to the emerging demands. It has caused businesses to re-invent and adapt to changes. Such as QR codes scanning for checking the menu at a restaurant, mobile key to enter guest's room in a hotel, audio guide at museums, installing robotic devices to prepare as well as serve food, contactless payments, and experiencing a site through virtual and augmented reality. All innovations and incorporation of technology in tourism and related sectors are directed towards enhancing the tourists' experience. For instance, Kerala tourism introduced a QR code aided tourism search that offers hundreds of geotagged pictures of destinations in the state and enhances their experience. Travelers have the facility to upload pictures of a destination they toppled upon during their visit (Pawar, 2020). Likewise, on the occasion of World Tourism Day on September 27, 2020, Google introduced a collection of 37 cultural and historical sites that can be accessed and experienced through augmented reality (AR) by people (Chatterjee, 2020).

The adoption and sophisticated implementation of technological tools are considered an appropriate alternative to resolve unsustainable tourism practices in both the pre and post-COVID-19 era. Several destination authorities have implemented advanced technology for better management, as shown in Table 20.1. Similarly, in post-COVID-19 situations, sites are depending on technology to manage tourist flow in a safe environment and enhance their experience. One such example is the Umrah pilgrimage. To manage the crowding issues, the government of Saudi Araba is promoting religious tourism with complete caution. Umrah has developed a mobile application app that allows visitors to register in advance and make bookings (Globaldata Travel and Tourism, 2020). Indeed, technology is emerging as a potent mediator between travelers and service providers after the crisis. Croce (2018) forecasted that even after crises would continue to rise and negatively impact the development of the tourism industry, the sector would witness steady growth in the long-term owing to innovation, technology, and increasing demand for travel in the global population.

Conclusion

A phenomenon as complex as overtourism cannot be viewed simply as a result of tourism. It requires a holistic understanding and scrutiny in the broader context of predicaments related to the management of destinations. Emphasizing that the issue of overtourism requires a multidisciplinary approach (Gowreesunkar & Seraphin,

2019), this chapter aimed to establish arguments related to the ambidextrous role of technology in overtourism. The upsurge in the unsustainable pattern of consumption in tourism is correlated with the evolution of information systems and technology on a global scale. The dark side of technology involves the ingenious use of social media and the self-complied nature of sharing economy that has hinders authentic tourist experiences at destinations. While on the other side, the technological advancements and emergence of smart cities have proven an asset to the tourism industry. It has made connectivity better and tourism more convenient than ever. The growth in the global tourism sector is linked with technological advancements that ensure a better tourists experience. However, smart technologies alone cannot tackle the situation of overcrowding, tourism-phobia or related unsustainable forms of tourism (UNWTO, 2018).

This chapter highlighted the technological strategies undertaken by different destinations concerning three approaches of managing overtourism (i.e., marketing, management, and regulation). The adaption of technology in the third approach "regulation" was least noticeable in comparison to the other two. Certainly, overtourism and related issues raised concerns related to unsustainable tourism practices on both the demand and supply side. The relationship between technology and overtourism related issues was rather complicated and paradoxical. However, the outbreak of COVID-19 brought a sudden transition in the global tourism scenario, and overtourism turned into a situation of zero-tourism over the period.

In short, overtourism needs cooperation amongst all the stakeholders and the use of multiple complimenting measures for sustainable development. It could take place with the help of a centralized system and technical innovations that facilitate coordination at all levels. Networking amongst the stakeholders appeared as a challenging task for destinations. However, the pandemic proved to be a blessing in disguise for the tourism industry, and technological tools played a key role in bringing stakeholders. Nevertheless, overtourism or undertourism, or zero-tourism are all forms of unsustainable tourism. They are considered as social concerns than just physical. Therefore, relying alone on technology cannot help destinations to overcome adversities. Besides, it is equally valid that digitalizing tourism and cognate sectors are inescapable in the post-pandemic era. Thus, destinations need to update their operations and management for sustainable resurrection.

References

Addo, P. C., Jiaming, F., Kulbo, N. B., & Liangqiang, L. (2020). COVID-19: Fear appeal favoring purchase behavior towards personal protective equipment. *Service Industries Journal, 40*(7–8), 471–490.

Agyeiwaah, E. (2020). Over-tourism and sustainable consumption of resources through sharing: The role of government. *International Journal of Tourism Cities, 6*(1), 99–116.

Alonso-Almeida, M.-d.-M., Borrajo-Millán, F., & Yi, L. (2019). Are social media data pushing overtourism? The case of Barcelona and Chinese tourists. *Sustainability, 11*(12), 3356.

Arias Sans, A., & Quaglieri Domínguez, A. (2016). *Unravelling Airbnb: Urban perspectives from Barcelona*. Retrieved from: https://www.degruyter.com/document/doi/10.21832/9781845415709-015/html. Accessed: 12 Dec 2020.

Ayeh, J. K., Leung, D., Au, N., & Law, R. (2012). Perceptions and strategies of hospitality and tourism practitioners on social media: An exploratory study. In M. Fuchs, F. Ricci & L. Cantoni (Eds.), *Information and communication technologies in tourism 2012* (pp. 1–12). Springer.

Barbhuiya, M. R. (2020). Overtourism in Indian cities: A case study of Nainital. *International Journal of Tourism Cities*. https://doi.org/10.1108/IJTC-08-2019-0148

Bourliataux-Lajoinie, S., Dosquet, F., & del Olmo Arriaga, J. L. (2019). The dark side of digital technology to overtourism: The case of Barcelona. *Worldwide Hospitality and Tourism Themes, 11*(5), 582–593.

Camatti, N., Bertocchi, D., Carić, H., & van der Borg, J. (2020). A digital response system to mitigate overtourism. The case of Dubrovnik. *Journal of Travel & Tourism Marketing, 37*(8–9), 887–901.

Cesarani, M., & Nechita, F. (2017). Tourism and the sharing economy. An evidence from Airbnb Usage in Italy and Romania. *SYMPHONYA Emerging Issues in Management, 3*, 32–47.

Chatterjee, A. (2020). Google's virtual tour lets you join heritage sites online. *The Hindu*. Retrieved from: thehindu.com: thehindu.com/sci-tech-technology/googles-virtual-tour-lets-you-join-heritage-sites-virtually/article32707550.ece. Accessed: 12 Dec 2020.

Cheung, K. S., & Li, L. H. (2019). Understanding visitor–resident relations in overtourism: Developing resilience for sustainable tourism. *Journal of Sustainable Tourism, 27*(8), 1197–1216.

Choi, K., Wang, Y., & Sparks, B. (2019). Travel app users' continued use intentions: It's a matter of value and trust. *Journal of Travel and Tourism Marketing, 36*(1), 131–143.

Croce, V. (2018). With growth comes accountability: Could a leisure activity turn into a driver for sustainable growth? *Journal of Tourism Futures, 4*(3), 218–232.

Dai, T., Hein, C., & Zhang, T. (2019). Understanding how Amsterdam City tourism marketing addresses cruise tourists' motivations regarding culture. *Tourism Management Perspectives, 29*(December 2018), 157–165.

Dichter, A., & Gloria, G. M. (2017). *Coping with success: Managing overcrowding in tourism destinations* (pp. 1–56). McKinsey & Company.

Dodds, R., & Butler, R. (2019). The phenomena of overtourism: A review. *International Journal of Tourism Cities, 5*(4), 519–528.

Falcao, R. P. Q., Ferreira, J. B., & Carrazedo Marques da Costa Filho, M. (2019). The influence of ubiquitous connectivity, trust, personality and generational effects on mobile tourism purchases. *Information Technology and Tourism, 21*, 483–514.

Globaldata Travel and Tourism. (2020). *Saudi Arabia's opt for religious tourism should be noted with caution*. airport-technology.com. Retrieved from: airport-technology.com/comment/saudi-arabia-religious-tourism-caution/. Accessed: 10 Dec 2020.

Goodwin, H. (2017). The challenge of overtourism. *Responsible Tourism Partnership*. https://doi.org/10.1016/S0140-6736(84)91114-0

Gowreesunkar, V., & Dixit, S. (2018). Consumer information-seeking behaviour. In S. K. Dixit (Ed.), *The Routledge handbook of consumer behaviour in hospitality and tourism*. Routledge.

Gowreesunkar, V., & Seraphin, H. (2019). Introduction: What smart and sustainable strategies could be used to reduce the impact of overtourism? *Worldwide Hospitality and Tourism Themes, 11*(5), 484–491.

Gretzel, U., Fuchs, M., Baggio, R., Hoepken, W., Law, R., Neidhardt, J., Pesonen, J., Zanker, M., & Xiang, Z. (2020).). e-Tourism beyond COVID-19: A call for transformative research. *Information Technology and Tourism, 22*(2), 187–203.

Gutiérrez, J., García-Palomares, J. C., Romanillos, G., & Salas-Olmedo, M. H. (2017). The eruption of Airbnb in tourist cities: Comparing spatial patterns of hotels and peer-to-peer accommodation in Barcelona. *Tourism Management, 62*, 278–291.

Hardy, A., Birenboim, A., & Wells, M. (2020). Using geoinformatics to assess tourist dispersal at the state level. *Annals of Tourism Research, 82*, 102903.

Hospers, G. J. (2019). Overtourism in European cities: From challenges to coping strategies. *CESifo Forum, 20*(3), 20–24.

Joppe, M. (2019). The roles of policy, planning and governance in preventing and managing overtourism. In R. Dodds & R. Butler (Eds.). *Overtourism: issues, realities and solutions* (pp. 250–261). De Gruyter.

Koens, K., Postma, A., & Papp, B. (2018). Is overtourism overused? Understanding the impact of tourism in a city context. *Sustainability, 10*(12), 4384.

Kraus, N. (2018). City of Dubrovnik fights overtourism Amid industry growth. *Tourism review*. Retrieved from: https://www.tourism-review.com/dubrovnik-dealing-with-overtourism-news10529. Accessed: 10 Dec 2020.

Martín, J. M. M., Martínez, J. M. G., & Fernández, J. A. S. (2018). An analysis of the factors behind the citizen's attitude of rejection towards tourism in a context of overtourism and economic dependence on this activity. *Sustainability, 10*(8), 2851.

McKinsey & Company, & World Travel and Tourism Council. (2017). *Coping with success: Managing overcrowding in tourism destinations*. Retrieved from: https://www.mckinsey.com/industries/travel-transport-and-logistics/our-insights/coping-with-success-managing-overcrowding-in-tourism-destinations. Accessed: 12 Dec 2020.

Milano, C. (2017). Overtourism and tourismphobia: Global trends and local contexts. *Ostelea School of Tourism*. https://doi.org/10.13140/RG.2.2.13463.88481

Milano, C., Cheer, J. M., & Novelli, M. (2018). *Overtourism a growing global problem film and contemporary representations of the exotic other: Gazing through filmic eyes view project*. Retrieved from: https://www.researchgate.net/publication/326573468. Accessed: 12 Dec 2020.

Milano, C., Novelli, M., & Cheer, J. M. (2019a). Overtourism and degrowth: A social movements perspective. *Journal of Sustainable Tourism, 27*(12), 1857–1875.

Milano, C., Novelli, M., & Cheer, J. M. (2019b). Overtourism and tourismphobia: A journey through four decades of tourism development, planning and local concerns. *Tourism Planning and Development, 16*, 353–357.

Nathan, R. J., Victor, V., Tan, M., & Fekete-Farkas, M. (2020). Tourists' use of Airbnb app for visiting a historical city. *Information Technology and Tourism, 22*(2), 217–242.

Neuhofer, B., & Buhalis, D. (2018). Understanding and managing technology-enabled enhanced Torurist experiences. *2nd Advances in Hospitality and Tourism Marketing and Management Conference, 53*(9), 1689–1699.

Niewiadomski, P. (2020). COVID-19: From temporary de-globalisation to a re-discovery of tourism? *Tourism Geographies, 22*(3), 651–656.

Nilsson, J. H. (2020). Conceptualizing and contextualizing overtourism: The dynamics of accelerating urban tourism. *International Journal of Tourism Cities, 6*(4), 657–671.

Nomadic Matt. (2020). *Overtourism: How you can help solve this worldwide problem*. nomadicmatt.com. Retrieved from: https://www.nomadicmatt.com/travel-blogs/overtourism-solutions/. Accessed: 12 Dec 2020.

Oklevik, O., Gössling, S., Hall, C. M., Steen Jacobsen, J. K., Grøtte, I. P., & McCabe, S. (2019). Overtourism, optimisation, and destination performance indicators: A case study of activities in Fjord Norway. *Journal of Sustainable Tourism, 27*(12), 1804–1824.

Pasquinelli, C., & Trunfio, M. (2020). Reframing urban overtourism through the Smart-City lens. *Cities, 102*, 102729.

Pawar, B. (2020). *Travelling in Kerala gets easier with launch of QR-coded signboards*. india.com. Retrieved from www.india.com/travel/articles/travelling-in-kerala-gets-easier-with-launch-of-qr-coded-signboards-3240785/. Accessed 10 Dec 2020.

Peeters, P., Gössling, S., Klijs, J., Milano, C., Novelli, M., Dijkmans, C., Eijgelaar, E., Hartman, S., Heslinga, J., Isaac, R., Mitas, O., Moretti, S., Nawijn, J., Papp, B., & Postma, A. (2018). *Research for TRAN Committee – Overtourism: impact and possible policy responses*. European Parliament, Policy Department for Structural and Cohesion Policies.

Perkumienė, D., & Pranskūnienė, R. (2019). Overtourism: Between the right to travel and residents' rights. *Sustainability, 11*(7), 1–17.

Rajasekaran, R., Tayal, M. A., & Kanimozhi, D. (2019). Opportunities and challenges in E-marketing with reference to tourism industry in India. *International Journal of Research in Business Studies, 4*(1), 49–62.
Renau, R. (2018). Touristification, sharing economies and the new geography of urban conflicts. *Urban Science, 2*(4), 104.
Sanchez, P. M., & Adams, K. M. (2008). The Janus-faced character of tourism in Cuba. *Annals of Tourism Research, 35*(1), 27–46.
Seraphin, H., Gowreesunkar, V., Zaman, M., & Lorey, T. (2019). Limitations of Trexit (tourism exit) as a solution to overtourism. *Worldwide Hospitality and Tourism Themes, 11*(5), 566–581.
Seraphin, H., & Ivanov, S. (2020). Overtourism: a revenue management perspective. *Journal of Revenue and Pricing Management, 19*(3), 146–150.
Seraphin, H., Sheeran, P., & Pilato, M. (2018). Over-tourism and the fall of Venice as a destination. *Journal of Destination Marketing and Management, 9*, 374–376.
Tiwari, P., Kainthola, S., & Chowdhary, N. R. (2020). Demarketing: A marketing framework for overtourism. In C. R. de Almeida, A. Quintano, M. Simancas, R. Huete & Z. Breda (Eds.). *Handbook of research on impacts, challenges, and policy responses to Overtourism* (pp. 94–114). IGI-Global.
Topham, G. (2019). *Uber loses London licence after TfL finds drivers faked identity*. The Guardian. Retrieved from https://www.theguardian.com/technology/2019/nov/25/uber-loses-licence-london-tfl. Accessed 10 Dec 2020,
Tscheu, F., & Buhalis, D. (2016). Augmented Reality at Cultural Heritage sites. . In A. Inversini and R. Schegg (Eds.), *Information and communication technologies in tourism 2016* (pp. 607–619). Springer.
UNWTO. (2018). *'Overtourism'? – Understanding and managing urban tourism growth beyond perceptions, executive summary*. UNWTO.
UNWTO. (2019). *International tourist arrivals reach 1.4 billion two years ahead of forecasts*. Retrieved from: https://www2.unwto.org/press-release. Accessed: 12 Dec 2020.
Vargas-Sánchez, A. (2020). Overtourism and smart cities: Present and future. In *The overtourism debate* (pp. 267–278).
Veríssimo, M., Moraes, M., Breda, Z., Guizi, A., & Costa, C. (2020). Overtourism and tourismphobia: A systematic literature review. *Tourism, 68*(2), 156–169.
Zmyślony, P., Kowalczyk-Anioł, J., & Dembińska, M. (2020). Deconstructing the Overtourism-related social conflicts. *Sustainability, 12*(4), 1695.

Chapter 21
What Could We Really Say About Overtourism?

Maximiliano E. Korstanje

Abstract The present chapter synthesizes an interesting discussion revolving around the problem of overtourism. From its outset, the tourism research has hotly debated furtherly on the dichotomies of tourism consumption and sustainability. While tourism tends to be considered a desirable activity –if not a right of liberal democracy- which brings prosperity and stability to countries, no less true is that the industry generates some collateral damages. Over recent years, overtourism exhibits not only the resentment and hostility of locals against foreign tourists but also the rise of digital technologies which escape from governmental regulation. Technology offers a fertile ground towards new solutions and discussions in the fields of overtourism. Some voices have eloquently alerted on the problems and limitations that Information and Communications Technology (ICT) has brought for tourist destinations. Nonetheless, ICT offers a fertile ground for planning more sustainable destinations while re-socializing tourists. The introduction of new ethical codes about tourism consumption seems to be vital to placate the negative effects of tourism. Said this, scholars should differentiate overtourism which is a direct effect of the industry and touristphobia. While overtourism entails practical solution, touristphobia denotes a slippery matter originated in the incapacity of West to tolerate the "Non-Western Other" who have become in an "undesired guest". This radical change means the end of hospitality, as least as we know it. Doubtless, the decline of hospitality has made more acute after 9/11 and the War on Terror.

Keywords Hospitality · Overtourism · Human rights · Ethics · Education

M. E. Korstanje (✉)
University of Palermo, Buenos Aires, Argentina
e-mail: mkorst@palermo.edu

A. Hassan, A. Sharma (eds.), *Overtourism, Technology Solutions and Decimated Destinations*, https://doi.org/10.1007/978-981-16-2474-2_21

Introduction

As a buzzword, which captivates the attention of academic circles, overtourism denotes overcrowding or congestion exceeding the carrying capacity of destination (Alexis, 2017; Milano et al., 2019). Overtourism affects negatively not only the quality of life of lay-citizens, some of them resistant to receive foreigner tourists, but also the tourist experience (Seraphin et al., 2018). From its inception, the tourism industry debated around the trend to develop consumption to revitalize economies and the urgency to protect destinations for future generations. The concept of sustainability has invariably played a leading role in the constellations of tourism fields. As Jost Krippendorf (2010) puts it, the uses of tourism are multiple and subject to a complex dynamic. Notwithstanding this fact, tourism should be seen as an instrument which fosters or destroys destinations according to the ways of exploitation. Sustainable destinations coincide with the mainstream cultural value of society. Having said this, the specialized literature rests on a philosophical quandary which is very hard to grasp. Since tourism is considered a peaceful activity which leads economies to prosperity and democracy, policy-makers and authorities endorse to the economic success of tourism to legitimate their administrations. At the same time, tourism is fostered as a desirable priority, local resources are notably exhausted. Another secondary effect of the stimulus of tourism rests on overcrowding and the excess of tourists visiting a destination (Weaver, 2001, 2017). Classic destinations as Venice, Rome, Barcelona or Amsterdam are hostile to receive tourists. Under lemmas like "tourists, you are the real terrorist" or "tourists go home," these global cities become the epicenter of much deeper disputes oriented to discourage mass-tourism consumption (Sharpley, 2000). What is more important, Koens et al. (2018) discuss critically the myths over the notion of overtourism. Beyond reaching a consensus, academicians have developed different conceptions revolving around overtourism. Scholars should adopt alternative solutions for a problem which is not caused by the tourism industry alone. Over-crowding associates often to higher levels of mobilities, poor infrastructure and global commerce. In fact, not only overtourism effects are not citywide but it is not the same than masstourism. This happens because though overtourism is caused by the increase of visitors, not all destinations keep the same carrying capacity to deal with increasing demand. In some cases, specialists strongly believe technology resolves the problems generated by mass consumption and overtourism, a point that Koens et al. (2018) amply refute. In some cases, emerging technologies or digital platforms as Airbnb or Couchsurfing –far from solving the problem- aggravates the condition of congested cities. In this respect, alternative digital technologies struggle to make more sustainable destinations but paradoxically they escape to tax and legal regulations which unless carefully controlled accelerates a rapid decline of the destination (Kadi et al., 2019; Volgger et al., 2019). Amid these controversies, Information and Communication Technologies (ICT) offers a fertile ground to decentralize congestion mitigating the negative effects of overtourism. Because of this, the present (short) concluding chapter deals with the dichotomies of overtourism and ICT, a new emerging field of research in the discipline. In the first section of the chapter, we discuss critically the causes and consequences of overtourism today. The specialized literature, in so doing, is placed under the critical lens of scrutiny laying the foundations towards a new understanding of the issue. The second section centers on the opportunities to employ technology to alleviate the negative effects of overtourism. Lastly, we draw some practical points for tourism research contributes to sustainable tourism in the next decades.

Overtourism: Causes and Consequences

At a closer look, the term overtourism denotes a local sentiment of hostility against tourists due to the compromise of local resources. Per this viewpoint, a planned program of sustainability would be enough to reverse the social maladies caused by mass-tourism. It is safe to say that these maladies or problems range from inter-ethnic conflict to inflation or real estate speculation (exhibited by the gentrification process) (Koens et al., 2018; Blanco Romero et al., 2018) without mentioning local violence, alcoholism and drug addiction (Thomas et al., 2019). To some extent, the host-guest encounter generates unexpected changes in local communities (Seraphin et al., 2018; Korstanje & George, 2020a, b). As Giang Phi eloquently observes, the support or rejection of tourism is given neither by its economic benefits nor hosting infrastructure. In some cases, anti-tourist discourses mark the pace in global cities while cultivating a negative image of foreign tourists in locals (Phi, 2020). These perceptions seem to be traversed by diverse and complex factors as historical exposure to visitors (Phi, 2020) or long-dormant geopolitical disputes between neighboring countries (Korstanje, 2011; Korstanje et al., 2016). In this vein, it is important not to lose the sight on the fact the term overtourism should be conceptually dissociated from tourist-phobia. While the former signals to negative effects perceived –or imagined- by the locals because of tourism activity, the latter refers to much complex emotional factors organized around chauvinist or racist expressions (Milano et al., 2019; Zerva et al., 2019; Volo, 2020). As Korstanje (2017) noted, the turn of the century brought unparalleled changes for the industry and West. The upsurge of global risks as terrorism, natural disasters and today virus outbreaks have prompted long-dormant racist discourses oriented to demonize the "Non-Western Other". In view of this, Islamophobia, tourist-phobia, travel bans and other restrictions to human mobilities are placed into the foreground. All these phenomena rest on the incapacity of Western culture to welcome the "strangers", what Korstanje dubbed as "the decline of hospitality, at least as we know it".

As the previous backdrop, Seraphin et al. (2020) alert that the future of tourism, probably lest the cultural value of community is changed, seems to be grim. In this token, education plays a crucial role in long-term return to the necessary investment to reverse and placate the effects of tourism at saturated destinations. For some reason, as experts said, education has been overlooked form the interests of tourism scholars. As Serena Volo adheres, the excess of carrying capacity at the destination is only one factor of overtourism but the problem is not limited to it. The quality of life in local communities is often degraded by improper behavior or consumption by the visitors (Volo, 2020). The point has been widely reported in several studies as Seraphin et al. 2018), Becken and Simons (2019), and Goodwin (2017). Hence it is vital to distinguish the meaning and reasons beyond tourist-phobia and overtourism as well as the utilization of digital technologies to sanitize the discontent and resentment of locals. Low cost travels associated with the recent technological revolution applied on mobilities are offering cheaper all packaged tours which notably affect – if not saturate- the targeted destination. Unlike other epochs where tourism has been limited to privileged elite, no less true is that nowadays the activity is seen as a global right all citizens should enjoy. Doubtless, these commented points lead policy-makers to cope with new challenges and problems to avoid overcrowding and massconsumption. As discussed, overtourism is not determined by urban life or the excess of visitors. Some rural areas unprepared to receive outsiders are in decline too. This suggests two important assumptions. On one hand, technology usefully disposes of a rational planning to allocate tourists according to the capacities of each destination (Volo, 2020; Butler, 2020). On another, digital apps can help policymakers to regulate informal activities of lodging which are outlined by the rise and expansion of rental online companies like Airbnb, Couchsurfing or Uber (Guttentag, 2015; Ioannides et al., 2019; Volgger et al., 2019).

Information and Communications Technology: Part of the Cure or a Symptom?

In a seminal book, which entitles *The technological society,* French philosopher Jacques Ellul argues convincingly that modern technology disposes of efficiency as a psychological need to be exploited. To put the same in other terms, technology – far from emancipating peoples- standardizes the reasoning into the channels of automatized forms of organizations. The rationality of the technique not only affirms the labor division but also a mechanical re-production of standards (stereotypes) organized to efface the "natural world" (Ellul, 1964). In sharp opposition to the critical viewpoint of Ellul, other optimist voices have emphasized on the dual nature of technologies. Manuel Castells overtly says that technology can be defined as a vehicle towards prosperity and development. As Castells recognizes, even if technology generates some material asymmetries as inequality or poverty, no less true seems to be that technology triggers some positive changes oriented to the "unleashing of the power of the mind" increasing the productivity, tolerance and environmental consciousness of "lay citizens" as never before (Castells, 1999; Carnoy & Castells, 2001). In tourism and hospitality fields, detractors and supporters of technology applied to development issues go in the same opposed poles (McCabe et al., 2012). Whatever the case may be, ICT may be very well defined as a double-edged sword. It potentiates while transforming the nature of tourism giving efficient tools to enhance local economies, but at the same time, posing serious challenges for the tourist system (Buhalis & O'Connor, 2005). This begs a more than a pungent question, is ICT aggravating or solving the problem of overtourism?

To respond this question, Vanessa Gowreesunkar and Tan Vo Than (2020) explain that digital technologies help to keep competitive the image of tourist destinations, as well as describing the potential niches to decentralize the affluence of tourists in the territory. Technology, so to speak, has played a leading role in enhancing the connectivity with the site while creating emancipated and unique experiences. The fluidity of travels and tourists are shaping new more complex (if not personalized) forms of consumption such as virtual tourism, e-marketing, and tailor-made tourism. In these mobile days, experiences are simply posted at digital platforms in the web in forms of opinions or reviews. These stories are carefully read and followed by other potential tourists. Under some circumstances, when these technologies move in a chaotic way, the probabilities a destination to be saturated turns higher. In fact, overtourism and technology are inextricably intertwined. The opposite is equally true since technology may accelerate the decline of some destinations, above all when the online-reviews are not favorable. Having said this, Jim Butcher (2020) calls attention to the function of media in the construction of overtourism as a social malady. Based on Barcelona, he toys with the belief that beyond the rhetoric of tourist-phobia lies the operations of some nationalist or pressure groups which struggle to confront the central administration. These groups allude often to the injustices of capitalism, technologies or mass-migration while forgetting the economic benefits historically amassed by tourism and global capitalism. The fear of losing culture or heritage before the advent of the tourist alluvium corresponds with the first point of entry in the voice of locals. Butcher, finally, dubs these local actors as "claims-makers". Of course, despite the philosophical quandary to regulate tourist behavior through technology because it directly contradicts the liberal right to move freely, we must understand that tourism-phobia is a political question (not a problem of carrying capacity). As Rodanthi Tzanelli (2020) brilliantly stressed, late-modernity is significantly transforming the re-signification of the interplay between the order and chaos. The allegory of western reasoning which is associated with a stable sense of reality is setting the pace to the advance of an egoist primitive drive. To put things in a straight, the postmodern consumption speaks us of a clear dissociation between the gains and costs. While locals surely enjoy the benefits of mass-tourism they develop some resistance to welcome tourists. This happens because the emotional aesthetics and landscape of global tourism are digitalized, packaged and sold in the lines of a fabricated memory which mediates among the gazers and the gazed actors.

Challenges for the Industry in the Next Years

Without any doubt, some pressure groups –probably supported by digital technology- has made public their aversion to mass-tourism in classic urban cities, which have historically lived from tourism. This point leads to thinking in the adoption of alternative instruments and programs to alleviate the carrying capacity of the tourist destination. Over the recent years, some scholars claimed that behind overtourism, the tourist-phobia evinces a complex phenomenon oriented to "demonize" or avoid the "stranger" (Korstanje, 2017). Some nationalist groups recently emerge to make public their discontent with mass-tourism which threatens the national identity. As Butcher noted, these discourse exert a caustic critique not only on tourism but also to technology, globalization and free-trade (Butcher, 2020). What is equally important, some broad strokes illustrate the ways hospitality is in decline (Tzanelli & Yar, 2020; Korstanje, 2017, 2019). To some extent, education would allow the reconstruction of bridges to connect locals with foreign visitors (Korstanje & George, 2020b). It is noteworthy that tourism-related policies are inapplicable with the exception to an education process which ignites positive changes in the paths of reversing overtourism. Tourists may be socialized to adopt more emphatic and conscious behavior about their effects in the environment. Over tourism, education process consists in the implementation of incentives and punishments organized to change unsustainable forms of consumptions (San Tropez, 2020). However, some voices lamented this stick and carrot system is not enough to change old habits or behavior culturally enrooted in society. Hence, education should be accompanied by cultural transformations that introduce ethical values to tourism consumption. American Anthropologist Dean MacCannell coins the term "the ethics of sightseeing" to the moral regulation to deal with the hegemony of capital. Per his viewpoint, global modernity has emptied not only the geographical landscapes, blurring the borders of time and space, but has initiated radical shifts towards anomie. Today, social institutions and personal behavior are subordinated to the capital. Tourists are not looking genuine experiences rather they are in quest of idealized landscapes to bear their own frustration (MacCannell, 2011). Ethics serve as an instrument of re-socialization embracing humanist values respecting the "Non-Western Other" (Korstanje & George, 2020a, b; Cakar, 2020; Maingi, 2020).

Conclusion

As discussed in the present concluding chapter, overtourism seems not to be new. From its outset, the discipline has debated furtherly on the dichotomies of tourism consumption and sustainability. While tourism tends to be considered a desirable activity which brings prosperity and stability to countries, no less true is that the industry generates some collateral damages. Over recent years, overtourism exhibits not only the resentment and hostility of locals against foreign tourists but also the rise of digital technologies which escape from governmental regulation. Some voices have eloquently alerted on the problems and limitations ICT has brought for tourist destinations. For experts, technology plays sometimes the role of a bogyman saturating destinations because of mass-consumption. Nonetheless, ICT offers a fertile ground for planning more sustainable destinations while re-socializing tourists. The introduction of new ethical codes about tourism consumption seems to be vital to placate the negative effects of tourism. Said this, scholars should differentiate overtourism which is a direct effect of the industry and tourist-phobia. While overtourism entails practical solution, tourist-phobia denotes a slippery matter originated in the incapacity of West to tolerate the "Non-Western Other" who have become in an "undesired guest". This radical change means the end of hospitality, as least as we know it. Doubtless, the decline of hospitality has made more acute after 9/11 and the War on Terror (Korstanje, 2017).

References

Alexis, P. (2017). Over-tourism and anti-tourist sentiment: An exploratory analysis and discussion. *Ovidius University Annals, Economic Sciences Series, 17*(2), 288–293.

Becken, S., & Simmons, D. G. (2019). Stakeholder management: Different interests and different actions. In R. Dodds & R. Butler (Eds.), *Overtourism: Issues, realities and solutions* (pp. 234–249). De Gruyter.

Buhalis, D., & O'Connor, P. (2005). Information communication technology revolutionizing tourism. *Tourism recreation research, 30*(3), 7–16.

Butcher, J. (2020). The construction of over tourism: The case of UK media coverage of Barcelona's 2017 tourism protests and their aftermath. In H. Seraphin, T. Gladkikh, & T. Vo Than (Eds.), *Overtourism: Causes, implications and solutions* (pp. 69–88). Palgrave Macmillan.

Butler, R. (2020). Overtourism in rural areas. In H. Seraphin, T. Gladkikh, & T. Vo Than (Eds.), *Overtourism: Causes, implications and solutions* (pp. 27–43). Palgrave Macmillan.

Blanco-Romero, A., Blázquez-Salom, M., & Cànoves, G. (2018). Barcelona, housing rent bubble in a tourist city. Social responses and local policies. *Sustainability, 10*(6), 2043–2055.

Cakar, K. (2020). Education as a way of tackle over tourism: The application of the principles of responsible management education. In H. Seraphin, T. Gladkikh, & T. Vo Than (Eds.), *Overtourism: Causes, implications and solutions* (pp. 361–382). Palgrave Macmillan.

Castells, M. (1999). *Information technology, globalization and social development* (no. 114). UNRISD.

Carnoy, M., & Castells, M. (2001). Globalization, the knowledge society, and the network state: Poulantzas at the millennium. *Global Networks, 1*(1), 1–18.

Ellul, J. (1964). *The technological society* (Vol. 303). Vintage books.

Goodwin, H. (2017). *The challenge of overtourism. Responsible tourism partnership*. Retrieved from: https://haroldgoodwin.info/pubs/RTP%27WP4Overtourism01%272017.Pdf. Accessed 1 Dec 2020.

Gowreesunkar, V., & Vo, T. T. (2020). Between over tourism and under tourism: Impacts, implications and probably solutions. In H. Seraphin, T. Gladkikh, & T. Vo Than (Eds.), *Overtourism: Causes, implications and solutions* (pp. 45–68). Palgrave Macmillan.

Guttentag, D. (2015). Airbnb: Disruptive innovation and the rise of an informal tourism accommodation sector. *Current Issues in Tourism, 18*(12), 1192–1217.

Ioannides, D., Röslmaier, M., & Van Der Zee, E. (2019). Airbnb as an instigator of 'tourism bubble' expansion in Utrecht's Lombok neighbourhood. *Tourism Geographies, 21*(5), 822–840.

Kadi, J., Plank, L., & Seidl, R. (2019). Airbnb as a tool for inclusive tourism? *Tourism Geographies*, https://doi.org/10.1080/14616688.2019.1654541

Koens, K., Postma, A., & Papp, B. (2018). Is overtourism overused? Understanding the impact of tourism in a city context. *Sustainability, 10*(12), 4384.

Korstanje, M. E. (2019). *Terrorism, technology and apocalyptic futures*. New York, Springer.

Korstanje, M., George, B. P., & Amorin, E. (2016). Chile Decime que se Siente: Sports, conflicts, and chronicles of miscarried hospitality. *Event Management, 20*(3), 457–462.

Korstanje, M., & George, B. (2020a). Demarketing overtourism. In G. Seraphin & A. Yallop (Eds.), *Overtourism and tourism education: A strategy for sustainable tourism futures* (pp. 81–95). Routledge.

Korstanje, M., & George, B. (2020b). Education strategy to tackle over tourism for overtourism and inclusive sustainability in the twenty-first century. In H. Seraphin, T. Gladkikh, & T. Vo Than (Eds.), *Overtourism: Causes, implications and solutions* (pp. 341–360). Palgrave Macmillan.

Korstanje, M. E. (2011). Influence of history in the encounter of guests and hosts. *Anatolia, 22*(2), 282–285.

Korstanje, M. E. (2017). *Terrorism, tourism and the end of hospitality in the west*. Palgrave Macmillan.

Krippendorf, J. (2010). *Holiday makers*. Taylor & Francis.

Maingi, S. W. (2020). Mainstream overtourism education for sustainable behavioural change in Kenya's tourism industry context. In H. Seraphin, T. Gladkikh, & T. Vo Than (Eds.), *Overtourism: Causes, implications and solutions* (pp. 383–400). Palgrave Macmillan.

McCabe, S., Sharples, M., & Foster, C. (2012). Stakeholder engagement in the design of scenarios of technology-enhanced tourism services. *Tourism Management Perspectives, 4*, 36–44.

MacCannell, D. (2011). *The ethics of sightseeing*. University of California Press.

Milano, C., Cheer, J. M., & Novelli, M. (Eds.). (2019). *Overtourism: Excesses, discontents and measures in travel and tourism*. CABI.

Phi, G. T. (2020). Framing overtourism: A critical news media analysis. *Current Issues in Tourism, 23*(17), 2093–2097.

San Tropez, M. (2020). ABC of overtourism education. In H. Seraphin, T. Gladkikh, & T. Vo Than (Eds.), *Overtourism: Causes, implications and solutions* (pp. 287–318). Palgrave Macmillan.

Seraphin, H., Gladkikh, T., & Vo Than, T. (2020). Introduction. In H. Seraphin, T. Gladkikh, & T. Vo Than (Eds.), *Overtourism: Causes, implications and solutions* (pp. 1–8). Palgrave Macmillan.

Seraphin, H., Sheeran, P., & Pilato, M. (2018). Over-tourism and the fall of Venice as a destination. *Journal of Destination Marketing & Management, 9*, 374–376.

Sharpley, R. (2000). Tourism and sustainable development: Exploring the theoretical divide. *Journal of Sustainable Tourism, 8*(1), 1–19.

Thomas, T. K., Mura, P., & Romy, A. (2019). Tourism and the 'dry law' in Kerala–exploring the nexus between tourism and alcohol. *Journal of Tourism and Cultural Change, 17*(5), 563–576.

Tzanelli, R. (2020). *Magical realist sociologies of belonging and becoming: The explorer*. Routledge.

Tzanelli, R., & Yar, M. (2020). Atmospheres of the inhospitable in staged kidnappings. *Consumption Markets & Culture*, 24(5), 439–455.

Volo, S. (2020). Overtourism: Definitions, enablers, impacts, and managerial challenges. In H. Seraphin, T. Gladkikh, & T. Vo Than (Eds.), *Overtourism: Causes, implications and solutions* (pp. 11–26). Palgrave Macmillan.

Volgger, M., Taplin, R., & Pforr, C. (2019). The evolution of 'Airbnb-tourism': Demand-side dynamics around international use of peer-to-peer accommodation in Australia. *Annals of Tourism Research, 75*, 322–337.

Weaver, D. B. (2001). Ecotourism as mass tourism: Contradiction or reality? *Cornell Hotel and Restaurant Administration Quarterly, 42*(2), 104–112.

Weaver, D. (2017). Sustainability and mass tourism: A contradiction in terms. In *Mass tourism in a small world* (pp. 63–74).

Zerva, K., Palou, S., Blasco, D., & Donaire, J. A. B. (2019). Tourism-philia versus tourism-phobia: Residents and destination management organization's publicly expressed tourism perceptions in Barcelona. *Tourism Geographies, 21*(2), 306–329.

Index

A. Hassan, A. Sharma (eds.), *Overtourism, Technology Solutions and Decimated Destinations*, https://doi.org/10.1007/978-981-16-2474-2

The manufacturer's authorised representative in the EU is Springer Nature Customer Service Centre GmbH, Europaplatz 3, 69115 Heidelberg, Germany. If you have any concerns regarding our products, please contact ProductSafety@springernature.com

Printed and bound by CPI Group (UK) Ltd, Croydon, CR0 4YY
30/09/2026
02225248-0011